T0280737

Strategic Computing

History of Computing
I. Bernard Cohen and William Aspray, editors

Strategic Computing
DARPA and the Quest for Machine Intelligence,
1983–1993

Alex Roland
with Philip Shiman

The MIT Press
Cambridge, Massachusetts
London, England

©2002 Massachusetts Institute of Technology

All rights reserved. No part of this book may be reproduced in any form or by any electronic or mechanical means (including photocopying recording, or information storage and retrieval) without permission in writing from the publisher.

This book was set in New Baskerville by Achorn Graphic Services, Inc.

Library of Congress Cataloging-in-Publication Data

Roland, Alex, 1944–
    Strategic computing : DARPA and the quest for machine intelligence, 1983–1993 / Alex Roland with Philip Shiman.
        p. cm. — (History of computing)
    Includes bibliographical references and index.
    ISBN 978-0-262-18226-3 (hc. : alk. paper) — 978-0-262-52926-6 (pb. : alk. paper)
    1. High performance computing.   2. Artificial intelligence.   I. Shiman, Philip.   II. Title.   III. Series.
QA76.88 .R65 2002
004.3—dc21
                                                            2001056252

The MIT Press is pleased to keep this title available in print by manufacturing single copies, on demand, via digital printing technology.

Contents

10
Conclusion   319

Illustrations

Preface

This story has its own story. Philip Shiman and I were commissioned by the Defense Advanced Research Projects Agency (DARPA) to prepare this study. Most of the research and writing were done under subcontract between the Naval Warfare Systems Center (NWSC), an agent of DARPA, and Duke University, which employed Shiman and me. Execution of this contract was fraught with difficulty, and the contract was terminated short of completion. The book that follows is best understood in the light of this relationship between the historians and the object of their study.

On the recommendation of I. Bernard Cohen and Merritt Roe Smith, members of a DARPA panel overseeing this book project, Keith Uncapher contacted me in March of 1992. I then submitted a proposal to serve as principal investigator, with Research Associate Philip Shiman, to prepare a 400-page history of the Strategic Computing Program (SCP). The three-year project would be funded by the Information Science and Technology Office (ISTO) of DARPA. On 11 November 1992, Shiman, Uncapher, and I met with Steven Squires, director of ISTO. At that meeting Squires displayed a remarkable intelligence, a boyish enthusiasm for computers, a petulant imperiousness, a passion for Isaac Asimov and Star Trek, and a permanent membership card to Disneyworld, which he wore on a chain around his neck.

Prudence dictated that Shiman and I walk away from this undertaking, but we succumbed to a siren song of curiosity. Squires claimed that the world was in a Seldon crisis, a reference to the Asimov *Foundation* trilogy. This classic of science fiction, which may be seen as a rumination on the collapse of the Roman empire or a metaphor for the Cold War, envisioned two "foundations" chartered at opposite ends of a galactic empire, which visionary Hari Seldon knew was coming to an end. Seldon tells one of the foundations, analogous to the United States or the free world, that it must survive a series of crises. Only then will it endure the millennium of chaos before the birth of galactic civilization. Squires clearly implied that he and his colleagues in computer research were solving critical problems, thereby ushering in a new era of civilization.

Shiman and I both wanted to know more about Squires and the strange world he inhabited. We sparred with Squires over the terms of the contract and our method of operation. He wanted us to work primarily from interviews with him and his colleagues and to write up the story as they saw it. We insisted that we would need other sources of evidence to validate whatever interpretation we settled on. We left the interview, I now think, each believing that the other had agreed to do it his way.

As we set about our work, it quickly became apparent that working with DARPA would prove more difficult than we had anticipated. It took eight months just to get the terms of the contract settled. The authorization chain led from Squires as director of ISTO, through a contracting officer at DARPA, the NWSC, a subcontracting agent (at first, Meridian Corporation, then Applied Ordnance Technology [AOT], and then Meridian again [now a subsidiary of Dyncorp]), through Duke University, and finally to the historians. The main issue in dispute was intellectual control over the findings of the study. Duke insisted that the historians function as researchers, delivering in the final report their "best effort" to understand and explicate SCP. DARPA, through its agents, insisted that the historians deliver an acceptable manuscript. When no resolution was achieved after months of negotiation, AOT replaced Meridian as the contracting agent and a contract was quickly signed along the lines proposed by Duke. Eighteen months later, however, when the contract came to be renewed, DynCorp replaced AOT and the same dispute resumed. The second contract was signed on 19 February 1995.

This contract called for us to work "on a best efforts basis," but required "approval of Buyer prior to . . . publication of any nature resulting from this subcontract." Squires had been relieved as ISTO director long before the second contract was signed, but his permanent successor, Howard Frank, did not take office until the new contract was in place. He quickly decided to cancel the contract. From his perspective, the history made sense, but DARPA sponsorship did not. If the history was favorable, it would appear to be a whitewash; if negative, it would be used against the agency by its enemies. This hard political calculus reflected the visibility and controversy that SCP had helped to attract to DARPA.[1]

After Shiman and I met with Frank, he agreed to continue funding. Then he changed his mind again and withdrew support. Nevertheless, I was told that sufficient funds were already in the pipeline to carry the project to completion. When that oral commitment was not confirmed in writing, I terminated the project, short of the completion date. Subsequently I learned that the money in the pipeline had already dried up; Duke and I absorbed the shortfall.

With funding terminated, Philip Shiman withdrew from the project at this point. In almost four years of work on this study, he had done the majority of the research, conducted most of the interviews, either alone or with me, helped to conceptualize the work, and drafted four of the ten chapters (5, 7, 8, and 9). Additionally, he contributed to the penultimate revision of the manuscript for publication.

As Shiman and I conducted our research, we developed a painful empathy for the computer researchers we were studying, learning firsthand how layers of bureaucracy can distract and suffocate researchers and distort the purposes they thought they were about. DARPA prides itself on being a lean organization relatively free of red tape; in practice it has simply contracted out its bureaucratization to other institutions.

Nowhere was the frustration of dealing with DARPA more salient than in gaining access to its records. The "Note on Sources" (pages 397–403) catalogs the advantages and drawbacks of having special access to an agency's records. Suffice it to say here that we never achieved the unfettered access to records that was promised to us at the outset. We saw more and better records that an outsider would have seen using the Freedom of Information Act, but we did not see many files to which we sought access. Some documents and information we had to obtain through the Freedom of Information Act when the project was done, though by then we had the advantage of knowing what to ask for.

This was not, as best we could tell, a cover up. Seldom did we discern that anyone at DARPA wanted to hide anything. Rather, they simply could not understand why we wanted to see the materials we requested. We appeared to be on a fishing expedition, and we appeared to be fishing for trouble. They believed they knew what the story was, and the documents we requested did not always seem to pertain to the story as they understood it. We were never explicitly denied access to records controlled by DARPA; we just never gained complete access.

Still, we were able to piece together a document trail that finally correlated rather closely with our reading of the interviews we conducted. We got more and better documents from the garages, attics, and basements

of former DARPA employees than we ever got from DARPA itself, but
the end result was similar. In the modern age of photocopying, documents that are withheld in one venue often turn up in another. The one
sampling we got of e-mail files whetted our appetite for more, but requests for access to these were most often met with laughter. Historians
in the future who want to pursue this topic further will be fortunate if
they can break into that cache.

We learned in our research that Strategic Computing (SC) was a house
of many rooms. Even DARPA veterans of the program disagreed about
its goals, its methods, and its achievements. Yet there is a fragile collegiality among the veterans of SC that makes them reluctant to confront one
another with their differences. They will disagree on technical matters,
often in the strongest possible terms, but they will seldom discuss for
attribution their equally strong beliefs about each other. The tension we
noted in our first interview with Squires never lessened.

Most veterans of SC believed that the program made a historic contribution to computer development. They wanted us to capture that
achievement and to convey the excitement they felt. They did not, however, want us to talk about people and politics. For them the thrill was
in the technology. Therefore, they resisted our efforts to pursue avenues
we found interesting, and they attempted to channel us into certain
paths of inquiry. They never lied to us, as best we can tell, but neither
did they tell us all they knew. We were privileged and enriched to have
access to them and their records, but it was always a mixed blessing.

I came to believe that the veterans wanted us to present a consensus
view of SC, one that they could all get behind but that would nonetheless
reflect their different opinions about what it was. I never found such a
view. Instead, I have tried to capture and present fairly the multiple possible interpretations of SC, then to let readers judge for themselves. I
reached my own conclusions on the program, but I hope the story is
presented here in such a way as to allow readers to make their own
judgments.

Alex Roland
Durham, North Carolina

Acknowledgments

Many people contributed to this book. Philip Shiman researched, conceptualized, and drafted significant parts of it. My wife Liz proofread, copyedited, and indexed the book. Keith Uncapher recruited me to conduct this study and remained throughout its production the most stalwart of supporters. My good friends I. Bernard Cohen and Merritt Roe Smith recommended me for the project and provided moral and technical support as the full measure of their "disservice" became manifest. Two abiding friendships sprang from the project, one new and one revived. J. Allen Sears reentered my life after a long absence since our college days together at the U.S. Naval Academy. Clinton Kelly and I became good friends based initially on our connection to Strategic Computing. I am grateful to them both for their help with the book and for their fellowship.

Many institutions contributed as well. Duke University provided the ideal scholarly environment in which to conduct such a study, and it contributed in many other ways as well. Its Office of Research Support persevered in prolonged and frustrating negotiations with DARPA's agents to develop contracts that would serve DARPA's interests while preserving our right to present the story as we found it. The late Charles E. Putman, Senior Vice President for Research Administration and Policy, and William H. Chafe, Dean of the Faculty of Arts and Sciences, backstopped this project with their own discretionary funds when DARPA support was withdrawn. Perkins Library and its exceptional staff supported the project from beginning to end, either maintaining the requisite information in its extensive holdings or finding it in other repositories. Duke's Engineering Library was equally efficient and accommodating. The Department of History at Duke challenged many of my early conceptions of this story and allowed me to try out the ideas from which this study finally emerged. Duke colleagues Alan Biermann and Hisham Massoud provided helpful insights from their respective disciplinary perspectives in Computer Science and Electrical and Computer Engineering. Robert A. Devoe of Duke's Educational Media Services turned sketchy drawings and faded photocopies into crisp and revealing illustrations.

Other institutions contributed as well. I spent a stimulating and productive year (1994–1995) as a fellow of the Dibner Institute for the History of Science and Technology at MIT. The directors of the Dibner, Jed Buchwald and Evelyn Simha, created an ideal atmosphere for scholarship and collegiality, and the other fellows offered valuable criticisms and suggestions as the book took shape. Helen Samuels and her staff at the MIT Institute Archives and Special Collections took pains to help me understand MIT's many intersections with the Strategic Computing Program. The late Michael Dertouzos, director of the MIT Laboratory for Computer Science, and his successor, Victor Zue, shared with me their own experiences with SC and their perspectives on the program as it appeared from MIT. Professor Charles Weiner allowed me the opportunity to try out my ideas on his undergraduate history class.

Arthur Norberg, co-author of the prequel to this volume, *Transforming Computer Technology: Information Processing for the Pentagon, 1962–1986,* welcomed me to the rich holdings of the Babbage Institute for the History of Computing at the University of Minnesota and offered his own considerable insight into Strategic Computing. Kind invitations from colleagues allowed me to test my ideas before informed and discriminating audiences. I am indebted to Eric Shatzberg for inviting me to speak to the history of science community at the University of Wisconsin and to David Hounshell for organizing a presentation to the history and computer science communities at Carnegie Mellon University. The library and records staffs of DARPA have struggled heroically to resist the centrifugal forces at work on the agency's documents; they provided invaluable assistance and commiseration. Alexander Lancaster, himself a veteran of Strategic Computing, brought the considerable resources and services of Meridian Corporation to our aid.

David Hounshell also read the entire manuscript for MIT Press and provided thoughtful, informed, and incisive suggestions for its improvement. I am equally indebted to the other referees for MIT Press, including William Aspray. Among the others who also read the manuscript, I am particularly grateful to I. Bernard Cohen, Ronald Ohlander, Philip Shiman, John Toole, Keith Uncapher, and M. Mitchell Waldrop. Most of my obligations to individuals are acknowledged in the list of the interviews conducted for this project. For less formal personal communications that added to my understanding of Strategic Computing, I am indebted to Gordon Bell, Donald MacKenzie, Arthur Norberg, Garrel Pottinger, and G. Paschal Zachary.

Larry Cohen and Deborah Cantor-Adams at MIT Press were exceptionally patient and tolerant in helping me to get a delinquent manuscript into publishable form. Errors that remain in the text in the face of the invaluable aid and good advice of these many people and institutions are my wife's fault.

Alex Roland
Annapolis, Maryland

Chronology: Key DARPA Personnel during the Strategic Computing Program

Directors of Darpa

Robert Cooper (July 1981–July 1985)

James Tegnalia (Acting) (July 1985–January 1986)

Robert Duncan (January 1986–February 1988)

Raymond S. Colladay (February 1988–July 1989)

Craig Fields (July 1989–April 1990)

Victor Reis (April 1990–June 1991)

Gary Denman (June 1991–1995)

Directors of Computer Technology Offices

Information Processing Techniques Office (IPTO)
Robert Kahn (August 1979–September 1985)
Saul Amarel (September 1985–June 1986)

Information Science and Technology Office (ISTO)
Saul Amarel (June 1986–November 1987)
Jack Schwartz (November 1987–September 1989)
Stephen Squires (Acting) (September 1989–November 1989)
Barry Boehm (November 1989–June 1991)

Computer Systems Technology Office (CSTO)
Stephen Squires (June 1991–September 1993)
John Toole (September 1993–1995)

Engineering Applications Office (EAO)
Clinton Kelly (November 1984–June 1986)

Software & Intelligent Systems Office (SISTO)
Barry Boehm (June 1991–August 1992)
Duane Adams (August 1992–December 1992)
Edward Thompson (February 1993–1995)

Acronyms

AAV	Autonomous Air Vehicle
ACCAT	Advanced Command & Control Applications Testbed
ADRIES	Advanced Digital Radar Imagery Exploitation System
ADS	Advanced Data Systems
AGVT	Advanced Ground Vehicle Technology
AI	Artificial Intelligence
ALBM	AirLand Battle Management
ALCOA	Aluminum Company of America
ALV	Autonomous Land Vehicle
AFWAL	Air Force Wright Aeronautical Laboratory
AOT	Applied Ordnance Technology
ARPA	Advanced Research Projects Agency
ARPANET	Advanced Research Projects Agency Network
ASIC	application-specific integrated circuit
AUV	Autonomous Underwater Vehicle
BAA	Broad Agency Announcement, Broad Area Announcement
BBN	Bolt, Beranek & Newman
BMS	Battle Management System
CAD	computer-aided design
CAE	computer-aided engineering
CAMPSIM	Campaign Simulation Expert System
CAP	computer-aided productivity

CASES	Capabilities Requirements Assessment Expert System
CASREPS	casualty reports
CAT	Combat Action Team
CBD	Commerce Business Daily
CDC	Control Data Corporation
CINCPACFLT	Commander-In-Chief, Pacific Fleet
CISC	complex instruction set computers
CM	Connection Machine
CMOS	Complementary Metal Oxide Semiconductor
CMU	Carnegie Mellon University
CNRI	Corporation for National Research Initiatives
CPSR	Computer Professionals for Social Responsibility
CPU	central processing unit
CSTO	Computer Systems Technology Office
CT	Component Technology
DARPA	Defense Advanced Research Projects Agency
DART	Dynamic Analysis & Reprogramming Tool
DDR&E	Director of Defense Research & Engineering
DEC	Digital Equipment Corporation
DoD	Department of Defense
DoE	Department of Energy
DRAM	digital random access memory
DSO	Defense Sciences Office
DSP	Digital Signal Processor
EAO	Engineering Applications Office
ECL	emitter-coupled logic
ERDA	Energy Research & Development Administration
ERIM	Environmental Research Institute of Michigan
ETL	Engineering Topographical Laboratory
FCCBMP	Fleet Command Center Battle Management Program
FCCSET	Federal Coordinating Council on Science, Engineering & Technology
FETs	field effect transistors

FLASH	Force Level Alerting System
FLOPS	floating point operations per second
FRESH	Force Requirements Expert Systems
GaAs	gallium arsenide
GAO	Government Accounting Office
GE	General Electric
GFE	government furnished equipment
gigaops	billion operations per second
gigaflops	billion floating point operations per second
GM	General Motors
GPP	general-purpose processor
HDTV	high-density television
HMMV	high mobility medium-utility vehicle
HPC	High Performance Computing
HPCC	High Performance Computing & Communications
IBM	International Business Machines
IC	integrated circuit
IM	intelligent munition
IMPs	Interface Message Processors
IP	Internet Protocol
IPL	Information Processing Language
IPTO	Information Processing Techniques Office
ISI	Information Sciences Institute
ISTO	Information Science and Technology Office
IT	information technology
IU	image understanding
JJ	Josephson Junction
KBS	knowledge-based system
KEE	Knowledge Engineering Environment
LISP	LISt Processing
LMI	Lisp Machine Inc.
MAC	Machine-Aided Cognition; Multiple Access Computer

MCC	Microelectronics & Computer Technology Corporation
MIMD	Multiple Instruction stream/Multiple Data stream
MIT	Massachusetts Institute of Technology
MITI	Ministry of International Trade & Industry
MOAs	memoranda of agreement
MOSAIC	Molecular Observation, Spectroscopy and Imaging Using Cantilevers
MOSFET	metal oxide semiconductor field effect transistor
MOSIS	Metal Oxide Semiconductor Implementation Service
MOUs	memoranda of understanding
MPP	massively parallel processor
MSAI	Master of Science in Artificial Intelligence
MSI	medium-scale integration
NASA	National Aeronautics and Space Administration
NAVELEX	Naval Electronics System Command
NavLab	Navigation Laboratory
NBS	National Bureau of Standards
NGS	New Generation System
NIST	National Institute of Science and Technology
NOSC	Naval Ocean Systems Center
NREN	National Research and Education Network
NSA	National Security Agency
NSF	National Science Foundation
NWSC	Naval Warfare Systems Center
NYU	New York University
OMB	Office of Management and Budget
OPGEN	Operation Plan Generation Expert System
OPLAN	operations plan
OSS	Operations Support System
OSTP	Office of Science and Technology Policy
PA	Pilot's Associate
PARC	Palo Alto Research Center

PC	personal computer
PERT	Program Evaluation and Review Technique
PM	program manager
PSC	programmable systolic chip
RADIUS	Research and Development Image Understanding System
RFP	request for proposals
RFQ	request for quotation
RISC	reduced instruction set computer
SAGE	Semi-Automatic Ground Environment
SAIC	Science Applications International Corporation
SAPE	Survivable, Adaptive Planning Experiment
SAR	Synthetic Aperture Radar
SC	Strategic Computing
SCI	Strategic Computing Initiative
SCORPIUS	Strategic Computing Objected-directed Reconnaissance Parallel-processing Image Understanding System
SCP	Strategic Computing Program
SDC	System Development Corporation
SDI	Strategic Defense Initiative
SDIO	Strategic Defense Initiative Office
SDMS	Spatial Data Management System
SIMD	Single Instruction stream/Multiple Data stream
SIOP	Single Integrated Operations Plan
SISTO	Software and Intelligent Systems Technology Office
SKAT	Smart Knowledge Acquisition Tool
SPAWAR	Space and Naval Warfare Systems Command
SPP	special-purpose processor
SRC	Semiconductor Research Corporation
SRI	Stanford Research Institute
SSI	small scale integration
STRATUS	Strategy Generation and Evaluation Expert System
SUR	speech understanding research

TCP	Transmission Control Protocol
TEMPLAR	Tactical Expert Mission Planner
TeraOPS	trillion operations per second
TeraFLOPS	trillion floating point operations per second
TI	Texas Instruments
TMC	Thinking Machines Corporation
TRW	Thompson Ramo Woolridge
TSR	Technology Status Review
TTL	transistor-transistor logic
UAV	unmanned air vehicles
UGV	unmanned ground vehicles
ULSI	ultra large scale integration
VHPC	Very High Performance Computing
VHSIC	Very High Speed Integrated Circuits
VLSI	very large scale integration

Strategic Computing

Introduction

Between 1983 and 1993, the Defense Advanced Research Projects Agency (DARPA) spent an extra $1 billion on computer research to achieve machine intelligence.[1] The Strategic Computing Initiative (SCI) was conceived at the outset as an integrated plan to promote computer chip design and manufacturing, computer architecture, and artificial intelligence software. These technologies seemed ripe in the early 1980s. If only DARPA could connect them, it might achieve what Pamela McCorduck called "machines who think."[2]

What distinguishes Strategic Computing (SC) from other stories of modern, large-scale technological development is that the program self-consciously set about advancing an entire research front. Instead of focusing on one problem after another, or of funding a whole field in hopes that all would prosper, SC treated intelligent machines as a single problem composed of interrelated subsystems. The strategy was to develop each of the subsystems cooperatively and map out the mechanisms by which they would connect. While most research programs entail tactics or strategy, SC boasted grand strategy, a master plan for an entire campaign.

The goal of connecting technology had a long tradition at DARPA. Psychologist J. C. R. Licklider introduced the concept when he created DARPA's Information Processing Techniques Office (IPTO) in 1962. Licklider's obsession was the man-machine interface, the connection between humans and computers.[3] A visionary and a crusader, Licklider campaigned tirelessly to develop computers as human helpmates. He sought to connect machines to people in ways that would enhance human power. To realize this dream, computers had to become extensions of human intention, instruments of human will.

Licklider's successors at DARPA played out his agenda during the Strategic Computing Program (SCP); IPTO director Robert Kahn focused

on switches and how to connect them. He first conceived the SCI as a
pyramid of related technologies; progress, in his scheme, would materialize as developments flowed up the pyramid from infrastructure at the
bottom through microelectronics, architecture, and artificial intelligence to machine intelligence at the top. The goal of SC was to develop
each of these layers and then connect them.

DARPA Director Robert Cooper focused on other connections. He
tied Kahn's vision to the Reagan defense buildup of the early 1980s and
then connected both to congressional concerns about the economic and
technological challenge posed by Japan. He connected university researchers to industry, ensuring that the new knowledge developed in
the program would "transition" to the marketplace. And he connected
researchers and users, who were in this case the military services.

Steven Squires, who rose from program manager to be Chief Scientist
of SC and then director of its parent office, sought orders-of-magnitude
increases in computing power through parallel connection of processors. He envisioned research as a continuum. Instead of point solutions,
single technologies to serve a given objective, he sought multiple implementations of related technologies, an array of capabilities from which
users could connect different possibilities to create the best solution for
their particular problem. He called it "gray coding." Research moved not
from the white of ignorance to the black of revelation, but rather it
inched along a trajectory stepping incrementally from one shade of gray
to another. His research map was not a quantum leap into the unknown
but a rational process of connecting the dots between here and there.

These and other DARPA managers attempted to orchestrate the advancement of an entire suite of technologies. The desideratum of their
symphony was connection. They perceived that research had to mirror
technology. If the system components were to be connected, then the
researchers had to be connected. If the system was to connect to its environment, then the researchers had to be connected to the users. Not
everyone in SC shared these insights, but the founders did, and they
attempted to instill this ethos in the program.

At the end of their decade, 1983–1993, the connection failed. SC
never achieved the machine intelligence it had promised. It did, however, achieve some remarkable technological successes. And the program leaders and researchers learned as much from their failures as
from their triumphs. They abandoned the weak components in their
system and reconfigured the strong ones. They called the new system
"high performance computing." Under this new rubric they continued

the campaign to improve computing systems. "Grand challenges" replaced the former goal, machine intelligence; but the strategy and even the tactics remained the same.

The pattern described in this study has been discerned by others. In their analysis conducted for the French government in 1980, Simon Nora and Alain Minc invoked the metaphor of "thick fabric" to describe the computer research and development infrastructure in the United States.[4] The woof and warp of their metaphor perform the same function as connections here. The strength of the fabric and the pattern in the cloth depend on how the threads are woven together, how they are connected.

"Connection" as metaphor derives its power in part from conceptual flexibility. James Burke has used the same term in his television series and his monthly column in *Scientific American* to explore the ways in which seemingly disparate technological developments appear to connect with one another in the march of material progress. Many historians of technology believe that the connections Burke sees have more to do with his fertile imagination than with the actual transfer of technology, but his title nonetheless demonstrates the interpretive flexibility of the term. In another instance, a research trajectory called "the new connectionism" intersected with SC in the late 1980s. This concept sought to model computer architecture on the network of synapses that connect neurons in the human brain. "The new connectionism" lost salience before SC ran its course, but it influenced SC and it demonstrated yet another way of thinking about connecting computer systems.

An attempt has been made in this study to apply the term "connection" consistently and judiciously, applying it to the *process* of technological development. It operates on at least three different levels. SC sought to build a complex technological system, of the kind conceptualized in Thomas P. Hughes's defining study, *Networks of Power*.[5] The first level of connection had to bring together the people who would develop the components of machine intelligence technology, what Hughes calls the system builders. These were the DARPA administrators themselves, the researchers they supported, and the developers who would turn research ideas into working technologies. If these individuals were not coordinated, they would be unable to transfer—"transition" in DARPA jargon—the technology from the laboratory to the marketplace, from the designer to the user. This development community had to share some common vision of what the enterprise was about and how it would achieve its goals.

The second sense in which the technology of SC had to connect was systems integration. The components had to work together. The chips produced in the infrastructure program had to work in the machines developed in architecture, which in turn had to host the software that would produce machine intelligence. The Apollo spacecraft that carried Americans to the moon in the 1960s and 1970s consisted of more than seven million parts; all of these had to be compatible. The machines envisioned by SC would run ten billion instructions per second to see, hear, speak, and think like a human. The degree of integration required would rival that achieved by the human brain, the most complex instrument known to man.

Finally, this technology had to connect with users. Paper studies from laboratory benches would avail nothing if they could not "transition" to some customer. SC first sought to connect university researchers with industry producers, not just to hand off ideas from one to the other but to harness them in a joint project to bring designs to fruition. Then both had to be connected with some end user, either the military services or the commercial marketplace. Just as Thomas Hughes's system builders had to invent customers for electricity in the late nineteenth century, so too did DARPA have to invent uses for intelligent machines in the 1980s. Whether the program pushed ripe technology into the hands of users, or users drew the technology along in a demand pull, the process had to be seamlessly connected all the way from the researcher's vision to the customer's application.

Connection, then, became the desideratum of SC. At every level from research through development to application, the workers, the components, the products, and the users had to connect. Unlike most such research projects, SC set this goal for itself at the outset and pursued it more or less faithfully throughout its history. Connection provides both the template for understanding SC and the litmus paper for testing its success. It will be applied to both purposes throughout this study.

Three questions dominate the history of SC. Why was it created? How did it work? What did it achieve? Each question has a historical significance that transcends the importance of the program itself.

The rationale behind SC raises the larger question of what drives technology. Many veterans and supporters of the program believe that it was created in response to a compelling technological opportunity. In the early 1980s, advances in microelectronics design and manufacturing held out the promise of greatly improved computer chips. New concepts of computer architecture could conceivably exploit these chips in machines of unprecedented speed and power. Such computers might finally have the muscle necessary to achieve artificial intelligence—machine capabilities comparable to those of human thought.[6] This image of Strategic Computing credits the technology itself with driving the program. Ripe technology imposed an imperative to act. Context did not matter; content was all. Technology determined history.

In contrast, many scholars believe that technology is socially constructed. They look to external, contextual forces to explain why people make the technological choices they do. SC offers some support for this view. Rather than a technological moment, the early 1980s presented a political moment. The Cold War was being warmed by the belligerent rhetoric and hegemonic ambitions of the new Reagan administration. Out of that caldron would come the Strategic Defense Initiative and an unprecedented acceleration of the arms race. The Falklands War between Britain and Argentina revealed the vulnerability of major industrialized powers to new "smart" weapons. Japan was challenging United States preeminence in computer technology with its Fifth Generation program, a research initiative that fueled concerns over the ability of the United States to remain competitive in the production of critical defense technologies.

The contextual forces shaping SC were personal as well as political. Key individuals within DARPA and the Department of Defense (DoD) joined with allies in Congress and elsewhere in the executive branch of government to conceive, package, and sell the SC Initiative. On a different stage with a different cast of characters, the story would have been different. These particular actors took existing technology and invented for it a new trajectory.

Scholars in recent years have debated the relative merits of these explanatory models. Technological determinism has been discredited, but it refuses to go away.[7] Social construction of technology has been extolled without, however, escaping criticism.[8] The best contemporary scholarship looks for explanatory power in both technology and context. But no formula suggests how these forces might be balanced in any particular case. This study seeks to find its own balance by casting a wide and impartial net.

The management of SC—how did it work?—raises two larger issues. First, is DARPA special? Supporters of SC claim that there is a DARPA style, or at least a style within those DARPA offices that promote computer development. People in these programs spend the taxpayer's money more wisely and get a higher payoff than other funding agencies, or so the apologists say. In the case of SC, they explicitly claim superiority to the National Aeronautics and Space Administration, the National Science Foundation, the National Security Agency, the DoD VHSIC program, and the Japanese Fifth Generation program. This study examines those claims, explores the nature of DARPA culture (or at least the information processing culture within DARPA), and compares it with other research and development programs both within government and outside it.

Second, SC nominally took on an unprecedented challenge. It purported to survey from Washington the entire field of computer development. It identified the most promising and important subfields, discerned the nature of the relationship between those subfields, and then orchestrated a research and development program that would move the entire field forward in concert. All the subfields would get the support they needed in timely fashion, and each would contribute to the advance of the others. In Hughes's military metaphor, it played Napoleon to the *Grande Armée* of U.S. computer development. Overlooking the battlefield, the program ascribed to itself the emperor's legendary *coup d'œil*, the ability to comprehend at a glance. It would marshall all the connections, vertical and horizontal, necessary to advance the army up the high ground of machine intelligence. It was ambitious beyond parallel to imagine that relationships this complex could be identified and understood; more ambitious still to think that they could be managed from a single office. This study explores how that remarkable feat was attempted.

But did it succeed? In many ways it is too soon to tell. The technologies supported by SC will not come to fruition for years, perhaps decades, to come. Nevertheless, some consequences of SC are already manifest; others are on predictable trajectories. This study attempts to evaluate those accomplishments and trajectories and set them in the context of overall computer development in the period from 1983 to 1993. This was the period during which computers entered the public domain and the public consciousness at a rate invoking the term "computer revolution." This study considers the nature of that revolution (if it was one) and explores the contribution of SC. In the end one hopes that the rationale and conduct of SC can be related to its consequences in such a way as to shed light on the complex question of how and why great technological change takes place. The answer to that question has enormous consequences for the future.

By focusing on these questions, this study joins a growing literature in several different fields. First, in the history of technology, it complements those studies that have sought to explain the nature of technological change. Thomas Hughes's *Networks of Power* holds a distinguished place in that literature. Other pivotal works have contributed as well. Edward Constant, for example, has adapted Thomas Kuhn's pathbreaking study of *The Structure of Scientific Revolutions* to explain quantum changes in technology.[9] His concept of presumptive anomaly looms large in the current study. Hugh G. J. Aitken has offered a powerful model of the relationships among science, technology, and market forces.[10] Donald MacKenzie has examined technological advance in the modern United States military from the point of view of social constructivism.[11] Michael Callon, Bruno Latour, and other social constructivists have proposed an actor-network theory that maps well with the SC Program.[12] These and other contributions in the history of technology inform the current analysis.

This study also joins the literature on the nature of government-funded research and development. During the Cold War, the U.S. government became a major sponsor of research and development. Within the government, the Department of Defense was the largest single contributor. These topics have a well-developed literature.[13] Within the Department of Defense, however, DARPA's history is less well documented. This lacuna belies the fact that DARPA is widely considered to be a model agency with an unparalleled record of success.[14] Indeed, as the Cold War wound down in the late 1980s, calls were heard for a civilian DARPA that could do for U.S. economic competitiveness what the old DARPA had done for military competitiveness.[15] These calls arose during the SC Program and contributed to its history.

Third, institutional histories of research establishments have sought to capture the styles of research that evolve within particular cultures and settings. Some focus on the institutions themselves,[16] while others target specific individuals who shaped institutions.[17] This study attempts to do both. It explores DARPA culture, especially the information processing culture within DARPA, and it also considers the ways in which individuals shaped the program. This too will help reveal the nature of DARPA.

Fourth is a literature on the relationship between civilian and military development. The "dual-use" enthusiasm of the late 1980s and 1990s is only the latest manifestation of a long-standing tension between military and civilian development. Though seldom recognized or acknowledged, military developments have long provided the basis for many civilian technologies.[18] In few areas is this more true than in computer development in the United States.[19] During the Cold War, however, this phenomenon raised concerns about the growth of a military-industrial complex.[20] DARPA functioned at the eye of this storm, and it still does. For decades it pursued both military and civilian technologies, choosing those that seemed to be in the best interests of the country. When forced by the armed services or the Congress, DARPA justified these programs strictly in terms of their contribution to national security. In practice, however, it cared little for the distinction. Long before it became fashionable nationally, DARPA understood that technologies such as computer development contribute to national security in ways that transcend rigid distinctions between military and civilian.

Fifth, economists and economic historians have attempted to understand the role of technology in economic development. This question has intrigued economists since Joseph Schumpeter introduced innovation as an independent variable in economic development.[21] Early attempts, such as John Jewkes's *The Sources of Invention,* suggested that technological change sprang from multiple sources, indeed from more sources than scholars could hope to discern and weigh.[22] More recently scholars have shifted to evolutionary models of technological change.[23] Eschewing the neoclassical economic model of rational choice by isolated actors, these scholars see change emerging from technological paradigms, a concept borrowed from Kuhn. Technical communities share a common understanding of the state of the art. They seek innovation by advancing the art in the direction of their economic goals, that is, in the direction in which they believe the market will move. These lines of development are called trajectories.[24]

In the case of SC, for example, the plan was shaped by Robert Kahn's understanding of the state of research and his goal of achieving machine intelligence. The path a development follows varies according to the technical results achieved and the evolving environment in which the technology must operate. Some developments are path-dependent, meaning that the outcome is determined, at least in part, by the trajectory of the development.[25] In Thomas Hughes's terms, a technology builds momentum as it proceeds, developing cumulative inertia and generating its own infrastructure. At some point it reaches what social constructivists call "closure"; the market selects one technological option at the expense of others. In computers, for example, the world chose digital processing and von Neumann architecture in the 1950s; thereafter analog machines and neural architecture had to overcome the weight of historical experience to challenge the resultant paradigm.

Most of the economic studies have focused on companies operating in the marketplace. Furthermore, they address product development more than basic research.[26] This is not the environment in which DARPA specialized. Still, some of the findings of these studies map closely with DARPA experience and with literature from the history of technology. They will be introduced here where they have explanatory power.

Finally, the computer itself boasts a well-developed history.[27] Its literature situates the SC Program in a period of remarkable and dramatic growth of computer capability and use in the United States. Participants in SC are inclined to see themselves at the heart of that growth. Others are inclined to see them as peripheral or even not to see them at all. One purpose of this study is to place SC within this larger context of computer development and to measure its impact. It is hoped that this study will connect with all this literature.

Part I

1

Robert Kahn: Visionary

Robert Kahn was the godfather of Strategic Computing (SC).[1] He conceived the program as a source of patronage for his professional family. He developed a rationale that gave the program the political respectability it needed to sell in Washington. He promoted it quietly and effectively behind the scenes. And he handed it over to his lieutenants when it was up and running.

Kahn's epithet was inspired by Mario Puzo's best-selling novel *The Godfather*. Within his community, he enjoyed the paternalistic authority of a mafia don. One colleague observed, only half facetiously, that when Kahn arrived at a DARPA research site, the contractors knelt and kissed his ring.[2] But Kahn presided over a family of a different sort. Its "cosa" is computers. In these machines, family members have a faith and an enthusiasm bordering on the religious. Computers, they believe, will define the future: who controls them, controls the world.[3] "The nation that dominates this information processing field," says Kahn, "will possess the keys to world leadership in the twenty-first century." He and his colleagues intend it to be the United States. Unlike the mafia described in Puzo's novel, they play by the rules, but they are nonetheless zealous in pursuit of their goals. About them hangs the air of true believers.[4]

Kahn presided over this family from the conception of the Strategic Computing Initiative (SCI) in 1981 through its baptism in 1983. He was then director of the Information Processing Techniques Office (IPTO) of the Defense Advanced Research Projects Agency (DARPA).[5] Kahn did not invent SC de novo; the idea was in the air.[6] But Kahn was the first to articulate a vision of what SC might be. He launched the project and shaped its early years. SC went on to have a life of its own, run by other people, but it never lost the imprint of Kahn.

To understand SC, then, it is necessary to understand Robert Kahn. That is no easy task, for part of his power over people and events derives from his reserved, understated, avuncular persona. Forty-four years of age when SCI began in 1983, of average height and comfortable girth, he favors the gray suits, white shirts, and muted ties that efface personality. His quiet demeanor and impassive face—square but comfortably rounded at the corners—suggest nothing of the burning enthusiasm that drives him. The only hint one gets of his galloping metabolism is the quiet intensity and relish with which he takes on food, apparently storing calories for the task ahead. Kahn is in it for the long haul, committed to selling the country on the power and potential of computers and ensuring that the United States has the best computer establishment in the world. As this book went to press, he was running the Corporation for National Research Initiatives, a private think tank outside Washington that quietly promotes his vision and agenda for the future.[7] To understand how and why he invented the Strategic Computing Initiative, however, it is necessary to go back to his professional roots and follow the path that led him to IPTO and DARPA.

Switching and Connecting

Kahn came to IPTO through switching and networking, through connecting electrical signals. He came not from computers per se but from signal processing of the kind that the military would use for command and control. He was part of what Paul Edwards has called "the closed world," a technical realm revolving about computer technology and shaped by military imperatives.[8] In this realm, government, industry, and academia intersect.

Kahn entered through academia. He earned a Ph.D. in electrical engineering at Princeton University in 1964,[9] specializing in applied mathematics and communications theory. His dissertation explored sampling and representation of signals and optimization of modulating systems. Before and during his studies at Princeton he worked as a member of the technical staff at Bell Laboratories at West Street (the old Bell Laboratories headquarters) in New York City and at nearby Murray Hill, New Jersey. Following graduation, he joined the electrical engineering faculty at MIT, where he worked in the Research Laboratory for Electronics. This fabled organization was an outgrowth of the MIT Radiation Laboratory, the center of World War II radar research.[10] It was paradigmatic of the growth of university research centers funded primarily by government defense agencies.

In 1966 Kahn took a year's leave from MIT to join Bolt, Beranek & Newman (BBN), a Cambridge consulting firm. The goal was to complement his mathematical and theoretical work with some practical experience in real-world problems. BBN, according to Kahn, felt like "a super hyped-up version" of Harvard and MIT combined, "except that you didn't have to teach."[11] It had made its reputation in acoustics, but it was just then reorganizing to move into computer research. Kahn quickly chose computer networks as his specialty.

The choice is hardly surprising. State-of-the-art computers were then being developed at MIT, as was the first major network to link these machines on a national scale. In the early 1950s, as the United States began to worry about the threat of Soviet bombers carrying nuclear weapons, MIT had initiated Project Charles, with support from the United States Air Force. The goal was nothing less than a network of digital computers that would link air defense radars with a central command structure to orchestrate fighter defense. The project, ultimately renamed Semi-Automatic Ground Environment (SAGE) soon moved off campus and became the core of MIT's Lincoln Laboratory in Lexington, MA, outside Boston. Lincoln would come to be seen by two journalists as the "perfect incubator for the kind of genius ARPA [Advanced Research Projects Agency] would need to push computing into the interactive, and interconnected, age."[12] It absorbed MIT's Whirlwind computer project, whose delays and cost escalations had threatened its continuance.[13] The addition of magnetic core memory salvaged Whirlwind and made it an appropriate instrument for the grand designs of SAGE, which was nothing less than an integrated, centralized command and control network.[14]

One of the first experiments in that grand design was the Cape Cod Network. It went on line on October 1, 1953, to test the ability of the revived Whirlwind to receive data from a radar network on the Massachusetts peninsula and convert it in real time into directions for deploying fighter aircraft. The success of the Cape Cod Network spurred commitment to a national system. To manufacture Whirlwind II, MIT turned to International Business Machines (IBM), which was just then following the advice of its new president, Tom Watson, Jr., son of the corporation's founder and chairman. The young Watson had spearheaded the move to take the company into the comparatively new field of digital computers.[15] SAGE offered an ideal opportunity to have the government underwrite the necessary research and development.

In what journalist Robert Buderi calls "a memorable clash of cultures,"[16] MIT and IBM combined to turn the Whirlwind into the FSQ-7 and FSQ-8, monstrous machines of unprecedented computing power. When the first of these went on line July 1, 1958, it established a benchmark in computer development, networking, and government-industry-university cooperation. Among other things, says Buderi, it consumed "the greatest military research and development outlay since the Manhattan Project" of World War II.[17]

In a pattern that foreshadowed the age of the mainframe then dawning, each pair of FSQ-7s filled an entire floor of the typical direction centers in which they were installed; supporting equipment filled three more floors. The pair of machines weighed 275 tons and consumed 1,042 miles of wiring, two million memory cores, 50,000 vacuum tubes, 600,000 resistors, and 170,000 diodes. They drew enough power to serve a town of 15,000. In all, twenty-seven pairs of FSQ-7s and FSQ-8s would go into service before Robert Kahn joined MIT; six remained in operation when Kahn created the SCI in 1983. Along the way, the SAGE program pioneered time-sharing, computer networking, modems, light guns, and computer graphics. It also allowed IBM to spin off its hard-won expertise into the 700-series computers that would inaugurate its dominance of the commercial computer manufacturing market.[18] It is small wonder that Kahn found himself drawn into the dawning world of computer networking that was opening up around MIT in the middle of the 1960s. If he was the godfather of SC, MIT was Sicily.

For all its technical innovation, the SAGE program had not solved all the problems associated with sending computer data over great distances. For one thing, it was not an interactive system between terminals; data went from the radars to the command center and from the command center to the air bases. Furthermore, it used dedicated land lines; no other traffic passed through the wires on the system. The full potential of interactive computing would be achieved only when obstacles such as these were overcome.

One institution interested in that challenge was ARPA's Information Processing Techniques Office. If it could connect the seventeen different computer research centers it then had under contract, resources and data could be shared instead of duplicated.[19] To pursue that goal, IPTO recruited Larry Roberts, who had first encountered computers as a graduate student at MIT. While working on his dissertation, he was a research assistant at the Research Laboratory for Electronics, where Kahn had spent two years, and then a staff associate at Lincoln Laboratory. His first contact with ARPA came when he sought support for a computer graphics program at Lincoln Laboratory. But his later research interest was sparked by a November 1964 conference at Homestead, Virginia, addressing the future of computing. Roberts came away committed to networking. With ARPA funding, he collaborated with a colleague at Lincoln Laboratory to design an experimental network linking computers in Boston and California. This project, however, produced little more than a published report.[20]

Like Kahn and others who would follow him, Roberts felt stifled by academic research and longed to put his knowledge into practice in the "real world." "I was coming to the point of view," he later recalled

that this research that we [the computing group at Lincoln] were doing was not getting to the rest of the world; that no matter what we did we could not get it known. It was part of that concept of building the network: how do we build a network to get it distributed so that people could use whatever was done? I was feeling that we were now probably twenty years ahead of what anybody was going to use and still there was no path for them to pick up. . . . So I really was feeling a pull towards getting more into the real world, rather than remain in that sort of ivory tower.[21]

In short, he wanted to connect his research to real applications. With this frustration in mind, and at the urging of both ARPA and Lincoln Laboratory, Roberts had gone to Washington in 1966 to advance computer networking. His contribution led on to the ARPA network (ARPANET) and ultimately the Internet. It also led to Robert Kahn's joining ARPA.

Soon after his arrival, Roberts presented his plans for a computer network at a ARPA PI [principal investigator] meeting. He encountered a cool reception. ARPA contractors feared that such a network would put demands on their time and their limited computer resources. PI Wesley Clark suggested to Roberts that some of those concerns could be addressed by layering a network of nodes—"interface message processors" they came to be called—on top of connected computers. In this way, message processing burdens would fall on the node layer, not on working computers. This neat solution allowed Roberts to cobble together an "ARPA network circle."[22] In October 1967 Roberts reported his plans at the Operating Systems Symposium of the Association for Computing Machinery. In response, he learned of the work of Donald Davies of Britain's National Physical Laboratory (with which he already had some acquaintance) and that of Paul Baran (of the RAND Corporation), which was less familiar to him. The two men had independently developed what Davies called packet switching, an alternative to the U.S. telephone system's message switching.

One way to send digital data streams is to load up a complete message and ship it in a single mass down a single communication track, much the way a train might carry freight. This has the advantage of bulk carriage, but it puts all of one's eggs in a single basket. Any delay or disruption along the line and the whole message will be held up. An alternative scheme is to break the data cargo into small batches and ship these separately, by different routes. This might be likened to a cargo divided among trucks, which take different highways to their destination. The cargo can then be reassembled at the terminal. Each truck can find its way by the most open available route, taking detours if necessary to avoid bottlenecks and outages.

Packet switching, as this technique came to be called in the field of communications, was the subject Kahn chose to explore at BBN.[23] Simple in theory, packet switching is complicated in practice. Messages must be broken into packets and tagged front and back with markers indicating where they are going and how they are to be reassembled once they get there. The markers also provide a means to test if the packets got there correctly. Protocols are needed at sending and receiving stations to ensure they speak the same language. Routing procedures must allow packets to move through nodes expeditiously, making the best choice at each node. And switches must be available at each node to handle streams of packets passing through from diverse sources to diverse destinations. Networks and gates, connecting and switching, were therefore indivisible parts of the same problem. To take up one is to inherit the other.

In his new position at IPTO, Roberts actively expanded upon what he already knew about the work of Baran and Davies. From Baran, he learned of layering, a technique that separated the component levels of a network to facilitate communications between levels. At the heart of the network was a communication subnet composed of switching nodes that came to be called interface message processors (IMPs). These routed packets around the network. A second layer was composed of hosts, server computers that packaged messages for delivery into the network and unpacked messages received from the network. Individual computer terminals, the end points of the system, dealt only with their host server. The subnet, where messages were being routed, was invisible to the terminal.

From these discussions and further work with his ARPA colleagues, Roberts developed a working concept of how a network might function. And he had a faith that such a network could benefit ARPA researchers. Instead of ARPA supplying expensive computing equipment, including mainframes, to all its sites, it would link those sites via networks to remote computers. They could use those assets by time-sharing, another advance recently promoted by ARPA. Instead of duplicating resources, he would connect them.

In 1967 Roberts set up his "ARPA network circle," advised by Paul Baran, to formulate specifications for the network. He targeted UCLA, the Stanford Research Institute, the University of California at Santa Barbara, and the University of Utah as the sites for the first IMPs of a prototype network. For overall system design, Roberts elected to award a contract through competitive bidding. Instead of simply contacting appropriate researchers in the field and enlisting them to work on some part of the program, Roberts advertised by a request for quotation (RFQ) for a contractor to integrate the whole network.

Coincidentally, Kahn contacted Roberts at about this time to tell him of his work on computer networks at BBN. Roberts encouraged Kahn and BBN to submit a proposal. With his academic and commercial background, his theoretical and practical experience, and his knowledge of communications networks, Kahn was ideally prepared for such an undertaking. What is more, he liked the challenge. BBN was also fortunate to have on its staff Frank Heart, who had recently joined the company from Lincoln Laboratory. Having worked on Whirlwind and SAGE, Heart was experienced in the development of complex, real-time systems using computers and digital communications. Under his leadership, Kahn and Severo Ornstein prepared the technical parts of the BBN proposal that won the prime contract for what would become ARPANET.[24]

Kahn and Heart also led the BBN team that laid out the conceptual design of the ARPANET. For Kahn this was "the epitome of practical experience," just what he had joined BBN to get.[25] When the job was done, he considered returning to MIT, his excursion in the real world successfully completed. But the problems of reducing his concept to practice soon captured his attention. "Most of the really interesting issues to me about networks," he later recalled, "were how they were going to work . . . : how did routing algorithms work, how did congestion control algorithms work, how did flow control work, how did buffering work."[26] To answer these questions he stayed on at BBN, working in the engineering group to solve the myriad problems that surrounded such a complicated prototype.

By 1972 the ARPANET was up and running as a computer network. The first four IMPs were connected in 1969 and linked to ARPA sites on the East Coast in 1970. The following year, fifteen ARPA sites across the country and a handful of non-ARPA hosts were also on line. Users complained of IMP failures and phone-line outages, but they kept on using the new network. One by one the bugs were flushed from the system and communications became more reliable, but the network still did not inspire great interest or enthusiasm within the computer research community. Only when Kahn planned and staged a demonstration at the first International Conference on Computer Communications in October 1972 did the initial take-off begin. Even then no one realized that the real driver of the ARPANET would be e-mail (electronic mail), not the resource sharing Roberts had envisioned.[27] But so it is with new technologies; they often find uses their inventors never anticipated.

Earlier that year, while Kahn was preparing the demonstration that launched ARPANET, he agreed to move to ARPA. Ironically, he and Roberts changed places. Roberts, who had succeeded Robert Taylor as IPTO director in 1969, recruited Kahn, not to work on networking but to take up a new ARPA program in manufacturing technology. Before Kahn left BBN he worked with Steve Levy, setting up Telenet, a "non-sanctioned spin off" company design to commercialize computer networking.[28] Turning down an offer to help run Telenet to go to ARPA, Kahn suggested that Roberts would be a good candidate to recruit. Kahn arrived at ARPA in November 1972; Roberts left in September 1973 to head Telenet.

This circulation of key personnel illuminates not just SC but the entire field of information processing technology. Key people moved freely among positions in industry, government, and academia. They provided a critical vector for the transfer of technology, transplanting both ideas and cultural values among the positions they held. The frequency of movement eroded institutional loyalty. They were part of a brotherhood of information technology (IT) professionals, committed to advancing their field, comparable to what anthropologist Anthony Wallace has called the nineteenth-century "international fraternity of mechanicians."[29] A given job might be at a work station, one of many sites on the great shop floor of IT. They had worked for other divisions and at other stations; they had friends working there now. They might compete with those friends, but they could just as easily trade places with them. In such an environment, ideas moved quickly and values converged. Most importantly, the disparate parts of an enormous technological system connected with one another and with their environment. They formed what French investigators Simon Nora and Alain Minc called the "thick fabric" of American computer development.[30]

The Information Processing Techniques Office (IPTO)

IPTO was created in 1962 to accelerate computer development. The first large electronic calculating machines in the United States had been developed in World War II to compute ballistic trajectories. The same machines were then turned to complex scientific calculations, such as those required for nuclear weapons development.[31] In the 1950s, the major focus turned to command and control, a reflection of the growing speed and complexity of modern war. The air force was the primary patron of computer development in this decade, complemented by ARPA after its creation in 1958.[32]

In the 1960s and 1970s IPTO had been led by a succession of visionary and powerful leaders. J. C. R. Licklider set the tone as the first IPTO director from 1962 to 1964; he returned to succeed Larry Roberts for a second tour from 1974 to 1975. Like Kahn he was an academic whose appreciation for practical applications led him out of the laboratory and into the corporate world. Trained in experimental psychology at the University of Rochester in the 1930s, Licklider spent World War II at Harvard, studying man-machine interaction, and patenting a fundamental new concept that facilitated radio communication with aircraft.[33] No militarist himself, he nonetheless appreciated the military need for advanced technology and he continued to work on military projects after the war. He joined the MIT Department of Acoustics in the late 1940s and transferred to Electrical Engineering in the 1950s. As a member of the Research Laboratory for Electronics he participated in Project Charles, the air defense study that led to the creation of Lincoln Laboratory and SAGE. In 1957 he became a vice-president of BBN.[34] From there he went on to ARPA, completing a career trajectory that Kahn would mimic a decade later.

Licklider brought to ARPA strong views on the need for connected, interactive computing such as shared computer resources and real-time machine response to input from operators. His vision of man-machine interaction became one of the guiding conceptualizations of the 1960s and 1970s, and his missionary zeal for the work made him a spectacularly effective fund raiser. In an influential 1960 paper entitled "Man-Computer Symbiosis," Licklider wrote:

The hope is that, in not too many years, human brains and computing machines will be coupled together very tightly, and that the resulting partnership will think as no human brain has ever thought and process data in a way not approached by the information handling machines we know today. . . . Those years should be intellectually the most creative and exciting in the history of mankind.[35]

When ARPA needed to move into computer research, Licklider was the ideal recruit to run the program. He had stature and credibility in his field. He had the ardor of a true believer. He had ideas and an agenda. He had experience in the laboratory and in the field. He believed in bending the power of science and technology to the purposes of national defense. And he thought that ARPA would be a bully platform. He went to ARPA to do good, to leverage his field with government money, and to make a difference.

ARPA's institutional culture proved fertile ground for his ambitions. The agency, or at least some parts of it, was a model for institutional support of innovation. Licklider drew upon that culture and transformed it in significant ways. He pioneered what came to be seen as the IPTO style.[36]

First, as with all of ARPA, IPTO restricted itself to a lean administrative structure. The staff consisted of an office director, a few program managers, an administrative manager, and one or two secretaries. The contracting functions that were the main business of the agency and that mire other government agencies in debilitating paperwork were themselves contracted out to private firms in the Washington area or to agents elsewhere in the DoD bureaucracy. Second, IPTO managers maintained close personal contacts with their colleagues in the computer research community, providing an informal circle that was commonly known among computer researchers as the "ARPA community." IPTO directors and some of the program managers came to ARPA from that community and usually returned to it after a few years. They prided themselves on knowing what was possible and who was capable of doing it. ARPA program managers like to repeat the quip that they are seventy-five entrepreneurs held together by a common travel agent.

Third, they often turned around proposals in weeks, sometimes days, as opposed to the months and even years that other agencies might take to launch a research undertaking. Fourth, they were willing to take risks. Indeed, risk was their niche. Other branches of DoD and the rest of the federal government were prepared to support the development work that led from good ideas to operational equipment; IPTO was to find the good ideas and distill from those the ones that might be made to work. Most failed; the small percentage that paid off made up for all the rest. Fifth, IPTO created eight "centers of excellence," university computer research organizations that had first-class researchers in sufficient numbers to create a critical mass, both for pioneering new developments and training the next generation of researchers. At least three of these centers were funded generously with sustaining contracts at three-year intervals.

Finally, IPTO subjected itself to self-scrutiny. Its program managers competed among themselves for resources and constantly tested the projects they were funding against others that they might fund. By staying in touch with the community, they collected feedback about the wisdom of their investments. By their own accounts, they were ruthless in cutting their losses and moving their funding from deadwood to live wires.

Soon IPTO became a main source of funding for computer research in the United States, a distinction it held through most of the 1960s and 1970s. J. C. R. Licklider revolutionized computing by aggressively supporting his twin hobby horses of time-sharing and man-machine symbiosis, and by proselytizing for his vision of an "intergalactic computer network."[37] He recruited Ivan Sutherland to do the same with graphics. Strong successors such as Taylor and Roberts ensured that artificial intelligence (AI) received the support it needed well into the 1970s. The payoff from AI was slower to come than in time-sharing, graphics, and networking, but IPTO kept faith with the field nonetheless.

In retrospect, by the standards of the 1990s, the pattern of IPTO funding in the 1960s and 1970s appears controversial. Most of its money went to a few institutions, including MIT, Carnegie Mellon, and Stanford. Some other schools, such as the University of Illinois, the University of Utah, and Columbia University, were also supported as centers of excellence, but they soon fell by the wayside. The goal was clearly to support the best people and their graduate students in a context promising some positive reinforcement among the different programs. Few strings attached to this funding; there was virtually no accountability.[38] Allen Newell, the AI legend and computer researcher at Carnegie Mellon University, said that there was no accounting for ARPA money in the 1960s; ARPA asked no questions:

They [ARPA] didn't have any control over the money they'd given us; that was our money now to go do what we wanted with. We didn't tell people in ARPA in the '60s what we were going to do with their money. We told them in sort of general terms. Once we got the money, we did what we thought was right with it.[39]

Furthermore, the support was generous by any standards. Project MAC at MIT was named for both "Machine-Aided Cognition" (i.e., artificial intelligence) and "Multiple Access Computer" (i.e., time sharing), but wags joked that it really meant "More Assets for Cambridge."[40]

This pattern gives cause for concern.[41] First, it concentrates the resources of the federal government in a few research centers. Those centers are likely to boast the best people and equipment, and ARPA money is likely to attract still more of both. The best students and the best researchers will want to work in that environment, and good people tend to attract even more money. These tendencies were even stronger than normal in the computer field of the 1960s and early 1970s, when the great cost and imposing bulk of powerful mainframe computers meant that computing power simply had to be highly centralized.

Critics maintain that such a pattern is unhealthy on several counts. First, it is elitist in its distribution of a democracy's money.[42] The best starve out the good by drawing the lion's share of resources. Centers that aspire to and might achieve excellence never get the chance, and the many institutions across the country where the great bulk of tomorrow's citizens are educated can offer only second-rate staff and equipment. Second, the diversity that is at the heart of innovation and creativity is starved when a handful of institutions, each with its peculiar style and agenda, pipes the tune for the whole national research community. *Mentalités* can take hold at such institutions and stifle new ideas not in keeping with the conventional wisdom.[43] Access to journals, professional meetings, fellowships, and the other levers of power in the research community is likely to be granted to those who subscribe to the views held at the centers of excellence, further endangering the new ideas that might spring up around a more diversified and heterogeneous national environment.

A second major area of controversy attached to the early years of IPTO was a potential or apparent conflict of interest. First, IPTO was inbred in a way that further stifles diversity. Rather than conducting national searches to identify the most qualified candidates, IPTO most often filled its vacancies by unilateral invitation to members of what has come to be called the "ARPA community," the way Roberts recruited Kahn. This community consists of those researchers who regularly get ARPA contracts and sit on ARPA advisory panels. These, it may be argued, make up the galaxy from which ARPA must naturally pick, but a counter argument can be made. By choosing only among those who share its view of the field, IPTO ensured that different views of the field would never get a hearing. Just as inbreeding narrows the gene pool, so too can this kind of recruitment narrow the range of research activity that IPTO will have at its disposal. Other research trajectories can enter the IPTO domain, but they are likely to come over the transom rather than through the door.[44]

Furthermore, IPTO directors and program managers maintained relationships with the institutions and people they funded, relationships that would raise eyebrows in later decades. Kahn going to ARPA on Roberts' invitation while Roberts went to Telenet on Kahn's recommendation is just one example. At ARPA, Licklider worked with his former MIT colleague Robert Fano in developing the proposal for Project MAC. Licklider then apparently reviewed the proposal by himself, without outside advice, and submitted it for approval to ARPA Director Jack Ruina. When Licklider left ARPA in 1964, he went first to IBM and then back to MIT to direct Project MAC.

So too with Licklider's recruit, Ivan Sutherland, who promoted computer graphics in his years at IPTO, leveraging his own pioneering research in a way that permanently transformed information processing. Among the centers of excellence that he supported was a start-up program at the University of Utah. When Sutherland left IPTO, he did not return to MIT but went instead to Harvard and then to Utah. There he joined with other colleagues to set up a private company to commercialize their research advances.

The appearance of, or at least the potential for, conflict of interest pervaded the early years of IPTO. Surprisingly, however, little criticism along these lines has ever surfaced. For one thing, the rules appropriate and necessary for a mature field do not always apply in the embryonic years when there are a few people struggling to create a niche. Furthermore, the contributions of Licklider and Sutherland were so profound, so fundamental, that few in the field appear inclined to quibble about their methods. Nor have they been charged with feathering their own nests, for most recognize that they could have acquired far more wealth than they did had they chosen to maximize the commercial potential of their contributions. Like others who were drawn to ARPA before and since, they appear to have been motivated primarily by an interest in advancing the field. The income they sought was far more psychic than monetary. They did not even seek power in the traditional sense of influence over people; rather they sought the larger and more intoxicating power of shaping the future.

Nor have there been significant charges of favoritism, cronyism, or impropriety. Even computer researchers who are not in the ARPA community are inclined to allow that ARPA has had a reasonable agenda, that its methods are by and large effective, and that its results have been impressive and at least commensurate with the funds invested.[45] Especially in a field and an era where researchers in university laboratories were regularly launching commercial ventures to capitalize on their institutionally funded expertise and knowledge, the directors and program managers who passed through IPTO in the 1960s and 1970s appeared upstanding, even selfless citizens who gave more than they got. Kahn was in that tradition.

AI Autumn

Kahn's arrival at IPTO in 1972 began his education in the ways of Washington. Almost immediately, Congress eliminated the program he had been recruited to direct, impressing upon him the distinction between good ideas and viable ideas. In Washington the politics of technology was as important as its substance, a lesson that would be reinforced in the struggle for SC. Electing to stay on at DARPA, Kahn returned to computer networking to exploit the possibilities being opened up by ARPANET. Among these was the challenge of extending network capabilities to radio transmission. Was it possible to link computers that were not connected by wires?

This is an instance in which military needs helped to drive the IPTO agenda. The ARPANET had been developed on its own merits, without direct military support, let alone demand. Roberts and Taylor had seen the potential of networking to multiply the power of computers. They had pursued the goal because it made technological sense. Only when it began to pay off and military applications began to appear did the military services enter the picture. In time, the Defense Communication Agency took over responsibility for the operation of ARPANET and freed up IPTO time and resources to move on to other projects.[46] DARPA, however, retained policy control over the ARPANET.

The dependence on wires to connect the ARPANET nodes posed particularly troublesome problems for the military. How could military forces on the move remain in contact with their mainframe computers? How could a commander relocate his headquarters without going off line? How could ships at sea and planes in the air take advantage of the power of large mainframes ill-suited to such environments? For these reasons and more, IPTO began thinking in the early 1970s about networking without wires. Kahn became involved in this work and thus began a journey that would lead to the Internet.

Some related work was already under way. In Hawaii, for example, ARPA had been supporting a research effort, "the Aloha System," to send packets over radio channels. Succeeding in that effort, ARPA extended the research to send packets up to communications satellites and back. Most of the principles were the same as those in packet switching over wires. Some interesting technical complications arose, but in Kahn's view these were not as troublesome as the political and administrative difficulties in sending messages between different, wired networks. He contributed, for example, to the development of the ARPANET Satellite Network, which used the Intelsat IV satellite to connect the ARPANET to several European countries. No economic tariffs existed in international agreements to support such communication, and Kahn found himself before the international board of governors of Intelsat, working out the details.[47]

Communicating between networks did, however, raise a new set of technical problems that wanted addressing. For example, ARPANET was designed to connect computers. The packets that flowed over the network bore tags identifying the source computer and the destination computer. IMPs were programmed to recognize these addresses and send the packets on their way. If a network was connected with other networks, however, then new and more sophisticated protocols were necessary. Packets would have to contain information about both the target network and the target computer on that network. IMPs would require algorithms to quickly process all that information. Protocols within participating networks would have to be compatible with the protocols on the networks to which they were connected. In short, connecting systems with systems required a higher level of coordination than on ARPANET or on other networks then appearing.

To address this problem, Kahn joined forces with Vinton Cerf, a recent Ph.D. in computer science from UCLA. The two conceptualized and then documented the architecture and protocols for what became the Internet and TCP/IP. One element of the architecture was a "gateway" that connected different networks. The Internet Protocol (IP) set the rules by which packets would be transported by gateways from source to destination. The Transmission Control Protocol (TCP) is a byte-count-oriented protocol that uses checksums and end-to-end acknowledgments to ensure flow control and reliable transmission. Each packet in a message carries one number that is nothing more than a count of the total bits in the message and a marker that indicates the position of the packet in the overall bit string. The receiving host uses these numbers to determine if any bits were lost or corrupted in transmission. The Kahn-Cerf protocol proved so useful and robust that it was adopted as a standard by the military in 1980 and replaced the ARPANET host protocol in 1983.

Packet radio and TCP/IP were only part of Kahn's portfolio. In his early years at DARPA, he managed as many as twenty separate programs on various aspects of computer communication and networking. He invited Cerf to join DARPA in 1976. He watched Roberts depart and Licklider return. And he watched George Heilmeier take over as DARPA director in 1975.

Heilmeier's tenure profoundly altered DARPA culture and the IPTO style. It changed the institutional atmosphere in which SC was born. To the DARPA directorship Heilmeier brought a technical background in electrical engineering and a management technique he preached as a "catechism." Of any proposal he asked six questions:

- What are the limitations of current practice?
- What is the current state of technology?
- What is new about these ideas?
- What would be the measure of success?
- What are the milestones and the "mid-term" exams?
- How will I know you are making progress?[48]

Office directors and program managers who wanted Heilmeier's signature on their program authorizations had to be prepared to answer these questions.

By itself Heilmeier's catechism was not unreasonable, but he reportedly applied it with a draconian enthusiasm that reflected a chilling political climate settling over Washington in the first half of the 1970s. The Vietnam War dragged on to an unsatisfactory settlement in 1973 and the collapse of South Vietnam two years later. A whole generation of Americans turned sour on the military, believing that the armed services were at least partly responsible for the disaster. Watergate brought on the resignation of President Richard Nixon in 1974, a deepening cynicism in the public at large, a predatory press, and a Congress bent on cutting the imperial presidency down to size. Hearings by the Senate Intelligence Oversight Committee under the chairmanship of Senator Frank Church revealed a politicization of the intelligence community and deepened the cynicism already abroad in the land. Jimmy Carter won the presidency in 1976 in part by running against the "bloated bureaucracy" of the federal government that was widely perceived as having failed its citizens.[49]

This troubled political environment visited ARPA in the form of the Mansfield amendment. Senator Mike Mansfield of Montana succeeded in attaching to the Defense Appropriation Act of 1970 a provision that DoD conduct no basic research.[50] His motivation was the widely held perception that the DoD was out of control, as evidenced by its failed yet continuing policies in Vietnam. It might perform more satisfactorily, Mansfield believed, if it concentrated on its primary business. Other agencies of government, especially the National Science Foundation, were specifically designed to support basic research. DoD should support no work that did not lead directly to military applications. In this way, its budget might be better controlled by Congress and its power limited to those areas for which it was intended.

The Mansfield amendment was repealed the following year, but its impact lingered. The DoD adjusted its own policies to conform to the law, and it left the adjustments in place. Congress had given clear notice that it wanted to see direct military payoff for the research-and-development dollars spent by DoD. The same enthusiasm led Congress in 1972 to change ARPA's name to DARPA, the Defense Advanced Research Projects Agency—a reminder that DARPA was not at liberty to support any technology it might find interesting. Rather it was to support only those projects with clear and present application to the military mission of the armed forces.

This changing atmosphere carried enormous portent for computer research at DARPA. At the bureaucratic level, it greatly constrained the easy autonomy with which research programs at centers of excellence had been approved and funded in the 1960s. As Program Manager Larry Roberts later recalled:

The Mansfield Amendment . . . forced us to generate considerable paperwork and to have to defend things on a different basis. It made us have more development work compared to the research work in order to get a mix such that we could defend it. . . . The formal submissions to Congress for AI were written so that the possible impact was emphasized, not the theoretical considerations.[51]

This trend toward greater accountability and specificity would also mark the public attention drawn to DARPA's programs during the SCI.

The Mansfield Amendment failed, however, to shake the fundamental faith of many IPTO leaders that technologies benefitting the country at large contributed to its security as well. Work might originate in clear military projects such as SAGE and Project Charles, but the technology could take on a life of its own. Licklider, for example, developed an almost messianic faith in the potential of man-machine symbiosis. His zeal transcended any specific military mission, even though he clearly believed that military applications would surely flow from this technological capability. IPTO directors and program managers pursued computers and networks because they believed they were good for the country. It followed that they must be good for the military.

But Heilmeier saw it differently. He had entered government in 1970 as a White House fellow. Instead of returning to his company, RCA, in 1971, he moved to the DoD to be Assistant Director of Defense Research and Engineering (DDR&E) for electronics, physical sciences, and computer sciences. There he imbibed the hard ethos of late-Vietnam-era Washington. He brought that ethos to DARPA when he became director in 1975.

He chided DARPA with the title "NSF West." By this he meant that it was a mirror image on the west side of the Potomac River of the National Science Foundation (NSF) which was chartered by Congress in 1950 to conduct basic scientific research.[52] Its research agenda thus differed from those of the so-called "mission agencies" such as the National Aeronautics and Space Administration (NASA) and the Energy Research and Development Administration (ERDA), which supported research directly applicable to their responsibilities. DARPA, in Heilmeier's mind, was a mission agency. Its goal was to fund research that directly supported the mission of the DoD.

Heilmeier demanded practical applications in all DARPA programs. Those who could respond flourished under his directorship. Kahn, for one, had no difficulty justifying his programs to Heilmeier. The two had been graduate students together at Princeton, so they enjoyed a long-standing personal relationship. More importantly, Kahn's networking projects were demonstrably effective and they were having real and immediate impact on the military.

Artificial intelligence fared less well. Heilmeier "tangled" with Licklider and by his own account "sent the AI community into turmoil." A deputation of AI researchers visited him to assert, as he recalls it, that "your job is to get the money to the good people and don't ask any questions." This was the attitude cultivated at the centers of excellence in the 1960s. Heilmeier replied, "That's pure bullshit. That's irresponsible."[53] He was prepared to support AI, but it had to be able to justify its funding. Told that AI researchers were too busy to write proposals, he scoffed. Told that AI was too complicated to explain to non-experts, he studied the materials himself. Told that the director of DARPA need not read proposed ARPA Orders, the agency's basic contracting document, Heilmeier objected. He gave all ARPA orders a "wire brushing" to ensure they had concrete "deliverables" and "milestones." Often he came to the same conclusion: many AI programs were too unstructured, too unfocused to qualify for ARPA funding. Heilmeier believed that AI researchers wanted "a cashier's booth set up in the Pentagon—give us the money and trust us."[54]

Heilmeier replaced the booth with a turnstile. To pass through, researchers had to demonstrate the utility of their projects. He cut back funding for basic research in AI and drove the field toward practical applications. He forced researchers to get their projects out of the laboratory and into the marketplace. For example, he promoted the Advanced Command & Control Applications Testbed (ACCAT), a simulated Navy command center that was linked via network with other command centers and capable of modeling war games for the Pacific Fleet. He promoted the development of LISP (short for LISt Processing) machines, computers based on a powerful and flexible language written by AI pioneer John McCarthy. But he cut drastically the ARPA funding for speech research when it failed to achieve its benchmarks. He also drove J. C. R. Licklider from DARPA and made it almost impossible to find a successor.[55] Finally he promoted Army Colonel Dave Russell, a Ph.D. physicist who had been deputy director under Licklider. The AI research community resisted Russell, because he had neither the pedigree nor the experience of his predecessors, but they were powerless to block his appointment. In the end, they learned to work with him. Russell adapted to Heilmeier's goals and his style, and he found no contradictions in tying DARPA programs directly to DoD missions.

By all accounts Heilmeier's measures forced some positive reforms on AI research, but they were a bitter pill for the community. In the beginning of the 1970s, the distribution of IPTO funding was approximately 60 percent basic research and 40 percent applied research.[56] In the second half of the 1970s, during Heilmeier's tenure (1975–1977), these levels were essentially reversed. The FY 80 budget was 42 percent basic and 58 percent applied. In the vernacular of the federal government, funding for AI shifted from "research" to "exploratory development."[57] The DoD Research and Development Program occupies category six of the agency's budget. It is divided into subcategories:

6.1 Research

6.2 Exploratory Development

6.3 Advanced Development

6.4   Engineering Development

6.5   Management and Support

6.6   Emergency Fund

In the 1960s, all IPTO funding was 6.1. Attempts to "transition" some of these paper studies to applications brought an infusion of 6.2 funding in the early 1970s. George Heilmeier accelerated that trend, while holding the total IPTO budget stagnant during his tenure.[58] Though some recovery was made in 1978 and 1979, funding for basic research still trailed demand. Kahn succeeded Dave Russell as IPTO director in 1979 determined to shore up support for worthy projects.

An Agent of Restoration

Kahn's first goal as IPTO director was to grow 6.1 funding, both to increase DARPA's impact and to help reverse the funding cuts for basic research of the mid-1970s. These cuts had compounded a relative shrinking of government support for computer research since the 1950s. In that decade, approximately two-thirds of the national expenditure for computer-related research came from the federal government—both military and civilian branches. By 1965 the percentage was down to half, falling to one-quarter when Heilmeier took over DARPA, and 15 percent when Kahn took over IPTO. What is more, the government share of all computers in use in the United States had fallen even more precipitously, from more than 50 percent in the 1950s to 20 percent in 1960, 10 percent in 1965, and 5 percent in 1975.[59] This relative decline was driven in part by the growth of the U.S. commercial computer industry, which had its own research establishment. But the government funded a different kind of research, one on which the long-term health of the field depended.[60]

Thus it was that Kahn made growth of the IPTO budget a precondition of accepting the directorship of the office. Robert Fossum, the DARPA director who had succeeded George Heilmeier in 1977, readily agreed.[61] In congressional testimony in 1981, Fossum reported that DARPA funding in constant dollars had fallen steadily from 1970 to 1976; it had not yet recovered its 1970 level when Kahn became IPTO director in 1979.[62] The post-Vietnam heat of the mid-1970s was abating, Fossum did not share Heilmeier's catechism or his pessimism about artificial intelligence, and Fossum wanted Kahn to take the job. By 1982 the IPTO budget had climbed from its 1979 low back up to a respectable $82 million.[63]

Kahn had three reasons for wanting to increase the IPTO budget. First, he had the faith of a true believer in the importance of computers and communications. Though he had entered the field of computer research via communications, he had found himself drawn increasingly into the enthusiasm of his predecessors: Licklider, Sutherland, Taylor, and Roberts. Computers, they all believed, had the potential to revolutionize society. Advances in the field were beginning to follow one another with remarkable speed, and the country that stayed out in front of this wave would reap enormous economic and military benefits. To fall behind would threaten the very health and safety of the nation.

American competitiveness in this rapidly evolving field depended on a robust research community. It needed adequate resources and a steady stream of funded graduate students, the researchers of tomorrow. In short, the best people had to be drawn to this field and they had to be encouraged and funded to do the best research. DARPA-IPTO had been the principal agent of that support since the early 1960s; it seemed to be the logical locus for continuing support in the future.

Kahn's second reason for seeking increased funding for IPTO was the belief, widely shared in the computer community, that several new technologies were poised on the verge of major advances.[64] The most important of these were in the realms of microelectronics, multiprocessor computer architecture, and artificial intelligence. One produced switches, one arranged switches in space, and one sequenced the switches. All connected.

Microelectronics seemed particularly ripe. Advances in this area provide infrastructural support for all computer development, for they make all computer equipment faster, cheaper, and more reliable. In the late 1970s, gallium arsenide (GaAs) promised to improve upon silicon as the semiconductor in computer chips. While GaAs was more expensive and more difficult to work with, it promised speeds up to six times faster than silicon and it offered greater resistance to high levels of electromagnetic interference. The latter was especially important for military applications, in which components might have to survive nuclear attack.

In the late 1970s the first VLSI microprocessors began to appear. (VLSI stands for very large scale integration of the circuitry on a chip.) Microprocessors are individual semiconductor chips containing thousands of transistors. Techniques then available were taking that count into the five-digit range, from 10,000 to 100,000 transistors per chip.

This posed enormous research challenges in developing designs to exploit this manufacturing potential.

The Josephson Junction (JJ) also promised potential improvements in high-performance computers. This tunnel junction could form the basis of a switching device with transition times from one state to another, from on to off, of picoseconds (trillionths of a second) and power dissipations of less than two microwatts. In integrated circuits then running hundreds of thousands of instructions per second, this promised significant improvements in performance, though its need to be supercooled limited its foreseeable applications for the military.

Computer architecture also seemed to be embarking on a promising new line of research. Analysis of existing supercomputers suggested that a technological ceiling hung over their performance. Cray machines, then the most powerful in the world, achieved their high speeds by clever refinements of von Neumann architecture. John von Neumann, a Princeton mathematician and computer theorist, had participated in the 1940s and 1950s in redirecting American computer research from electromechanical analog machines to electronic digital machines. In recommending the step, he noted that signals in an electronic machine moved so fast that for all practical purposes one could ignore the time it took a signal to go from one part of the computer to another. He therefore suggested an architecture in which steps in problem solution followed one another serially. An instruction from the central processing unit (CPU) would travel to the memory, where it would prompt another signal to return the appropriate data from the memory to the CPU. Once returned, the data would be operated on by the CPU and the new piece of resulting data would then be returned to memory. Then the process would begin again.[65]

In von Neumann's day this reasoning was valid. For example, the ENIAC computer on which von Neumann worked in the late 1940s, had a clock speed of 100 Khz and contained only twenty "accumulators." Clock speed refers to how often all the switches in the machine change state or process a signal. 100 Khz meant that all the gates in the machine could operate on a signal 100,000 times per second. Accumulators were gate arrays capable of storing, adding, or subtracting a ten-digit number.

By the late 1970s, however, the problem had changed. The Cray-1, for example, a supercomputer first introduced in 1976, had cycle times of 112.5 nanoseconds (billionths of a second) and theoretical peak performance of 160 million floating point operations per second (FLOPS). Wires carrying tens of millions of signals per second must be short; therefore switches must be as close together as possible. The long, leisurely trip back and forth between the CPU switches and memory switches was a luxury the supercomputers could no longer afford. And the switches themselves had to change state as quickly as possible with the least consumption of energy. Energy consumption became a design limitation and energy transfer in the form of heat build-up became a dilemma, what historian Edward Constant would call a presumptive anomaly.[66] Opening space within the machine for coolant only separated the components still further from one another, increasing the time and energy necessary to move signals from switch to switch. A different design paradigm, a different research trajectory, would be required to lift the capability of these machines beyond a certain predictable level.

The most promising alternative appeared to be parallel processing. The principle is simple. Instead of a single, massive processor, through which all data must be routed serially, why not have multiple processors? Problems could then be broken into parts, each part assigned to a different processor to be run concurrently. In addition, each processor could be associated, logically and spatially, with the memory it would need to work its part of the problem, further reducing the travel time and energy that signals would require. In the end, the data would be gathered from the various processors and integrated. The limits on such machines would be established not by the total distances that signals moved and their net speed, but by the numbers of processors and the creativity with which designers positioned memory and created software programs.

Such a machine already existed. The ILLIAC IV, designed and built at the University of Illinois with IPTO support, had been completed in 1972. Consisting of four quadrants of sixty-four processors each, the ILLIAC IV was the world's fastest computer until it was shut down in 1981. But it had been enormously expensive to build and debug, and it had proven equally difficult to program—harbinger of a characteristic that continues to haunt parallel machines. What was needed in the late 1970s, as the ILLIAC IV played out its contribution, was new research to exploit the potential of parallel processing. Ideally, the new experiments should be scalable, that is, the results on a modest prototype of say 64 processors should be readily expansible to a massively parallel machine of say 64,000 processors.

In addition, some other promising ideas for non-von Neumann architectures then in the air wanted development and testing. Data flow architectures, for example, had been significantly improved in the late 1970s.[67] These ideas were being taken up by the Japanese without having been tried out in practice in the United States. As with VLSI, there was an urgent need to build prototypes of the new machines, get them in the hands of users, and begin the ensuing software development that would make them truly useful.

Artificial intelligence was also seen by many as ripe for exploitation in the later 1970s, Heilmeier's machinations notwithstanding. This line of research promised to produce machines that would appear to have human intelligence. Many of the optimistic early predictions about AI had failed realization because computers simply lacked the necessary power. It has been estimated that the human brain is capable of 10^{18} arithmetic operations per second and has available 10^{16} or 10^{17} bits of memory.[68] Computer capacity at the end of the 1970s was on the order of 2×10^7 arithmetic operations per second and 1.3×10^9 bits of memory.[69] But promising advances in microelectronics and computer architecture meant that computer speeds and memory would likely be enhanced dramatically in the near future. This tantalizing prospect convinced many that AI would likewise come into its own. For some, it had already demonstrated its potential; it only needed to be transitioned from the laboratory to the marketplace.

These two factors, faith in the revolutionary potential of computers and a belief that many computer technologies were now ripe for exploitation, shaped Kahn's thinking when he took over IPTO in 1979. He believed that the computer research community had to "get its hands on VLSI to survive," and that it needed large funding for basic research and more modest support for applied research.[70]

In retrospect, it is remarkable that Kahn did not see, or did not fully appreciate, another technical development in the field that would turn out to be of comparable, if not greater, importance. In just the years when he was taking over IPTO and planning its future, the personal computer (PC) burst upon the commercial scene. The watershed event was the introduction of the Apple II computer in 1977. IBM secured the revolution four years later when it introduced its own version, the PC. Most computer professionals, including the DARPA community, were contemptuous of the personal computer, but the research and development funded by the competition in this new and unanticipated market would eventually dwarf the billion dollars invested in SC. More importantly, it would take the research trajectory in directions that Kahn and his associates never dreamed of.

Of those factors that did influence Kahn, a third appeared in 1981. In October of that year, the Japanese Ministry of International Trade and Industry (MITI) announced a ten-year program to develop a "Fifth Generation" of computers, machines capable of human intelligence. The first generation machines had been vacuum tube devices, of the kind that John von Neumann worked with in the period during and after World War II. With the invention of the transistor in 1948, a second generation of computers, extant between 1959 and 1964, was made possible. The electronic components in these machines—transistors, diodes, and so forth—were physically placed on printed circuit wiring boards and connected with copper wires. The third generation, such as the IBM 360 system of the 1960s, consisted of integrated circuits (ICs) in which the components were fabricated on single silicon wafers. Each of these ICs or chips was then connected by wires to the other chips to make a circuit of small scale integration (SSI). By the late 1970s the United States was well into very large scale integration (VLSI), in which an entire processor, less memory and peripherals, could be fabricated on a single chip. The Japanese were now promising to make a quantum leap to ultra large scale integration (ULSI) and produce machines powerful enough to achieve AI.

At the time the United States was clearly ahead of Japan and all other countries in virtually every important aspect of computer research and development. It had the largest and most sophisticated infrastructure, the widest scale of operations, the broadest commercial base, and the most advanced research establishment.[71] Still, Japan was not to be ignored. It had already challenged once-dominant U.S. industries in such critical areas as automobiles and commercial electronics. It was then threatening to do the same in chip manufacture and memory technology. Some comforted themselves that the Japanese advantage in these areas was based on production technology, not on the fundamental research necessary to lead the world in computer science. But who could be sure? Having educated the armed forces of the United States in the military importance of computers, DARPA could hardly ignore the possibility of losing its lead in this field to another country. Japanese expansion in chip manufacture and memory technology was already causing alarm in military circles.[72] It behooved IPTO to ensure that the U.S. grip on world leadership in computer technology was secure.

For all these reasons, Kahn wanted to step up IPTO activities in two realms. First, he wanted to continue the growth of the IPTO budget, to provide adequate support for the computer research community. Second, he wanted to transition research advances out of the university laboratory and into industry. The budget restorations he had negotiated with

DARPA Director Robert Fossum had gotten things moving, but they had not begun to realize the levels to which Kahn aspired. Ripe technology was a necessary, but not a sufficient, precondition of the Strategic Computing Initiative. The new levels that Kahn sought would be achieved only when a new government and a new DARPA director took office in 1981.

2

Robert Cooper: Salesman

The Pull of Robert Cooper

Robert Cooper, who became the tenth DARPA director in July 1981, was the perfect foil for Robert Kahn. Without him, Kahn would have been hard-pressed to sell the ambitious program that became SC. Cooper packaged and presented Kahn's plan to the outside world and sold it in the marketplace of Washington's budget competition. But just as the medium can be the message, so too can the package become the product. Cooper not only sold Kahn's plan, he transformed it. Former Program Manager Paul Losleben said there was a growing "competition between the two Bobs as to whose program it was."[1]

Robert Cooper was born in Kansas City, Missouri, 8 February 1932.[2] Growing up in Cedar Rapids, Iowa, during the Depression and World War II, Cooper developed the work ethic that would drive his entire career. He went to the University of Iowa in Iowa City on a baseball scholarship, but changed over to an ROTC scholarship to concentrate on academics and "to stay out of the Korean War."[3] After graduation and two years in the air force at Eglin Air Force Base in Florida, Cooper went to Ohio State University on a Westinghouse Fellowship. There he received a master's degree in biotechnology and worked on x-ray cinematography. On graduation, he literally flipped a coin to choose between engineering and medicine. The coin toss directed him to MIT, where he earned a Ph.D. in electrical engineering, studying plasma and wave propagation.

Graduating in 1963, he stayed at MIT, joining the faculty of the Department of Electrical Engineering and the staff of the Research Laboratory for Electronics at Lincoln Laboratory.[4] Alienated by the politicization of the campus in 1968, Cooper gave up teaching and moved full-time to Lincoln Lab. In 1972, he took leave from MIT to accept a three-year assignment as Assistant Director of Defense Research and Engineering (DDR&E), managing high-energy lasers and the military space program. In that capacity, he worked closely with George Low, Deputy Administrator of the National Aeronautics and Space Administration (NASA). As Cooper neared the end of his term as Assistant DDR&E, Low asked him to assume the directorship of NASA's Goddard Space Flight Center in Greenbelt, Maryland, just outside Washington. Though it entailed cutting his ties to MIT, Cooper accepted the invitation; he found the size and diversity of the space projects under way at Goddard challenging and unique. After the four-year tour that he had promised George Low, Cooper entered the private sector in 1979 to earn the money to put his two sons through college. As Vice President for Engineering of Satellite Business Systems, Cooper worked on commercial satellite communications, a field not far from the area that Kahn was directing at DARPA. In 1981, Cooper's friend Richard DeLauer, Under Secretary of Defense for Research in the new administration of Ronald Reagan, invited Cooper to become Director of DARPA. Judging this to be "the best engineering job in the world," Cooper agreed to a four-year term.[5]

Cooper's route to DARPA resembled that traversed by Robert Kahn. Both had passed through MIT at about the same time on their way to Washington. Both had experience in the "thick fabric" of academia, industry, and government. And both believed that DARPA could contribute to the economic and military security of the United States. In addition to the tacit knowledge that they absorbed and transmitted from venue to venue, they also became ambassadors of culture. They spoke the languages of academia, industry, and government, and they learned how to mediate among those realms.

They were alike in other ways as well. In world view, they were both practical, no-nonsense, world-wise, principled men who had done well by doing good. In patterns of thought they were both goal-oriented, ambitious, inner-directed men who found satisfaction in achievement. They held similar places in society, well-off without being wealthy, respected by their peers, influential in the circles they frequented, and effective at what they did.

But there were also significant differences in their views on the development of new technology. Kahn came to DARPA from the computing community, by way of mathematics and communications. Cooper came to DARPA from the business community, by way of electrical engineering and the space program. Kahn's priority was basic research, a precursor to many useful applications. Cooper's priority was goal-directed, engineering research compatible with industry adoption. Though both wanted to see innovative developments find their way into commercial and military applications, they had different views of how to achieve that end. In this respect they shared the same differences that had separated J. C. R. Licklider and George Heilmeier.

But it was their differing styles, more than anything, that shaped their interaction and their joint contribution to SC. Kahn is quiet, soft-spoken, almost retiring in his manner. But his low-key demeanor belies a potent will and a cheerful, resigned tenacity. He harbors visions and ambitions, but he takes no offense when others do not yet share them. It is his mission to educate, persuade, and entice. Obstacles to him are not frustrations but challenges; he meets them not by frontal assault but by what Basil H. Liddell Hart would call the indirect approach.[6] Kahn considers the nature of the obstacle, the location of its vulnerability, and the possibility of co-optation. If one's position is right, then one need only find the way to allow superior ideas to triumph over misinformation or lack of education. Kahn works quietly, often behind the scenes, exercising a sweet persuasion that is powerful and relentless. If he does not bring the target around to his point of view, he retires to devise a new strategy and try again. One colleague said of Kahn that you might as well give him the money when he walks in the door because he will bring you around eventually. Even when he has the power to command acceptance of his views, he prefers to lead by conversion.

Cooper, by contrast, is more public and outspoken. He too can be charming and persuasive, but his style is more direct. While Kahn plays his cards close to the vest, Cooper lays his on the table. Where Kahn will assault the fortress by ruse, Cooper will charge the front gate. Where Kahn will examine the vulnerability in the enemy's array, Cooper will gather more battalions. In any given circumstance, Cooper and Kahn might well choose the same tactics, as they often did during their time together at DARPA. But when they chose different tactics, though their goals might be identical, the distance between them was significant.

That distance infused SC with an institutional schizophrenia it never shook off. Kahn's SC program was a research initiative designed to build up a technology base in university laboratories and transfer it to industry. Cooper's SC was a development program designed to engage industry in the military application of recent computer research. Kahn wanted to push new technology out of the laboratory and into production. Cooper wanted to pull new technology with the force of specific applications.

In theory their approaches complemented each other; in practice they sometimes worked at cross purposes.

The Defense Advanced Research Projects Agency (DARPA)

The DARPA that Robert Cooper inherited in 1981 was a small agency. Its budget of $576.2 million was managed by a staff of 147, of whom about half were program managers.[7] By comparison, the overall defense budget that year was $157.5 billion, of which $15.3 billion was spent on research and development.[8] The ratio of dollars to individuals was $3.92 million within DARPA, or $7.68 million per program manager, compared with $51 thousand per individual in the DoD.[9] Each DARPA employee on average controlled 77 times as much budget authority as his or her DoD counterpart; each program manager 150 times as much.[10] Program managers had good reason to consider themselves powerful forces within the defense establishment, even if they occasionally found that establishment perversely intransigent.

But DARPA's distinction was not simply that it had fewer people with a lot of money to distribute. Rather, DARPA had built up over the years an unparalleled reputation for efficiency and effectiveness in research and development. Senator Jeff Bingaman (D-N.M.) and Admiral Bobby R. Inman, the former Director of the National Security Agency and Deputy Director of the Central Intelligence Agency, have called DARPA "the crown jewel in the defense research establishment, and one of the most successful agencies in the entire federal government."[11] Jeffrey Rothfeder credits it with "efficiency and aplomb rare among Washington bureaucracies." While it has had its share of miscues over the years, it has been able, in Rothfeder's view, "to gracefully skirt the shifting tides of Congressional opinion and ride the waves of technological advance."[12] As former DARPA Director Jack Ruina put it, "we made more mistakes, but because of that we also had more accomplishments."[13]

ARPA's record could hardly have been predicted from its origins. It arose out of the political uproar following the Soviet Union's launch of Sputnik, the first artificial satellite of earth, in October 1957. The failure of the United States to launch its satellite, Viking, before the Soviets convinced many Americans that interservice rivalry in Washington was threatening national security. For years the military services had bickered over roles and missions in ballistic missile development, duplicating each other's programs and creating what came to be called the missile mess. Appointment of "missile czars" in 1956 and 1957 proved unavailing. After Sputnik, Eisenhower reorganized the DoD for the second time in his administration, creating this time the office of director of Defense Research and Engineering (DDR&E) within the office of the Secretary of Defense. He presented to Congress the legislation that became the National Defense Education Act, a measure to stimulate the training of a new generation of scientists and engineers. He created the National Aeronautics and Space Administration and vested it with most of the space activities that the services had been squabbling over.[14]

Before taking any of these steps, however, Eisenhower had created ARPA. The first priorities for this new agency had been to oversee space activities until NASA was up and running and then to act as an institutional damper on the technological enthusiasm that gripped the Pentagon. When a service could propose, as the Army did in Project Horizon, that it should establish a base on the moon because of the age-old military injunction to "take the high ground," then restraint was clearly in order.[15] ARPA came to life based on the premise that science and technology were too important to be left to the military. The second ARPA director, Army Brigadier General Austin Betts, suffered under the perception within the Pentagon that he favored his own service; on his recommendation, all subsequent directors have been civilians.

At the outset, ARPA was first and foremost a screening agency, a brake on the technophilia of the military services. It discharged its first responsibility with dispatch, handing over the civilian space program to NASA when that agency opened its doors on 1 October 1958. Thereafter, ARPA confined its space functions largely to the military realm, until that too was formally sorted out early in the Kennedy administration. In other realms, such as arms control verification, ARPA began a career of exploration and administration that continued into the 1990s. It contributed significantly to the 1963 Nuclear Test Ban Treaty by convincing Congress that atmospheric tests were detectable. In the Kennedy and Johnson years, it sought out technologies to assist American ground troops with the intractable partisan war in Vietnam. It contributed, for example, to testing of the AR-15, precursor to the M-16. It also played a role in development of surveillance technologies, such as the Camp Sentinel Radar, capable of penetrating jungle foliage, and the X-26B-QT-2 silent night surveillance aircraft.[16]

In general, ARPA's initial role in all these areas was to screen new technological possibilities that the services otherwise might have embarked upon themselves. It determined which ones had merit; those without potential were to be shut down. Those with possibilities were to be pursued to proof of concept; at that threshold, ARPA would identify the service most appropriate to take up the technology and carry it to application. Its investigations lacked the competitive passions and institutional self-interest that so often carried the military-industrial complex to wretched excess.

Over time, however, the role of ARPA shifted. Instead of a brake on technological enthusiasm, ARPA became a positive driver. Instead of a screen, it became a funnel. The transformation occurred in part because ARPA could never completely control the services in the quest for new technologies; these organizations were simply too powerful to circumscribe. But ARPA also chafed at its exclusively negative role. The technical personnel attracted to ARPA service had enthusiasms of their own; they were more anxious to promote good technology than simply to squelch weak technology. Surely this was true, for example, of J. C. R. Licklider and his colleagues in the Information Processing Techniques Office (IPTO).

In the course of the 1960s, ARPA took on two new roles. These have driven the agency ever since. First, ARPA has been promoting cutting-edge research that appears to hold significant potential for the nation. ARPA itself does not seek to carry technology to production, but rather to bring along ideas that other segments of the nation will not or cannot develop and carry them to proof of concept. At that juncture, ARPA tries to "transition" the technology out of the laboratory and into the hands of users or producers who will conduct it to full adoption and exploitation. Often this entails targeting potential manufacturers and users and convincing them to pick up the ball. Former DARPA Director Victor Reis calls this a "high technology paradigm shift."[17]

Furthermore, since 1976, DARPA has been working on its own development projects, a trend that disturbed Cooper. In areas such as stealth technology and forward-swept-wing aircraft, DARPA failed to convince the services to take up the technology. It therefore carried its development past proof of concept and into prototype development. Cooper lamented that neither the services nor the private sector were willing to take up such projects, preferring to leave the funding burden on DARPA until the expensive development work had been completed and all the risk in the technology had been removed.[18]

DARPA's mission may be understood as surprise prevention.[19] The agency constantly scans the technological horizon, looking for areas in which enemies or potential enemies might steal a technological march. Not all of these technologies will appear worth pursuing, but DARPA is responsible for knowing enough about them to make an accurate assessment of the potential threat they pose. DARPA blazes trails into the most promising technological wildernesses, to ensure that other nations do not wander alone into some prospect the United States has not explored. Former DARPA Director Craig Fields has argued that DARPA causes surprises rather than prevents them, but these are opposite sides of the same medal. He noted, for example, that stealth technology and phased-array radars caught other countries unawares even while assuring that the United States was not.

Faced with such responsibilities, ARPA took a broad view of the technological horizon. Ever mindful of its primary responsibility to the DoD, the agency nonetheless delved into technologies whose military applications were not yet clear. The space program with which ARPA began, for example, was itself an artifact of the Cold War. It was arbitrarily divided between NASA and the armed forces, but even the Apollo program was a continuation of the Cold War by other means.[20]

The same was true of computers. Originally the military services were the main supporters of computer research and the main consumers of computer manufacture.[21] But even when civilian imperatives began to drive technological development in these domains in the late 1960s and 1970s, it was apparent to most that domination of computer research and manufacture was as fundamental to the East-West competition as was the race to put multiple warheads on ICBMs. President Eisenhower had always insisted that the Cold War was fundamentally a contest between two economic systems and that it would be won and lost economically, not militarily.[22] In such a competition, the distinction between military and civilian technology was often blurred beyond recognition.

ARPA/DARPA honored the distinction in principle. In practice, however, it concentrated more on developing the technology than on determining its end use. Directors and program managers explored promising technologies in their chosen areas and directed them to applications that appeared most auspicious, be they military or civilian. The agency never lost sight of its primary obligation to the armed services, but it did not let that distract it from other applications as well. From time to time a new director came on board determined to keep ARPA research sharply focused on the military, as George Heilmeier did in 1975. And from time to time Congress reminded the agency of its primary mission, as it did with the Mansfield amendment in 1970, and the 1972 name change from ARPA to DARPA, the Defense Advanced Research Projects Agency. By and large, however, ARPA—and especially IPTO—followed its nose. It promoted the technology, such as ARPANET, that appeared best for the country and for the military. It was taken as axiomatic that military applications would follow naturally if the technology was good.

To carry out these responsibilities, ARPA organized itself in 1980 into a director's office, three administrative offices, and six program offices. Figure 2.1 shows how they were related and where IPTO fit in. Defense Sciences was the largest DARPA office, running programs in nuclear monitoring, materials sciences, and cybernetics. IPTO ran "the largest research program in the federal government" on information processing and computer communications.[23] The offices of Tactical Technology and Strategic Technology focused on specific weapons and on communications and reconnaissance systems. It was this organization that Cooper inherited in 1981.

An Agent of Transition

Robert Cooper entered DARPA with a schedule, an agenda, a background, and an open mind. All four would shape what he did at DARPA and how he did it. He was one of those directors who was going to leave his mark, but the mark awaited definition.

Of the four factors, his schedule was the simplest and the most concrete. He would stay for exactly four years. His previous two government assignments had lasted for exactly the terms agreed upon at the outset, three years as Assistant DDR&E and four years as Director of Goddard. At the pleasure of the president, he would spend four years at DARPA, no more.[24]

His agenda was to help his friend Dick DeLauer realize the research and development goals of the Reagan defense buildup. The defense budget had bottomed out in the Carter administration, driven down by antimilitary sentiment after the Vietnam War, by the disarray of the Republican administrations of Richard Nixon and Gerald Ford, and by the pacific leanings of President Carter. In constant dollars, it reached its lowest point since 1951 in 1977. As a percentage of the gross domestic product and total government expenditures, the low point came in 1979.[25] The reversal began in earnest following the Soviet invasion of Afghanistan in that year. President Carter's defense budgets grew significantly in 1980 and 1981; before leaving office in 1981 he recommended a $26.4 billion increase in the fiscal year 1982 defense budget. Shortly after taking office, President Reagan recommended an addi-

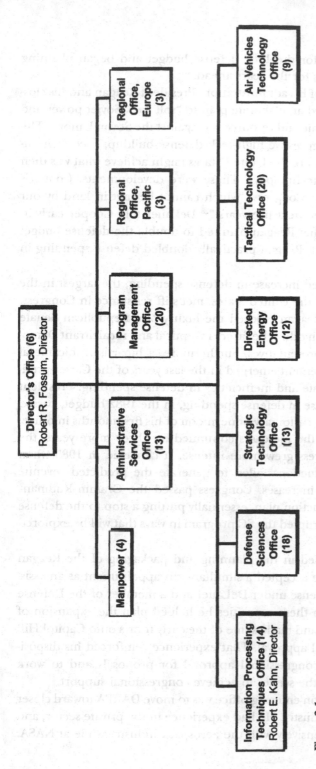

Figure 2.1
DARPA organizational chart, June 1980. *Source:* DARPA telephone directory.

tional $32.6 billion for the 1982 defense budget and began planning even larger increases for the years ahead.[26]

In the early years of his administration, President Reagan and his closest advisers developed an elaborate plan to "roll back Soviet power and engage in an economic and resource war against the Soviet Union." The plan included "an aggressive high-tech defense buildup,"[27] concentrating on those areas where the United States might achieve what was then called "asymmetric technology." These were developments, Cooper's predecessor had told Congress, "which cannot be met in kind by our adversaries within the next five years."[28] DeLauer told Cooper early in the administration that Reagan planned to double the defense budget in five years.[29] In fact, Reagan practically doubled defense spending in his first year.

The unprecedented increase in defense spending, the largest in the peacetime history of the United States, met stiff resistance in Congress. Though President Reagan enjoyed the luxury of a Republican Senate in his first two years in office, he faced a skeptical and recalcitrant Democratic House of Representatives. Furthermore, a bipartisan, bicameral defense reform movement emerged in the last years of the Carter presidency to oppose waste and inefficiency in defense spending. President Reagan's first increase in defense spending, in the 1982 budget, passed Congress during the customary honeymoon of his first months in office. Thereafter, though the increases continued for three more years, the struggle with Congress grew more intense. It climaxed in 1985 when the Reagan tax reductions failed to generate the predicted revenue to pay for defense increases. Congress passed the Gramm-Rudman-Hollings deficit reduction plan, essentially putting a stop to the defense buildup.[30] It also disrupted the SC program in ways that will be explored further on.

Cooper participated in the planning and packaging of the Reagan defense buildup. He accepted a simultaneous appointment as an assistant secretary of defense under DeLauer and a member of the Defense Resources Board. In these capacities he helped plan the expansion of the defense budget and made some of the early forays onto Capitol Hill to win congressional approval. That experience reinforced his disposition to seek early congressional approval for proposals and to work closely and behind the scenes to achieve congressional support.

Cooper's agenda on entering office was to move DARPA toward closer cooperation with industry. He had experience in the private sector, and he had worked extensively with the aerospace industry while at NASA.

While 80 percent or more of NASA's budget went to industry, only about 50 percent of DARPA's did.[31] DARPA's reliance on industry was on a par with the average for government agencies in the 1970s and 1980s,[32] but Cooper was imbued with the NASA model from his years at Goddard. Most of DARPA's investment in computers went to universities, with IPTO spending the lion's share. To Cooper's mind this meant too many paper studies in academia, not enough practical applications out in the marketplace. He polled industry early in his tenure and heard repeatedly that it was hard to get results from universities. Both he and DeLauer were determined "to give an enema to the pent up technologies that were ready, ripe."[33]

His experience at NASA's Goddard Spaceflight Center had taught him about clogged government bureaucracies. In January 1977 a group of NASA research engineers convinced him that conventional computers could not process the imaging data that would be transmitted from spacecraft in the near future. They wanted to develop a multiprocessor machine called the massively parallel processor (MPP). Within a week Cooper confirmed their judgment and promised to double their staff and quintuple their budget. This quick action, however, went unmatched by the system. A request for proposals (RFP) took a year to work its way through the bureaucracy. Another ten months elapsed before the study phase of the contract began. Only in December 1979, almost two years after Cooper's decisive action, was a fabrication contract let. The computer was finally delivered to NASA on 2 May 1983.[34] This was six years after Cooper's approval, four years after Cooper had left NASA, and almost two years after he arrived at DARPA. He was determined to do better.

He began by asking his staff at DARPA for new ideas, a reflection of the fourth and final characteristic of his early tenure at DARPA—an open mind. Robert Kahn was among those who stepped forward. Pleased with the restoration of funding for computer research that had been realized since he took over IPTO in 1979, Kahn nevertheless remained far from satisfied. He felt that the field needed a major infusion of new money to accomplish two goals: (1) exploit technologies that were now ripe by (2) transferring them to industry.

Both objectives appealed to Cooper, but he wanted to test the waters before diving in. He began a series of exploratory steps that would ultimately extend a year and a half into his four-year term at DARPA. This was not the one-week turnaround he had given to his researchers at Goddard when they had proposed developing a massively parallel processor, but then again this was not a comparable program. Kahn was proposing a major research initiative that could well cost hundreds of millions of dollars.

Coincidentally, Kahn's suggestion reached Cooper in the fall of 1981, just as the Japanese were unveiling their Fifth Generation program. In Cooper's view, the Japanese initiative was driven primarily by an imperative for machine language translation. Constantly at a disadvantage vis-à-vis Western countries, Japan was prepared to invest heavily in supercomputers and artificial intelligence in hopes that they could "breach the language barrier," as Cooper put it.[35] But for this reason, and also because of the technological path the Japanese had chosen, Cooper came to believe that the Fifth Generation did not pose a major threat to the United States. It did, however, provide a rationale and a potent political lever for attracting new money to U.S. computer research.

Cooper and Kahn traveled around the country speaking to computer researchers and businesspeople.[36] From the researchers they received enthusiastic support. Some, such as Michael Dertouzos, director of MIT's Laboratory for Computer Science (successor to Project MAC), were already alarmed by the Japanese. Dertouzos shared Kahn's perception that microelectronics, architectures, and artificial intelligence were ripe for development, and he feared that the Japanese would take advantage of American advances in these fields and get to market first.[37] He said that when he learned of the Fifth Generation plan, "I got on the horn and started screaming. I wrote to our computer corporation presidents. I went to visit the Defense Science Board. I started working with Cooper and Kahn."[38] Others within the research community, though less alarmed at the Japanese Fifth Generation program, were equally supportive. They respected Kahn and they realized that the tide of money being contemplated would raise all their boats.

Industry, however, was cool to Cooper and Kahn. Cooper heard nothing but reluctance and indifference. "They had no vision," he remembers, "no inclination to do anything other than plod along doing what they had been doing for years."[39] Cooper returned from his travels convinced that the transfer of new ideas from academia to industry was a difficult chore and that DARPA would have to seize the initiative and make it happen.

This perception is difficult to explain. Even while Cooper and Kahn were making their rounds, the microelectronics and computer industries were already engaged in new and imaginative responses to the Japanese challenge. For example, in 1982 forty U.S. firms founded the Semiconductor Research Corporation (SRC), a nonprofit research cooperative to fund basic semiconductor research in universities.[40] The subscribers clearly intended to transition the results of that research into commercial products. An even larger undertaking by the DoD in 1979 had brought together six prime contractors (Westinghouse, Honeywell, Hughes, Texas Instruments, IBM, and TRW) and many more subcontractors in very high speed integrated circuits (VHSIC), a program to enhance chip manufacture for military purposes. Gene Strull, vice president of technology at the Westinghouse Defense Electronics Center, said that at the time "there wasn't a single weapons system in the U.S. arsenal with electronics as sophisticated as what you could find then in video games."[41] The DoD was conceding that commercial development of computer chips, spurred by Japanese competition, had outstripped military development. It was prepared to spend $320.5 million over six years to bring military standards up to the state of the art.[42] A Defense Science Board summer study in 1981, chaired by former DARPA Director George Heilmeier, identified VHSIC as number one among seventeen technologies "that could make an order of magnitude difference in U.S. weapons capabilities"; machine intelligence ranked eighth and supercomputers ninth. Industry was expressing a willingness to cooperate in improving military chips, even while DARPA ran a parallel program in very large scale integration (VLSI).[43]

The largest cooperative undertaking, and the one most parallel to strategic computing, was forming as Kahn and Cooper made their tour. Partially in response to the call of alarm from Michael Dertouzos, William C. Norris, chairman of Control Data Corporation, convened a meeting of eighteen industry leaders in Orlando, Florida, early in 1982.[44] They formed a research consortium, the Microelectronics and Computer Technology Corporation (MCC), focusing on four problems: computer-aided design and manufacture (CAD/CAM), computer architecture, software, and microelectronics packaging. The latter three were very close to SC priorities. Member companies paid $150,000 (later $1 million) to join and then invested in those research programs of interest to them.[45] Researchers from the collaborating companies would work together instead of competing with one another. MCC held all patents from the research licensed for three years exclusively to those companies that supported the program producing the patent. Nineteen companies joined when MCC opened its doors in September 1983; over the next eight years, membership turned over as $300 million was invested in the enterprise. In the scale of the undertaking and the goals of the research, MCC parallels SC and will warrant comparison later in this study.

Meanwhile, still other cooperative programs were taking shape. The Stanford University Center for Integrated Systems attracted nineteen industry sponsors to join with government in supporting university research in VLSI systems and training of new researchers.[46] The Microelectronics Center of North Carolina took up an even broader research agenda in the Research Triangle Park.[47] Similar programs grew up around the University of Minnesota, Rensselaer Polytechnic Institute, and the University of California.[48] All of these had an industrial component.

These cooperative programs call into question Cooper's characterization of the microelectronics and computer industries as indifferent and lethargic. More likely he encountered a coolness to working with the government and a reluctance to engage in cooperative ventures that might subsequently be construed as violations of the antitrust laws. Harvey Newquist believes that IBM and AT&T declined the invitation to join MCC in 1982 because they were "gun shy" from their recent Justice Department scrutiny.[49] The passage of the Joint Research and Development Act of 1984 diminished the latter concerns, reducing by two-thirds industry liability for damages in antitrust cases.[50] In the same year, Congress passed the National Cooperative Research Act, protecting research consortia such as MCC from antitrust legislation. Between 1984 and 1991 more than 173 consortia registered with the Department of Commerce.[51] More importantly, perhaps, many of the charter members of MCC expressed concerns about accepting government funds, with all the red tape and interference that accompanied them.[52] Some companies appeared willing to join government programs such as VHSIC, where the goals were comparatively clear and specific. They were less willing, however, to share with government the management of long-range research.

The other reason for industry coolness to the SC proposal was research horizon. Industry tends to value research that will produce marketable results in the near term, say three to five years. This tends to be directed or applied research. Investigations at a more basic or fundamental level often have larger, more long-term payoff. But they seldom contribute to this year's balance sheets and they often produce results that profit industry generally, not the company that produced them. Industry looks to government and academia to support this research. In spite of Kahn's belief that long-term developments in microelectronics, architecture, and artificial intelligence were now ripe for application, industry—or at least the segment of it that Cooper consulted—appears to have been skeptical. SC would turn out to be a test of their skepticism.

In spite of industry reticence, Cooper remained convinced that Kahn's idea had merit. Indeed it could be just the right mechanism to promote transfer—what the DARPA computer community called "transition"— of university research to industrial practice. If industry would not come to the mountain, Cooper would take the mountain to industry. He asked Kahn to draft a proposal. The resulting document, "A Defense Program in Supercomputation from Microelectronics to Artificial Intelligence for the 1990's," appeared in September 1982. It is remarkable on several counts, not least of which is the perspective it offers on SC. Throughout the subsequent history of this program, participants and observers would argue about the strengths and weaknesses, accomplishments and shortcomings of SC. They always measured the program that came to be against the program they thought was intended. Always they tested what SC did or failed to do against what they thought it was supposed to do. Kahn's September proposal was the first written articulation of what he intended the program to be. It also laid out most of the component elements of SC, though many of them were going to change substantially before the program was adopted. With this document, Kahn took the first step toward justifying what would eventually be a ten-year, $1 billion investment.

As its title suggested, the program was to focus on supercomputation, microelectronics, and artificial intelligence. The last two, together with architecture, were the three areas that Kahn and others thought of as ripe or at least as having ripe developments within them. Interestingly, Kahn did not present these in the paper as necessarily related to or dependent upon one another, although he undoubtedly linked them in his mind.[53] Rather, each one deserved support on its own merits and each promised significant new developments.[54] This was not yet an integrated, connected program but rather a funding category to support three areas of research.

Kahn took pains to differentiate between fast machines and smart machines. Increases in machine speed, attributable to advances in microelectronics, had produced high-speed microcomputers, the report said. But "speed [alone] will not produce intelligent machines. . . . Software and programming will be central factors."[55] In other words, artificial intelligence would have to be mated with high speed to produce intelligent machines. Those intelligent machines were not yet the major focus of the plan, but they were a logical consequence of what might be achieved if Kahn's three areas of research concentration proved fruitful.[56]

In spite of there being three main components, Kahn nevertheless characterized the program as encompassing "four key areas": microelectronics, supercomputers, generic applications, and specific applications. Artificial intelligence, to Kahn, was subsumed under "generic applications." Kahn divided these "generic applications" into three categories based on ripeness. It is worth reproducing the lists in each of the three categories, for subsequently they would be pruned and distilled to form the goals for SC.

Category I
- Distributed data bases
- Multimedia message handling
- Natural language front ends
- Display management systems
- Common information services

Category II
- Language and development tools for expert systems
- Speech understanding
- Text comprehension
- Knowledge representation systems
- Natural language generation systems

Category III
- Image understanding
- Interpretation analysis
- Planning aids
- Knowledge acquisition systems
- Reasoning and explanation capabilities
- Information presentation systems

Kahn envisioned "Category I programs being initiated first, followed a year or so later by Category II and finally by Category III." Then an ultimate integration would "meld these capabilities into a single whole." Presumably that whole would be a thinking machine, a machine capable of human intelligence. The "main challenge," he said, was "to utilize AI technology to provide computer based capabilities that model human capabilities."[57]

From this "environment of generic applications," this implementation of artificial intelligence, Kahn foresaw the development of "specific applications" to serve needs of defense "or any other area." These might include photo interpretation, data fusion, planning and scheduling, a pilot's assistant, AI-based simulation systems, applied speech understanding, and robotics applications. "In each case," said Kahn, "the underlying components are simply used rather than rebuilt, addressed rather than reintegrated." In other words, the generic applications, the AI developments, would be tailored to suit a particular need. Once the general capabilities were developed, the specific applications would flow from them naturally.

The generic software envisioned here paralleled similar developments in computer hardware. The earliest electrical digital computer, the ENIAC of World War II, was hard-wired to perform certain calculations. Its switches were connected in a sequence corresponding to the algorithm it was processing. To work a different problem, the wires had to be reconfigured. Mathematician John von Neuman appraised that design and suggested instead that the sequence of switching be determined by a software program retained in computer memory. The computer processor would be wired to handle multiple algorithms. In any given problem, the program would determine which sequence of switches to engage. Thus was born von Neuman architecture and the modern, general-purpose computer. Kahn was seeking the same kind of generality in software applications. A program would be capable of sequencing certain kinds of operations, for example, data fusion. Instead of writing a separate software program for each set of data to be integrated, programmers would write one generic shell that would work for all data bases. One size fits all, a goal that promised standardization, economies of scale, and human-like versatility.

Kahn envisioned a six-year program, starting with funding of $15 or $30 million dollars in fiscal year 1983 and rising to $225 million in fiscal year 1988. Total cost would range from $648 million to $730 million. This amount would be in addition to the funding of computer science, communications, VLSI, artificial intelligence, and materials sciences already under way in IPTO and elsewhere in DARPA. The microelectronics portion of the program would be carried out by the Defense Sciences Office within DARPA, the supercomputer and generic applications (i.e., AI) by IPTO, and the specific applications by other DARPA offices.

Funding for those specific applications would mostly be follow-on to the technology development that was at the heart of Kahn's proposal; he expected that only one or two specific applications "could reach critical mass" during the six-year program.

The rationale for spending at this level was competition with Japan.[58] Noting that the Japanese planned to quintuple U.S. spending on "generic AI based technology," Kahn concluded that they were aiming for world leadership in the computer field by the 1990s. Given that the Japanese had overtaken the United States in memory chips in a decade, it was essential to prevent the same fate in artificial intelligence, "the last bastion of U.S. leadership in the computer field."[59]

It is easy to read this document as an AI manifesto, easy as well to understand why some observers came to see SC as an AI program. But the title and the text also suggest that AI was only one of three goals along with supercomputers and microelectronics. These goals are connected both to each other and to the possible environments in which they might work. Unlike subsequent versions of the SC charter, this one does not lay out a hierarchy of developments, with microelectronics making possible the revolutionary new architecture of machines that think. This plan does, however, culminate in AI, in "generic applications," in a way that suggests a hierarchy even if none is explicit.

In fact the unusual terminology of generic and specific applications appears to be an attempt to mediate between the strategies of Kahn and Cooper. One can read between the lines of this document a desire by Kahn to fund on their own merits and without strings the three areas of computer development that he considered ripe: microelectronics, computer architecture, and AI. Cooper, on the other hand, was pushing him to identify the payoff, the specific military applications, that might flow from that development. By labeling AI "generic applications," Kahn was simply putting Cooper's language on top of what Kahn wanted to do. Specific applications would come mostly in a follow-on program.

Testing the Market

Kahn's approach failed to satisfy Cooper, but it did convince him to move ahead. He charged Kahn to write another version of the proposal while Cooper laid the groundwork for selling the program. First, he consulted with his friend and boss Richard DeLauer, who was particularly exercised by the Japanese Fifth Generation.[60] Both agreed that Kahn's proposal had merit; both wanted validation. With funding from

Cooper's DARPA budget, DeLauer charged the Defense Science Board to investigate defense applications of computer technology. The study would be chaired by Nobel Laureate Joshua Lederberg, a geneticist noted for innovative use of computers in his research. The executive secretary for the study would be Ronald Ohlander, a DARPA program manager who would come to play an important role in SC. The study had the twin goals of garnering DoD support for the program and clarifying the issue of defense applications, which had been largely omitted in Kahn's first formulation.

Second, Cooper went to Capitol Hill to try out the idea in private with some influential congresspeople and their key aides. He found a warm reception, but he got strong advice as well. To get $600 or $700 million new dollars from Congress, in addition to the existing DARPA budget for computer research, he would have to come armed with a clear, hard-hitting proposal. This project was sellable, but it wanted selling. Cooper conveyed that message to Kahn.[61]

As Kahn worked on a new draft of the proposal, the fundamental difference between his strategy and Cooper's emerged with ever greater clarity.[62] They agreed on the overall goals of SC and the urgency behind the initiative. They both believed that this program was in the best interests of the country at large and that it was important to national security, in the broadest sense of the term. And they agreed that successful developments in what Kahn called the "technology base" would lead to important military applications. They disagreed only on how to sell that proposition.

Kahn is fundamentally cerebral. The whole proposal was clear in his mind. He could picture the computer research community with which he had now worked intimately for fifteen years. He knew its potential and he was convinced that curbed spending through the 1970s had starved the field to inefficiency. It had excess capacity in personnel and research opportunities. Other programs such as VHSIC and MCC targeted industry; they did little to nurture university research. SRC, which did provide support for universities, distributed comparatively little money. There was no doubt in Kahn's mind that increased funding for the technology base would lead to important new applications. What he could not say, what no one could say, was precisely what the results of this research might look like and when they would be achieved. Rather, one had to invest as an act of faith, an argument that George Heilmeier had rejected when Licklider advanced it seven years earlier.[63]

Kahn was arguing for what Cooper called a "level of effort activity."[64] Annually the centers of excellence supported by IPTO would submit budgets for the coming year. The budgets combined ongoing research programs with new initiatives. IPTO often critiqued these budgets and returned them for revision, suggesting how to label and characterize the research to make it salable in Washington. Sometimes IPTO rewrote the documents, couching them in terms that Congress wanted to hear.[65] Year after year, the centers of excellence received consistent support. This money sustained the researchers, trained graduate students, and supported research that the "ARPA community"—university researchers and their IPTO patrons—judged to be promising and potentially useful to the country and the DoD.

So close were the ties between IPTO and the universities that some observers wondered who was directing whom. Michael Dertouzos of MIT probably knew DARPA better than its own employees did; he studied the agency for years, watching directors and program managers come and go. After attending the principal investigator's meeting in Monterey, California, 12–22 April 1980, Dertouzos returned to MIT and wrote out a sixteen-page report for his colleagues, describing DARPA organization, funding trends, contracting and budgeting procedures, and current research interests. Attached to his report was a hand-written seating chart of the dinner in Monterey, indicating every place at a table held by an MIT representative, an ex-MIT faculty member, or an MIT alumnus (see figure 2.2).[66] When Robert Cooper visited MIT in October 1982 on his tour of the U.S. computer community, Dertouzos briefed his staff on Cooper's "most successful visit" [at Xerox PARC] and his "worst visit" [at Stanford], suggesting how MIT could present itself in the best light.[67]

Dertouzos claimed to have been one of those "sitting around the table ... putting together the ideas behind Strategic Computing."[68] Allen Newell of Carnegie Mellon University claimed even more. "The guys who ran IPTO were ourselves," he boasted, referring to the elites of the AI community. Kahn was their "agent," their man in Washington.[69] This characterization accords with the views of many that throughout his tenure Kahn was carrying water for the ARPA community, especially for the centers of excellence built up by his predecessors.[70] In return, recalled DARPA Deputy Director Charles Buffalano, Kahn was "canonized" by the favored university researchers. For five years, he was the patron saint of AI, a "cardinal." "When he got off an airplane," said Buffalano, "the locals got on their knees and kissed his ring."[71]

Kahn could hardly expect such reverence from Cooper, DeLauer, Reagan, and Congress, but he did want them to trust his judgment.

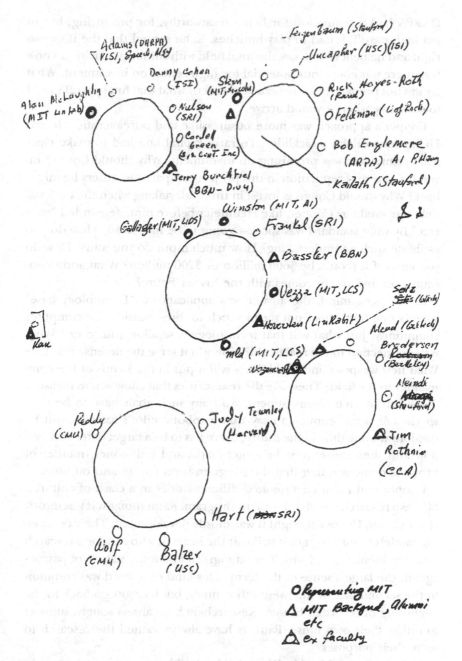

Feigenbaum (Stanford)

Uncapher (USC)(ISI)

Adams (DARPA)
VLSI, Speech, MSg

Danny Cohen
(ISI)

Skern
(MIT, Lincoln)

Rick Hayes-Roth
(Rand)

Alan McLaughlin
(MIT Lin lab)

Nilson
(SRI)

Feldman (U of Roch.)

Cordel
Green
(Sys. Cont. Inc)

Bob Englemore
(ARPA) AI P.Hay

Jerry Burchfiel
(BBN-Div 4)

Kailath (Stanford)

Winston (MIT, AI)

Gallager (MIT, LIDS)

Frankel (SRI)

Bassler (BBN)

Vezza (MIT, LCS)

Seitz
Estes (Caltch)

Klau

Hovenden (LinRabit)

Mead (Caltch)

Bredersen
Robinson
(Berkeley)

mld (MIT, LCS)

Alendi
Atkiasan
(Stanford)

Reddy
(CMU)

Judy Townley
(Harvard)

Jim
Rothnie
(CCA)

Hart (BBN SRI)

Wolf
(CMU)

Balzer
(USC)

Representing MIT

△ MIT Backgnd, Alumni
etc

△ ex faculty

Figure 2.2
Dertouzos seating chart. *Source:* MIT archives, records of the LCS, 268/19/7.

DARPA had a reputation for being trustworthy, for producing, but to get high payoff, it had to play hunches. Kahn sensed that the time was right and he believed he saw the total field with sufficient clarity to know that there was enormous potential for high return on investment. What he was not prepared to articulate was exactly what that final payoff would look like or when it would arrive.

Cooper's approach was more businesslike and bureaucratic. He was Heilmeier to Kahn's Licklider. To raise capital one had to make clear what return one was promising on investment. Why should Cooper invest his time and reputation in this project compared to others he might back? Why should Congress invest in this undertaking when the national debt was swelling? Cooper, like Heilmeier before him, demanded "metrics." By what standard was success going to be measured? How do you calibrate such an undertaking? How much input do you want? How do you know if it should be $600 million or $700 million? What additional return on investment comes with the higher figure?

In Cooper's mind, the answer was applications. "Technology base" sounds fine but it does not mean much to those outside the computer community. Just what will that technology base allow you to do? What is its connection to the real world? How will it serve the defense mission? What new weapons and capabilities will it put in the hands of the commander in the field? These are the real metrics that allow you to measure the worth of such an investment. And any such goals have to be time specific. It is not enough to say that a robotic pilot's assistant will be possible at some time in the future; there has to be a target date. Between now and then there must be benchmarks and milestones, metrics of progress demonstrating that the program is on course and on time.

Cooper and Kahn championed different sides in a clash of cultures. SC was a research and development program. Kahn thought it was mostly research and Cooper thought it was mostly development. Their competing models of funding goals strike at the heart of who supports research and development, and why. They disagreed about the nature of patronage, in the largest sense of the term. This kind of discord was common in the second half of the twentieth century, but its roots go back to the beginning of recorded history. Researchers have always sought support to follow their own muse. Patrons have always wanted the research to serve their purposes.[72]

Archimedes was hired by Dionysius I to build war machines. Leonardo and Galileo sold their services for similar purposes. Some few patrons support research out of pure altruism or love of science or admiration for the researcher or some other form of psychic income. King Frederick II of Denmark gave the sixteenth-century astronomer Tycho Brahe his own island and financial support to conduct observations of no practical worth to the crown. Albert Einstein was installed at the Institute for Advanced Study in Princeton, free to conduct whatever work he chose. DARPA, however, was no such patron. In the golden years from the mid-1960s to the mid-1970s the agency may have looked that way to some fortunate centers of excellence pursuing AI, but those days had ended with George Heilmeier, if not before.

Kahn did not object to Cooper's expectation of results, that is, applications, but rather to the demand for precise definition of what those results might be and when they would appear. He and his colleagues were conducting the most challenging kind of research, operating, as he liked to say, at the "cutting edge" of technology. Building up a "technological base" along that edge ensured that practical applications would follow. But because research at the boundary operates in unknown territory, it is always impossible to predict exactly what applications will follow and when.

Analogies are often invoked to describe this dilemma. One is logistic support for troops at the front. Logistics do not capture any ground or break through any lines, but without it, those advances are impossible. Kahn wanted to build up the logistics base while Cooper wanted to know what combat operations the base would support. Another analogy invoked in this debate is the pool of knowledge. Basic researchers gather knowledge for its own sake and store it in a pool of information that is available to all. Applied researchers then dip into that pool to draw out the information they need for their specific application. If the pool is not replenished regularly, the field of applications withers on the vine for want of irrigation. Yet another analogy, one of Kahn's favorites, is the expedition of Lewis and Clark; to demand prediction of specific applications in advance of fundamental research is like asking these two explorers to describe what they are going to find. Had they known that, they would not have needed to take the trip in the first place. If you knew specific answers, the research would not be necessary.

Kahn and his colleagues believed that Cooper was trying to impose on them a NASA style of operation, one with which he had presumably become familiar during his years as director of Goddard Spaceflight Center. The model comes from the Apollo program. You identify a challenging goal that you believe is within the range of current or foreseeable technological capabilities, then you delineate a series of steps to get from here to there. Because the project is goal-driven, you can identify objectives at the outset and you can also predict what milestones you will have to pass at what time to reach the objective on time.[73] Such a schedule draws on the techniques of program evaluation and review technique (PERT) management developed by the Navy for the Polaris submarine program and on the management technique of concurrency, developed by Air Force General Bernard Schriever for the Atlas missile program.[74] All of these programs were expensive, high-priority undertakings that had about them an air of public relations and gimmickry.

It was not the gimmickry that Kahn and his colleagues objected to; rather they believed that they were undertaking a fundamentally different enterprise. Apollo, Polaris, and Atlas were engineering and development programs. One knew at the outset where one was going and how one would get there. Kahn's objective was by definition less precise. It was to push out the bounds of known capabilities. The horizon opens up as the program proceeds. It requires an evolving, adaptive plan, one that exploits new knowledge and new discoveries to refocus the research agenda. Instead of a clear trajectory with a fixed destination, the project is a kind of feedback loop in which advances early on open up targets of opportunity, even while reverses and dead ends may close down some paths of development that appeared promising at the outset. Like Lewis and Clark, one adjusts one's itinerary as one goes along, based in part on what one learns as one proceeds.[75]

In the terminology of economics, Cooper and Kahn were debating whether or not machine intelligence was "path dependent." Economist Giovanni Dosi helped introduce the notion of developmental paths in the 1980s.[76] Building on the work of R. Nelson and S. Winter,[77] Dosi hypothesized "technological paradigms" of the kind that Robert Kahn embraced about computer development in the early 1980s. These paradigms have strong directional components; they suggest "normal" technological trajectories which can have an "exclusion effect," leaving researchers blind to alternative lines of development. Furthermore, the trajectories may develop their own momentum and carry the program toward a predetermined goal. Dosi's theory relates only tangentially to the "path dependence" concepts of scholars such as Paul David, who argues that lines of economic development are determined in part by "historical accidents" experienced on the way to rational goals.[78] Both theories recognize that "history matters," that events along the way alter the outcome of research programs and economic plans, but the later work of David and his colleagues accords more importance to historical contingency.

When applied to SC these theories help to illuminate the differences between Kahn and Cooper. If a research field begins from a clearly defined point in conceptual space—the current state of the art—how does it move to its research goal? In this case Kahn's goal, for example, was the realm of machine intelligence. Within that realm resided subordinate goals, such as machine vision, planning, and natural language understanding. Kahn believed that research should be launched on several trajectories at once. Perhaps there was an ideal path that led directly to the goal, but that path was knowable only in the case of an engineering project such as Apollo, which applied known capabilities to a new problem.

In the real world of research, success was path dependent, because not all paths would arrive at the goal, and those that did would get there more or less directly. For that reason, one should impose no a priori expectation that any given path would pass benchmarks or milestones at any fixed time. In other words, Heilmeier's milestones might place an unrealistic expectation on research trajectories. Only in hindsight would one know that a given path was the direct route to the goal. But then when would one abandon failed paths? Do all trajectories eventually arrive at machine intelligence? Or is achievement of machine intelligence path dependent? If some paths are better than others, what criteria will reveal the preferred path short of achieving the goal? In other words, how do researchers know when they are on course? How did Lewis and Clark know where to go next?

The Pyramid as Conceptual Icon

Cooper accepted in principle Kahn's faith in path flexibility, that the researcher knew best where to turn. In practice, however, he could not sell it on Capitol Hill. He charged Kahn to write a fuller version of his proposal, one that would include clear and compelling military applications. The resulting document, "Strategic Computing and Survivability," completed in May 1983, was a serious attempt to close the gap between Kahn's and Cooper's views of what SC should be.

The term "Strategic Computing" was itself a Cooper signature. As a key member of the Reagan defense team, Cooper was drawn into the discussions following the president's surprise announcement on 23 March 1983 that he intended to accelerate ballistic missile defense.

Cooper claims to have named that undertaking the Strategic Defense Initiative (SDI) and in the same period to have dubbed Kahn's proposal the Strategic Computing Initiative (SCI).[79] The coincidence in timing and terminology convinced many observers that the two programs were intimately connected, a perception that haunted SC for years.[80]

The evidence linking SCI to SDI is circumstantial but impressive. SCI was first announced to Congress just three weeks after President Reagan's "Star Wars" speech on 23 March, 1983. Its defining document identified missile defense as a military function requiring automated systems. The Fletcher Report on the feasibility of SDI, named for chairman and former NASA administrator James C. Fletcher, devoted a complete volume to the computer problem. Robert Cooper linked the two programs more than once in congressional testimony. For reasons that will be explored later, gallium arsenide, originally envisioned as a foundation of the SC microelectronics program, was actually transferred to SDI in 1984.[81]

These connections with SDI were not yet clear in the document that Robert Kahn produced in the spring of 1983. His working title, "Strategic Computing and Survivability," addressed Cooper's insistence on clear military relevance. The armed forces needed electronic components that could survive the electromagnetic pulse given off by a nuclear explosion, a need that Kahn proposed to meet by exploring gallium arsenide (GaAs) as a substitute semiconductor material for silicon. The subtitle of the plan might have been written by Cooper himself: "A Program to Develop Super Intelligent Computers with Application to Critical Defense Problems."

Other concessions to Cooper's point of view run through the document. It is now a ten-year, two-phase program, with the second phase focused on producing specific applications for the military: a pilot's assistant, a battle management system, and an autonomous underwater vehicle. The Soviet Union replaces Japan as the external stimulant of this development, even though the program is described as "a fifth generation activity." No longer an open-ended research program of the kind Kahn had first proposed, SC is now "an engineering development program," a label one might apply to Atlas, Polaris, or Apollo. Universities would receive only 25 percent to 30 percent of the funding in the program; most of the rest would go to industry, with small amounts for federally funded research centers and in-house government laboratories. Of the $689 million proposed for the first phase, only $130.5 million went to applications. But the implied promise of the program was that phase two would see enhanced focus on applications as the technology developed in phase one came on line.

Even in the technical discussion of the various research fields, Cooper's insistence on benchmarks and milestones is manifest. The three major areas—microelectronics, computer architecture, and artificial intelligence—appear as before but in much greater detail. An attempt is made to describe where each field is in 1983 and where Kahn expects it to go by the end of the decade. Some fields even have intermediate benchmarks.

Microelectronics, for example, focuses on GaAs technology, the area most directly responsive to military needs. Kahn proposes to establish one or more pilot production lines for GaAs computer chips, each line capable of producing 100 multichip wafers a week. Chip capacity would be targeted at 6K equivalent gate array and 16K memory by 1987, meaning the production chips by that time would have 6,000 electrical components (transistors, resistors, capacitors) that perform logical operations such as AND and OR, and they would have 16,000 bytes of memory. Kahn also proposed other improvements in microelectronics, such as development of wafer-level integration technology, so that entire wafers would be devoted to a single chip, thus eliminating postfabrication chip slicing. For these goals, however, Kahn offered no concrete time tables or benchmarks.

Architecture received the fullest articulation of goals and milestones. Kahn first cautioned that while the program would indeed aim for "a super intelligent computer," this should not be conceived as "a single box full of microelectronics which fills all needs."[82] Rather the program would seek progress in five modular areas: signal processing, symbolic processing, simulation and control, data base handling, and graphics. The super computer, in his mind, was "really a family of computing capability built up out of all the combinations of the basic modules and their communication system."[83]

Progress on all five fronts would be achieved by pursuing four different categories of computer structures or topologies: linear, planar, multidimensional, and random. These subsumed a total of seven configuration schemes then under consideration. Such topologies were being pursued to "maximize the concurrency of the computers," that is, to achieve greater speeds by designing machines with multiple processors that could perform operations in parallel. Some progress had already been made on this front, said Kahn, in numeric processing, the kind of arithmetic number-crunching demanded by many scientific applications. The symbolic processing necessary to realize artificial intelligence, however, did not lend itself to comparable techniques.

"The regularity of numeric computation," said Kahn, permits "a structure decomposition of the computation, *in advance of execution,* into concurrently executable components." But symbolic processing required "dynamic decomposition and concurrent execution as the computation is executed."[84] In other words, parallel processing lent itself to numeric computation more readily than to symbolic processing, to scientific calculation more readily than to AI. Kahn here was addressing with great clarity and candor a problem that challenged the development of multiprocessor architecture throughout SC. This prospectus, in the end, could serve as a post mortem on the achievements and failures of parallel processing.

Kahn went beyond mere articulation of the problem. He made predictions. Signal processing, he said, would improve under SC from 10–50 million operations per second (ops/sec) in 1983 to 1 billion ops/sec (giga ops) in 1986 and 1 trillion ops/sec (tera ops) in 1990. High-speed LISP machines would accelerate from 5 MIPS (million instructions per second) to 50 MIPS by 1986. Logic machines, of the kind then using Prolog architecture, would be improved from 10,000 inferences per second to 100,000 by 1986 and 100 million by 1990—four orders of magnitude. The data flow technology developed but not exploited in the United States and recently taken up by the Japanese for their Fifth Generation program would be simulated so as to emulate a 100-processor system by 1986. More importantly, other multiprocessor technology, then capable of few tens of processors only, would be benchmarked at a 100-processor system in 1985 and a 1,000-processor system in 1987.

In every one of the eleven areas of architecture research laid out by Kahn, he promised either orders-of-magnitude improvements in performance or clear, measurable goals for developments not yet attempted. And he promised software for these machines capable of delivering real-time response—the interactive computing that Licklider had espoused two decades earlier. Real-time response, while enormously difficult to achieve, was a sine qua non of the applications envisioned for the military. After all, a pilot's assistant that took hours or even minutes to process data would hardly be much use to a pilot closing with an enemy aircraft faster than the speed of sound.

AI garnered similar commitments. Kahn identified twelve areas that would be targeted by the program, down from the sixteen areas identified in the September 1982 plan. Seven of these—vision, natural language, navigation, speech, graphics, display, and DM/IM/KB ("a combination of data management, information management and knowledge base technology")—would be developed in "generic software packages . . . independent of particular applications." In other words, all these lent themselves to development in forms that would be universally suitable to different military applications—one size fits all. The other five—planning, reasoning, signal interpretation, distributed communication, and system control—were more application-dependent. The program would seek the general knowledge necessary to custom-design such systems for particular applications.

In both areas Kahn again committed the program to clear metrics. Speech, for example, would produce "real-time understanding of speaker-independent connected speech" with vocabularies of 10,000 words in an office environment and 500 words in a "very noisy environment" such as an airplane cockpit or a battlefield. Signal interpretation would be able to identify military hardware on the battlefield at a range of one mile with 90 percent accuracy and track targets of interest in successive images with 95 percent accuracy. Unlike the predictions made in architecture, Kahn attached no dates to these final goals, nor did he provide benchmarks. Still, all fit within the proposal's overall goal: "U.S. industrial production of intelligent computers, with associated programming technology, for achieving three to four orders of magnitude faster execution of functions needed for intelligent military systems."[85]

To these three major areas of development, Kahn added a fourth component to the program: support systems. Included here were access to VLSI-VHSIC fabrication facilities, access to ARPANET, deployment of DARPA's standard management techniques, availability of high-powered machines for researchers and students, and training of graduate students. Subsequently, this category of activities would come to be called "infrastructure," but for now it was appended apparently as an afterthought that did not find its way into the graphic designed to capture the overall program structure.

This graphic image was the first form of what came to be the defining iconography of SC—the SC pyramid. This diagram reified Kahn's vision of SC (see figure 2.3). The fundamental concept is of a hierarchy of research categories. Microelectronics, at the base, provides the technical components that will make possible the supercomputers on the second level. These in turn will be the platforms on which AI, the next level, will run. These three layers support the military applications, the fourth layer, which held Cooper's attention. To Kahn, however, these are not

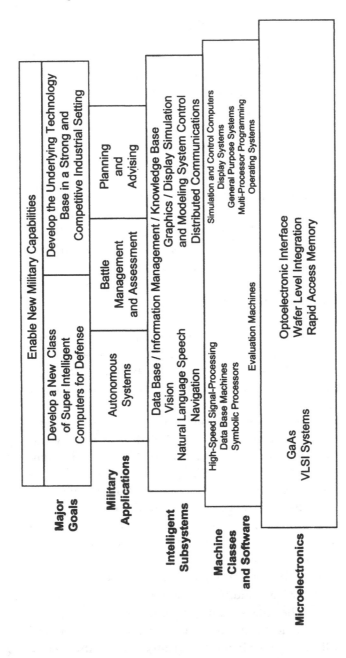

Figure 2.3

"Program Structure and Logic," the SC pyramid before it became a pyramid. Adapted from "Strategic Computing and Survivability" [May 1983], fig. II-1.

the pinnacle of the pyramid, not even the most important goal. To Kahn, the real objective, the "major goals" were "a new class of super intelligent computers for defense" (and, he might have added, for civilian applications as well) and "the underlying technology base in a strong and competitive industrial environment." These, in turn, would "enable new military capabilities" that could not yet be specified with any precision. They lay beyond a horizon that would be brought closer during the SCI program. Kahn's reluctance to adopt Cooper's applications early in the program revealed itself in the inversion of the pyramid in its top two levels. To him, these larger, more abstract, less concrete goals, outshone mere applications in significance and merit.

The SCI pyramid amounted to what Giovanni Dosi calls a technological paradigm.[86] This differs from a Kuhnian paradigm in that it is not necessarily shared across an entire field or discipline. Rather, it reflects the way a technological phenomenon is conceptualized within a particular community—a research laboratory, a company, a government agency. This one might be called the DARPA paradigm for computer development. Pyramidal in shape, resting on infrastructure, and culminating in "generic applications," the paradigm focused on "technological base" as the key to future development. This was fundamental research, largely in universities, which would percolate up from microelectronics through new architecture to artificial intelligence, making possible unimagined applications in machine intelligence. Kahn was proposing not just a plan but a paradigm, a way of conceptualizing the next decade of computer development and a plan for plotting what Dosi would call a "technological trajectory." As yet that trajectory was precise in neither direction nor terminus; rather the trajectory would be defined on the fly by evolving research results. It was not yet known if the goal—machine intelligence—was path dependent, or in Dosi's terminology, trajectory dependent.

In deference to Cooper's approach, Kahn included a second icon of strategic computing on the page next to the pyramid (see figure 2.4). This schema represented essentially the same information, but conceived as time lines, milestones, benchmarks, and metrics. It flowed, while the pyramid rose. It focused on movement toward goals. It introduced the notion of a "management timeline," going from "definition" through "experimental development" and "integration and test," to "demonstration" and "transition." Arrows bent upward along the way, suggesting movement and action. While the pyramid was static and monumental, this schema was dynamic and goal-directed. It looked for all

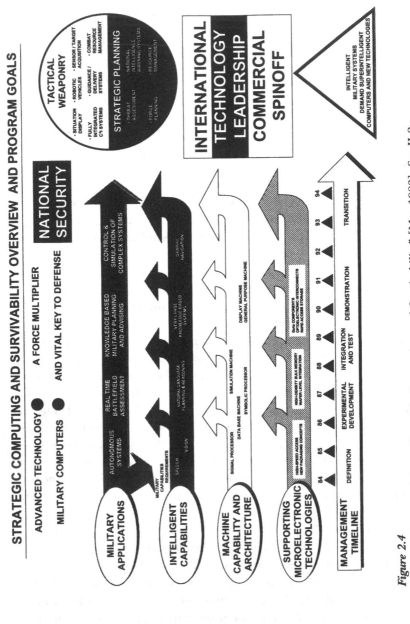

Figure 2.4
Time line. Adapted from "Strategic Computing and Survivability" [May 1983], fig. II-2.

the world like a PERT chart or a NASA time-line for getting to the moon by the end of the decade. The chart, generated by Cooper's subsequent deputy director, Charles "Chuck" Buffalano, working with Kahn and Craig Fields, finessed the differences between the two approaches. The computer community could embrace the pyramid and construe the program accordingly; Congress could fasten on the timeline and see what it wanted to see. It was a program for all seasons.

Unfortunately, the pyramid and the time line were not alternate models of the same reality. They were incompatible representations of fundamentally different paradigms. The SC plan embodied a cognitive dissonance that would preclude successful execution of the program. DARPA officials, from the director down through the program managers, had to agree on the goals and methods of SC. Unless they shared the same research paradigm, they could not hope to administer a coherent program in which the disparate elements connected with each other and reinforced each other. Those curved arrows in the time line and the implied percolation within the pyramid indicated how the various research projects were expected to cross-fertilize one another, to connect. To make that work, the community had to adopt one paradigm or the other. Kahn said build the technology base and nice things will happen. Cooper said focus on a pilot's associate and the technology base will follow. One paradigm is technology push, the other technology pull. One is bubble up, the other trickle down. Both can work, but could they work together? The tension between them stressed SC through much of its history.

Of Time Lines and Applications

Cooper remained dissatisfied with the SC plan. It was not so much the conceptual dissonance between his paradigm and Kahn's. Rather it was his perception that Kahn would not or could not grasp the imperatives of packaging the program for Congress. Cooper still believed in SC and he still believed in Kahn. He did not believe, however, that Kahn was willing or able to sell the program. In his view, Kahn's plan was still too cerebral, still too inner-directed, still too focused on what DARPA could do for computers, not yet focused enough on what computers could do for the military. No doubt Kahn believed that what was good for computers was good for the military. But Cooper needed a better argument to sell this program in the Pentagon, at the White House, and on Capitol Hill. He therefore sought help.

At this point, Kahn began to lose control of the program. One sympathetic program manager, Paul Losleben, later concluded that "the vision was larger than Bob Kahn was capable of—either capable of delivering or [being] allowed to deliver."[87] In the lore of SC, Kahn would always be "the godfather" of SC but he would never again exercise the exclusive control over his idea. When Kahn went into the hospital in September 1983 for elective surgery, Cooper moved decisively to reshape the program.

His instrument was Lynn Conway, a computer scientist recently recruited from Xerox PARC (Palo Alto Research Center) to work under Kahn's supervision in running the Strategic Computing Program within IPTO. Holding a 1963 Master of Science degree in electrical engineering from Columbia University, Conway had worked at IBM and Memorex before joining Xerox PARC in 1973.[88] She arrived at DARPA in August 1983, just in time to be caught up in the struggle over how to define and represent SC.

Lynn Conway had made her reputation in VLSI, the production of computer chips with 10^5–10^6 transistors. Her special contribution had been to help her colleague, Carver Mead of CalTech, reduce to practice the scheme he had developed for standardizing the design characteristics of VLSI chips. By establishing standardized protocols, Mead facilitated the production of chips and also made it possible to digitize and communicate chip design electronically. When he turned to Xerox PARC for help in reducing his ideas to practice, he teamed up with Conway. Together they wrote the standard textbook on the topic[89] and Conway tried out the idea during a year's leave at MIT. She taught a course there in which her students were allowed to develop their own chip designs. These were sent electronically back to Xerox PARC and then to Hewlett-Packard for fabrication. Within a matter of weeks, students had the chips they had designed, chips they could install, test, debug, and develop software for.[90]

Conway then turned to DARPA, seeking access to the ARPANET and underwriting of fabrication costs. Kahn approved the proposal and launched in 1979 what would become the metal oxide silicon implementation service (MOSIS) program at the Information Sciences Institute of the University of Southern California.[91] This was a key technology in the area that Kahn was calling "support systems" in his May proposal, but one that he would finally call infrastructure.

Kahn and Cooper admired Conway's organizational and entrepreneurial talents. Cooper had seen her in action at a recent conference and had come away impressed. She was smart and tough, she had done cooperative projects at Xerox PARC and elsewhere, and she had worked in both academia and industry. Perhaps most importantly for Cooper, she was clearly a process person, someone who could do for Kahn what she had done for Mead. Mead had had the idea for VLSI; Conway helped him reduce it to practice. "Bob was a content person entirely," said Cooper.

He managed things in detail and he had the structure of the arrangements, the goals and objectives, the tasks that needed to be done, down at the minute level of his brain. He didn't ever write things down. . . . He knew what needed to be done and he knew who he had to give direction to and how he had to interact with them and review what they were doing and so on. . . . There was not as much process in the execution of the plan as there was in the creation of the plan.[92]

Kahn assessed himself in similar terms. He said that it was his style "to actually formulate the project and run it, . . . because we couldn't even articulate what we wanted to do well enough to write it down."[93]

Cooper charged Conway to impose process on SC. While Kahn was in the hospital, he directed Conway to transform Kahn's plan into a document he could sell on Capitol Hill. Conway interpreted this command as a "clear and reset."[94] She would take Kahn's ideas, but begin afresh with a clean piece of paper. She organized a team of mostly IPTO principals and took them to a remote site, away from the distractions of their regular DARPA office. She chose the quarters of BDM, a commercial firm in business primarily to contract support services to government agencies such as DARPA. At that time, BDM was developing a contract proposal to provide support services for the SC program.[95] In BDM's McLean, Virginia, offices, some ten miles from DARPA headquarters in Arlington, Conway and her colleagues rewrote the SC plan. They narrowed the focus, strengthened the military applications, and enlivened the style.

By the time Kahn emerged from the hospital and returned to work, the revised plan was in its new and close-to-penultimate form. Kahn found it a "punchy" and "inflammatory" document, but Cooper appeared to like it. All it lacked, from Cooper's perspective, were time lines and milestones. Conway did not have the expertise to do these. Under pressure from Cooper, Kahn agreed to provide these, though on condition that the report would be "nameless."[96] Too embarrassed to claim authorship and too jealous to credit Conway, Kahn insisted that neither of them be credited. He called upon experts throughout the national computer community. Most were from the "ARPA community." He asked all of them to look into the future of their fields and predict where they might go in five or ten years with appropriate funding. Their charge was to be ambitious but realistic. Emphasize the possibilities, they were told, but do not promise the impossible. These predictions were then set out on elaborate time lines, which were coordinated with one another and integrated in the text.

The result was "Strategic Computing: New-Generation Computing Technology: A Strategic Plan for its Development and Application to Critical Problems in Defense," which appeared on 28 October 1983. Kahn's vision still permeates this document, but the approach is clearly Cooper's. It is a proposal to fund the computer research community, but it reads like a manifesto to transform the military. The consequences of the struggle between Cooper and Kahn to define strategic computing can be seen most clearly in the ways in which this final document differs from the one drafted by Kahn in May.

The most important change is the pervasive emphasis on applications. Instead of applications being a possible payoff of a computer research program, applications are now the stimulants that will pull technology to higher levels. In Kahn's view, microelectronics would make possible improved architectures on which AI would run. This in turn would make possible the intelligent machines that could do all sorts of wonderful things for the military. In the new dispensation:

- Applications drive requirements of intelligent functions.
- Intelligent functions drive requirements of system architectures.
- System architectures drive requirements of microelectronics and infrastructure.[97]

The impetus, in other words, flows down Kahn's pyramid, not up. Technology advances by being pulled from above instead of pushed from below. The two modes of technological development are not entirely incompatible, but the underlying goals and philosophies are profoundly different.

The way in which applications were to drive technology in this new rendering of strategic computing is apparent in the descriptions of the main applications. The "Autonomous Land Vehicle," successor to the autonomous underwater vehicle of the previous version, is exemplified by an unmanned reconnaissance vehicle that could travel 50 kilometers cross country at speeds up to 60 km/hr, conduct a reconnaissance mission, and report back its findings. This would entail planning a route, navigating by visual reading of topographical features and landmarks, seeing and dodging obstacles, reconnoitering the target, and transmitting the data. This, said the plan, would require an expert system of approximately 6,500 rules firing at 7,000 rules per second.[98]

Then current expert systems, software programs that mimicked certain domain-specific human behavior, contained about 2,000 rules firing at 50–100 rules per second. The vision system in such a vehicle would have to process 10–100 BIPS (billion equivalent von-Neumann instructions per second), compared with the 30–40 MIPS (million instructions per second) achieved by contemporary von-Neumann machines. Furthermore, to be practical on the battlefield, the onboard computer should occupy no more than 6 to 15 cubic feet, weigh less than 200 to 500 pounds, and consume less than 1 kw of power—1 to 4 orders of magnitude improvement over existing machines.

So too with the "pilot's associate." Envisioned as a kind of R2D2[99] to help an aircraft pilot with everything from communication and navigation to battle tactics and mission strategy, this computer would have to operate in real time. Only one metric was given: the monitoring of basic flight systems such as power, electrical, and hydraulic systems could require processing rates two orders of magnitude faster than currently achieved. But the associated time line for expert systems development gave a compelling example of what was entailed. By 1993, it was predicted, expert systems would be able to operate at five times real time. This meant that in 20 seconds the systems could simulate 100 seconds of real time.[100]

Expert systems would also be driven by the "battle management" application, which posed a category of problem known as "decision making under uncertainty."[101] In the naval battle management system envisioned for SC, the expert system would "make inferences about enemy and own force order-of-battle which explicitly include uncertainty, generate strike options, carry out simulations for evaluating these options, generate the OPLAN [operations plan], and produce explanations."[102] This would demand an expert system running some 20,000 rules at processing speeds of 10 BIPS. The natural language system to communicate with the expert system would require 1 BIPS by itself. The space-based signal processing requirements to support this system with intelligence and communications would require gallium arsenide components operating at speeds of 200 megahertz with tens of milliwatts of power radiating at the order of 10^7 rads [radiation absorbed dose, equivalent to 0.01 joules of energy per kilogram].

The driving force of applications on technology is only the most striking change in this document from the one Kahn drafted in May. Other changes were equally telling. The SC pyramid was refined (see figure 2.5). A new level, infrastructure, was added at the bottom. It incorporated what Kahn earlier had called support structure: networking, research machines, and other resources that the computer research community needed to do its work, resources that were common to the entire enterprise and benefitted everyone. The next three layers were the now-familiar areas in which the great breakthroughs were anticipated: microelectronics, architecture, and AI—the last rendered here as "intelligent functional capabilities." These three were now given the common label of "technology base." These were the focus of Kahn's interest. Above them now were two layers instead of three. Immediately above were the three specific applications, one approach for each military service, which were presented as driving development of the technology base. And above those was placed Kahn's version of what the appropriate goal of the program should be: "a broad base of machine intelligence technology." This resonated with the "technology base" which had all along been at the heart of Kahn's agenda and encompassed the great mass of the pyramid.

A narrowing and sharpening of focus also was present. Three applications were targeted for special attention. The broad range of architectural technology was narrowed to four: multiprocessing, signal processing, symbolic processing, and multifunction machines. Artificial intelligence was narrowed from the sixteen topics identified in September 1982 and the twelve of May 1983 to just four: natural language, vision, speech, and expert systems.[103] As with the other parts of the technology base, these were to be driven by applications. The pilot's associate would drive speech, the autonomous land vehicle would drive vision, battle management would drive natural language, and all three would drive expert systems.

If this was Kahn's pyramid with a bow to Cooper, then the accompanying time lines embodied Cooper's vision with a bow to Kahn (see figure 2.6). Appendices at the back of the document offered separate time lines for each component of the program, one for each level of the pyramid, and a separate time line for each of the three applications. They ran from 1984 through 1993, clearly indicating that this was meant to be a ten-year program. A disclaimer warned the reader that the "planning

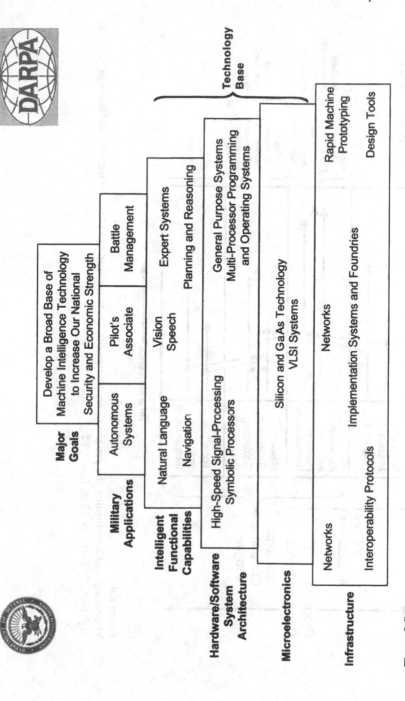

Figure 2.5
SCI pyramid. Adapted from "Strategic Computing" plan, 28 Oct. 1983, fig. 4.1.

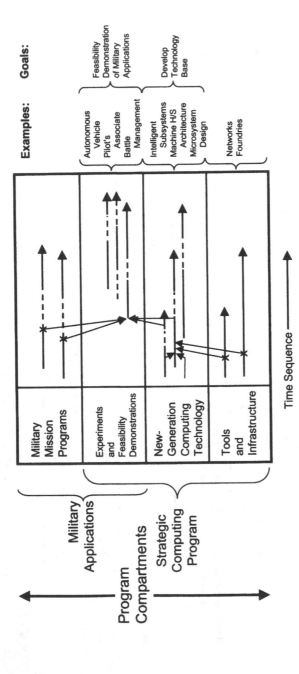

Figure 2.6
Visualizing and interpreting the program's compartments, elements, and time line. Adapted from "Strategic Computing" plan, 28 October 1983, figure 4.2.

framework" in the appendices "is not a closed 'PERT' system chart." The authors clearly wanted to distance themselves from the NASA style of management and program packaging that came to be associated with Cooper. These time lines were to be read instead as describing a "technology envelope," a conceptualization that was "open to accommodate available opportunities and competively [sic] selected technologies during appropriate time windows."[104] This apology echoed Kahn. Contingency must shape any research program that was truly pushing the bounds of the unknown.

In terms of path dependence, the new SCI plan imposed upon the program the burden of point solutions. Not only would the program research trajectories have to find their way to the realm of machine intelligence, they would have to intersect specific applications. Cooper appeared to envision an ideal trajectory that would proceed directly, predictably, and on time to an application. Kahn believed that there were many possible paths and that some would arrive not only in the realm of machine intelligence but also in a predictable subset, such as "vision." Such fields, he believed, were ripe. But he also suspected that some trajectories might never penetrate the realm of machine intelligence, let alone one of the ripe subfields. He believed it entirely unrealistic to define in advance a point solution that might be intersected by one of the program's research trajectories. Indeed, point solutions would come to be seen within SC as a classic conceptual trap. Point solutions worked as targets, but what the trajectory gained in direction it lost in energy. Time and resources invested in trying to stay exactly on line retarded movement. Projects ground to a halt for lack of resources.

The desideratum was informed flexibility. Research trajectories constrained early on by a point solution wasted resources. On the other hand, research trajectories that always followed the path of least resistance lacked direction. The former, though on line, would never reach its goal. The latter might whiz through the benchmarks ahead of schedule only to miss the entire realm of machine intelligence. Kahn knew how to avoid the two extremes, but he did not know how to get his judgment and his insight institutionalized in the plan.

The title announced that SC was a "New Generation" program, the U.S. version of the Fifth Generation. Like the Japanese program, it was clearly aimed at "machine intelligence technology" to be achieved by enhanced AI running on faster machines. As with earlier versions, it proposed to spend about $600 million in its first phase, but it envisioned getting up to speed faster—$50 million in the first year—but leveling out at $150 million a year instead of the $225 million Kahn had suggested in 1982. To do the research, the plan proposed the creation of approximately ten new computing communities and five to ten new "applications communities," each at about the size of a small research center, say around 100 professionals. These were reminiscent of the centers of excellence promoted by IPTO in the 1960s and 1970s.

A chapter on "Program Management" stressed bureaucratic issues but failed to come to grips with the technology integration that distinguished SC from other such research ventures. The ambitious goal of advancing an entire technological field in tandem begged elaboration of exactly how the various components would be integrated, but on this point the plan was silent. Within the government, DARPA would be the lead agency for SC, working with the White House Office of Science and Technology Policy, the Departments of Energy, Commerce, and Defense, the National Science Foundation, and the National Aeronautics and Space Administration. Within DARPA, IPTO would be the lead office, though "significant responsibilities [were] allocated to other offices."[105] In spite of Conway's penchant for process, this plan provided less detail on cooperation with other DARPA offices than did Kahn's earlier versions, a reflection, perhaps, of Conway's newness to the organization. Instead, Conway stressed panels, working groups, internal reporting, communication, budgeting, and progress reviews—connection by networking.

All of this was fine, but it failed to address the two most important issue of process in the whole undertaking. Both flowed from Cooper's emphasis on applications. First, how could managers of applications programs begin designing their systems when the new technology on which they were to be based had not yet been developed? Could they depend upon research at the lower levels of the SC pyramid to deliver the predicted capabilities on time? If so, could this research agenda be all that challenging? If not, what back-up technologies would they have to rely on? If they were going to design their systems with off-the-shelf technology, then they did indeed look like engineering development programs of the NASA style, not research programs of the DARPA style. Especially not a DARPA program as ambitious as this one.

The July 1983 version of the "Strategic Computing and Survivability" plan had admitted that success would depend on "technological breakthroughs."[106] By definition, a breakthrough is a leap into the unknown, a step up to a plateau of technological capability heretofore impossible. They may be anticipated, perhaps, but they cannot be predicted or foreseen; otherwise they would not be breakthroughs. Unknown and un-knowable, they left applications managers little to work with. A capability might come, but researchers could not say when it would appear or what it would look like.

Second, if applications were going to rely on developments in the technology base, how would the two realms be coordinated, connected? By what mechanism would demand-pull reach down through the layers of the pyramid to shape microelectronics research or computer architecture? By what mechanism would new research results flow up the pyramid into waiting applications? By concentrating on what was to be done, the SC plan had neglected how to do it. Kahn carried a plan around in his head, but Conway appears not to have known about it or not to have cared for it. Thus, the director of SC and her boss launched the program with differing views of how it would run. Soon, those views proved to be incompatible.

3

Lynn Conway: Executor

Connecting with the Customer

The Strategic Computing Initiative changed the entire landscape of information processing at DARPA. It changed the DARPA budget. It changed the visibility of the agency. It changed IPTO, its way of doing business, and its relationship with the computer research community. It changed the working relationship among DARPA offices. And finally it changed the computer research agenda in the United States. Not all of these changes were unprecedented. Not all of them were permanent. Not all of them were profound. But DARPA and computer research would never be quite the same again.

The magnitude of the change that was under way revealed itself in a curious artifact of the struggle to write an SC plan. Having overseen a critical phase of the plan's drafting, the revision of the document that took place while Robert Kahn was hospitalized, Lynn Conway displayed a pride of authorship that seldom surfaced at DARPA. She had a significant number of copies of the plan bound in expensive, hard cover. The result was a kind of presentation document, something more suited for a coffee table than a working government agency or a research laboratory. This was surely not Robert Kahn's style, nor the DARPA-IPTO style. They preferred the innocuous typescript and soft cover of the working paper, emphasizing content over packaging, substance over image.

This difference of style mirrored the larger conflict between Kahn and Cooper over technical base and applications. Kahn saw clearly in his own mind what SC might accomplish, but it was a vision of bits and bytes and cycles per second. It was all Greek to the generals and members of Congress who would have to approve the funding. When Kahn reduced his vision to a plan, it came across to the layperson as technical, impenetrable, otherworldly. Cooper insisted that he couch the argument in terms of products, concrete results such as R2-D2, robot tanks, battle management wizards. Kahn conceptualized the plan; Cooper packaged it. In this endeavor, Conway was Cooper's loyal aide. She spoke's Kahn's language but she put it in a form Cooper could sell at the Pentagon and on Capitol Hill. Pleased with her work, she bound it as she saw it, a glossy package that would attract customers.

The selling of SC differed little from the advocacy surrounding other big science projects funded by the government. The rhetoric escalated with the budget. The bigger the price tag, the more exaggerated the hype. Commercial nuclear power was going to provide electricity too cheap to meter. The space shuttle was going to make spaceflight routine and inexpensive. The breeder reactor was going to produce energy and new fuel at the same time. The superconducting super collider was going to unlock the secrets of the atom. Star Wars, the sister program to SC, was going to build a dome of security over the United States. When billions of dollars are at stake in scientific and technological undertakings of incomprehensible complexity, projects must be reduced to a language and an image that politicians and their constituents can understand. Scientists and engineers may shudder at the vulgarization of their work, but without this salesmanship the funding does not flow.

In his first year in office, President John F. Kennedy was looking for a bold technological initiative to distract public attention from the Bay of Pigs fiasco and to restore America's reputation for scientific and technological preeminence, recently tarnished by Soviet achievements in space. His science advisor, MIT President Jerome Weisner, suggested desalinization of sea water, an undertaking with enormous implications for health and demographics around the world. Kennedy replied that Weisner was a great scientist but a lousy politician. The astute president chose instead the Apollo moon mission, an undertaking of great political salience and little scientific substance.[1]

In 1983 DARPA-IPTO faced its moon decision. Kahn spoke for the desalinization school. Cooper and Conway were in the Apollo camp. First, get the money. Working with ten years and a billion dollars, DARPA would surely produce something of value. So long as there was payoff that the politicians could report to their constituents, no one would question if it was exactly the results that had been promised. Kahn preferred more truth in advertising. Conway's glossy, hardcovered prospectus would sell the program to DARPA's customers, but its claims might prove hard to defend ten years hence.

Those customers came in three categories: the computer research community, the Reagan administration, and Congress. All had been primed to expect and accept SC, but all needed to hear a formal pitch and scrutinize exactly what the plan entailed.[2] The pitch was a little different for each community.[3] Predictably, each community came away with a somewhat different understanding of what SC was and what it would do. Those differing views shaped people's expectations of the program and determined in the end their impression of whether or not it succeeded.

The easiest sale was to the computer research community. The reason was simple but nonetheless profound: this was new money. DARPA might have launched such an initiative by redirecting existing funds, but this would have produced a zero-sum game in which the new undertakings came out of someone else's hide. The allegiance of the newly supported researchers would have been bought with the enmity of those cut off. Instead, the SCI left the extant information processing budget intact and layered on top of that the $1 billion new dollars that would finally flow through this program. In Washington parlance, this is what "initiative" meant. No one's ox was gored; all boats rose on this tide.[4]

Cooper, Kahn, and Conway all participated in making this pitch, sometimes separately, often together. They attended meetings of ARPA principal investigators (PIs), because their program was going to cut across office lines and influence research agendas in fields outside of information processing. They met with IPTO PIs, many of whom would move into new SC fields or have their research shifted from existing budgets to SC funds. They met with computer professionals outside the DARPA community, both to win their support for the general concept behind SC and to encourage them to submit proposals for the new undertakings.

In these circles Kahn was well known and highly regarded; he had done important work in networking and packet switching, he had funded sound projects as IPTO director, he was smart and knowledgeable, and he was a straight shooter.[5] Conway was known by reputation for her work with Carver Mead on VLSI, but she had neither Kahn's network of personal relationships nor his stature as a central player of many years' standing. Cooper was known to most simply as the DARPA director smart enough to back a large investment in computer research, though he quickly won his own following with his firm grasp of the technical issues and his frank, unadorned, common-sense style of discourse.

The pitch to the computer community rested on the three pillars of the SCI argument—in the order that the audience wanted to hear.[6] First,

SC would underwrite infrastructure and build the technology base. Second, it would keep the United States ahead of the Japanese, about whom many in the computer community were worried. And third, the program would naturally provide useful products for the military.

Community response was at once positive and skeptical. Computer professionals appreciated the funding but doubted the claims. Willie Shatz and John W. Verity called the SCI "remarkably ambitious."[7] Dwight Davis, editor of *HT*, observed that the applications "pose[d] staggering technical challenges."[8] But commentators appeared to believe that exaggerated claims were DARPA's problem, not theirs. They showed no concern that a failure for SC might tarnish the reputation of the entire field and jeopardize future funding. Rather, they worried that the tidal wave of funding from SC would flow to established centers of excellence and not to the computer community generally.

DARPA had two answers. First, there were many new opportunities here and it was not clear that even the centers of excellence had the resources to bid on and win all the contracts that would flow from SC. Second, and more importantly, Cooper would insist that SC programs be put up for competitive bidding.[9] Even if the lion's share of IPTO funds had gone to centers of excellence in the past, this was no longer automatic. Competitors for SC funds would be judged on their merits. No doubt many in these audiences suspected that the centers of excellence had used past IPTO funding to accumulate more merit than less favored research centers, but in principle there was no arguing with the challenge.

More controversy arose over the military purposes to which SC research would be bent.[10] Leading the criticism were the Computer Professionals for Social Responsibility (CPSR), a newly formed, nonprofit organization whose early history paralleled the birth of SC. An electronic discussion group had formed at Xerox Palo Alto Research Center (Xerox PARC) in October 1981 to share concerns over the threat of nuclear war and the role of computers in strategic weapons systems. An organizational meeting the following June adopted the name CPSR and led to formal incorporation in March 1983, just two weeks before President Reagan announced the Strategic Defense Initiative (SDI).[11] CPSR focused on this one "main" issue over the next several years, naturally drawing SC into its purview as well.[12] For its first few years, recalls former CPSR chair Terry Winograd, "CPSR was basically a single-problem organization."[13]

Their critique took two forms. First, they insisted that SC was bent toward producing what they characterized as "battling robots" and "killer robots."[14] In their view applications drove the program. No computer researcher could work on chip design or software or even cognitive theory without contributing in the end to the production of autonomous killing machines. These machines were evil both because they were killers and because they were autonomous. The lethality of computer-based weapons was clear enough on the face of it. The question of autonomy was more oblique but nonetheless troubling, especially in an age of nuclear weapons.

Did we, they asked, want to design machines empowered to make life-and-death decisions? Would not the autonomous land vehicle, nominally a reconnaissance vehicle, soon turn into a robot tank? And who would program its rules of engagement? What computer would detect friendly forces in its gun sights, women and children in its path?[15] What artificial intelligence in any foreseeable future would allow a pilot's associate or a battle management program to make such fine distinctions? What did the battlefield of the future look like without human agency?[16]

The question of autonomous weapons systems took on horrendous proportions when applied to nuclear war. This issue was joined in President Reagan's SDI, which had been named by Cooper and was partially housed in DARPA. The timing of the two initiatives, the similarity in their names, and the coincidence of their location in DARPA all contributed to the widespread impression that SC was really just a cat's paw for Star Wars. The real cost of Star Wars, some believed, would be hidden by lodging some of the money in SC, but the goals were the same. Indeed, computing soon emerged in public debate as one of the pacing items of Star Wars. It was argued that no one knew how to write an error-free program complex enough to control a layered defense system.[17] Surely there would be bugs in any such program, and those bugs would have the potential to launch nuclear weapons and start World War III. This was tantamount to a doomsday machine.

Cooper was the primary target of CPSR. Members were comfortable with Kahn personally, with his emphasis on technology base, and with his vagueness about applications. Cooper was less well known to them, less attuned to their sensibilities, and less abashed in his enthusiasm for the Reagan defense buildup. They took particular alarm from a famous piece of testimony given by Cooper before the Senate Foreign Relations Committee on President Reagan's ballistic missile defense system. Appearing with other administration witnesses, Cooper found himself drawn into a testy exchange with senators Paul Tsongas (D-MA) and

Joseph Biden (D-DE). Cooper shared their distrust of the president's SDI, but he tried nonetheless to be a good soldier. The two senators worried about the same issues troubling CPSR, an autonomous system empowered to take military action against a perceived Soviet missile attack in its initial burn phase. How much time, the senators wanted to know, would the United States have to decide whether or not to fire defensive weapons in space? Could the president realistically be notified and make an informed decision? Cooper dismissed the question as hypothetical and became increasingly irritated as the senators continued to press it, calling Tsongas's argument spurious and suggesting that Biden was turning the hearing into a "carnival." In the end, Cooper appears to have raised more concerns than he calmed, prompting Senator Biden to say with uncharacteristic bluntness, "you have convinced me that I don't want the program [SDI] in the hands of a man like you."[18] CPSR members concluded from this exchange that they did not want SC in his hands either.[19]

Tellingly, the CPSR critics attacked SC on exactly the same grounds that Cooper advocated it—AI. Cooper took pains testifying before Congress to distinguish between mere supercomputers and the specialized supercomputers SC would develop. Existing supercomputers, he told Congress in June 1983, were designed for number crunching. They did complicated numerical calculations, mostly for scientific purposes. The military had little use for such machines. SC sought supercomputers capable of symbolic processing. These were the machines that would finally realize the potential of artificial intelligence.[20] But it was just these machines that most alarmed the CPSR. AI, said Severo Ornstein, had been "full of false promises and hype."[21] Computers, maintained Ornstein and AI pioneer Terry Winograd, were unreliable and unpredictable. To put them in control of nuclear weapons, as the Star Wars program promised to do, was to court disaster.

Ironically, Cooper himself opposed the SDI, as did his boss, Richard DeLauer. Like others in the Pentagon, they did not believe it could work. Once begun, they feared, it would drain resources from other, more worthy programs. Sensing this opposition within DoD, the Reagan White House elected to give SDI independent status, lest it be suffocated by a hostile bureaucracy. Cooper and DeLauer cooperated, moving resources from DARPA and other agencies to create the Strategic Defense Initiative Office (SDIO). "The SDI technologies," recalled Cooper, "were not going to be successful in my lifetime. . . . I'm a results kind of guy. I like to see what happens before I die, so I dumped as many of those [technologies] as I could into the Strategic Defense Program and got rid of them."[22]

SDI was not so much a new program as an amalgamation of existing research on ballistic missile defense. When that research was collected from agencies around the government, including DARPA, it was found to total $1.6 billion. Then $400 million in new money was added to cross the $2 billion threshold viewed as commensurate with the rhetoric of President Reagan's announcement. The funding was invested in SDIO.[23] Critics of SC were not dissuaded by these actions, but concerns about the program subsided over the years as the SCI and the SDI proceeded on different historical trajectories.[24]

Concerns of the critics notwithstanding, the computer community by and large embraced SC. Traditional members of the ARPA community, such as the IPTO centers of excellence, made proposals to conduct SC projects. Start-up companies such as Thinking Machines, Teknowledge, and Intellicorp submitted successful proposals that drew them into the expanded DARPA community.[25] And industries with no prior experience in this kind of research, such as aerospace giant Martin-Marietta, took advantage of the opportunity to expand their portfolios. SC was a gift horse whose teeth most researchers were loath to inspect.

Selling the SCI within the Reagan administration was almost as easy. Cooper and his boss, Richard DeLauer, had done their homework before the plan was drafted. They knew the Reagan defense agenda and they saw how SC could fit in. At meetings of the Defense Science Board and in summer studies in 1981 and 1982 they had taken the pulse of the Reagan defense team. Because DARPA had a good reputation within the DoD, and because its budget was seen as independent of anyone else's, there was little to fear from bureaucratic politics or turf struggles. The opposition of CPSR to the SCI had enhanced its credibility in the Pentagon.[26] Nonetheless, coolness or even indifference on the part of the armed services could still derail the program. If the services did not believe that SC would ultimately benefit them, then it might be difficult to convince others to fund it.

It was for this reason that Cooper had insisted on couching the program in terms of applications. Admirals and generals would not likely endorse VLSI or multiprocessors, but they might support a system that would help a pilot hit his target or help a fleet commander take on multiple targets at once or send an unmanned reconnaissance vehicle across a minefield. Just like CPSR, uniformed officers looked at SCI and saw walking war machines, but in their eyes, this was good.

To cement this perception, Cooper and DeLauer had funded a study of "military applications of new-generation computing technologies" by the influential Defense Science Board. When Chairman Joshua Lederberg reported in December 1984, well after SC was under way, he endorsed the applications identified by DARPA and appended some that had not yet been incorporated in DARPA plans.[27] Added to the autonomous vehicle, the pilot's associate, and battle management were warfare simulation, logistics management, and electronic warfare. Also included was an endorsement of ballistic missile defense, but this function had already been transferred from DARPA to SDIO.

Included also in the generally positive review of defense applications was a serious note of "dissension." Two of the panel members, Frederick P. Brooks, Kenan Professor and Chairman of the Department of Computer Science at the University of North Carolina, Chapel Hill, and Robert R. Everett, President of the Mitre Corporation, tended to "fall off the bandwagon" of SC at about the same point—around AI and applications. The two men brought enormous stature and credibility to the study. Everett had played key roles in development of the Whirlwind computer and the SAGE air defense system; he had formed MITRE Corporation in 1959 to do systems integration for SAGE. Brooks had led the software development for IBM's landmark System 360 computer in the 1960s, later collecting his insights in a book, *The Mythical Man-month*, that took on near-Biblical reputation in the computer community.[28] Both doubted that AI was as ripe as the plan suggested, and both had various reservations about the particular applications chosen to drive the program.

Rather than increased computer power, Everett thought the military needed "much better understanding of the problems we are trying to solve and of the ways to organize to solve them."[29] Brooks felt the plan "overemphasizes the glamorous and speculative *at the expense of the doable*," and feared "wasting a generation of our brightest young scientists, as they beat fruitlessly on unripe real problems or over-abstracted toy ones."[30] Their criticism revealed an appreciation of the issues that separated Cooper and Kahn. They scored Cooper's position for overselling the program and promising "glamorous" applications that were not demonstrably attainable. They doubted Kahn's belief that the field was poised to achieve AI. In both cases, they felt the SCI proposal concentrated too much on what the technology might do and not enough on what the military services might need. Kahn proposed to push the technology into new applications, while Cooper invented applications that he expected to pull the technology. Neither listened very carefully to what the services wanted.

In an appendix to the Defense Science Board report, Brooks offered an entirely different model of how computer development could best serve military needs. He recommended the promotion of "super microcomputers," which he believed were already being driven in healthy ways by market forces.[31] His recommendation prophesied an important technological trajectory that SC did not anticipate and did not lead. Neither Brooks nor Everett sought to derail SC, only to suggest that the applications it promised would be hard to realize. Their concerns were muted, however, embedded in the recesses of a report that addressed a fait accompli. They had little power to shape the program, let alone stop it.

Nor would Congress prove to be an obstacle. Once again, Cooper had already done his homework on Capitol Hill before the SCI plan was announced. In preliminary discussions with key members of Congress and staff aides, Cooper had learned that a project such as this seeking hundreds of millions of dollars would have to have a clear payoff. The applications, then, one of the three pillars of the SCI argument, were designed as much to satisfy Congress as to persuade the DoD. The second pillar, building the technology base, appealed on Capitol Hill because DARPA had a strong record of selecting the right technologies and investing in them wisely. Most members of Congress could not say with any specificity exactly what it was that DARPA funded, but they appear to share the conventional wisdom that DARPA's high-risk and high-payoff policy worked.

The pillar that did most to sell the program on Capitol Hill was the one that Kahn and Cooper had least conviction about: Japan. Congress was more exercised by Japan's Fifth Generation program than either the Reagan administration or the computer community. Much of the consternation flowed from a book conceived in 1982 and published in 1983. *The Fifth Generation* by Edward Feigenbaum and Pamela McCorduck argued that the Japanese were about to run away with yet another technology invented in the United States: artificial intelligence.[32] Having visited Japan and taken the measure of the Japanese program announced with such great fanfare in 1981, Feigenbaum and McCorduck concluded that the United States was blind to the danger on the horizon.

The computer community derided Feigenbaum as "chicken little," but Congress embraced him as a seer prophesying doom. In testimony before the House Committee on Science, Space, and Technology, he infected the legislators with his sense of alarm. "The era of reasoning machines is inevitable," he told them. "It is the manifest destiny of computing."[33] Long a beneficiary of IPTO funding, Feigenbaum recommended DARPA as one of the institutional responses the United States might call upon to ward off the Japanese challenge. He went so far as to compare DARPA to Japan's legendary Ministry of International Trade and Industry (MITI).[34]

Congress embraced Feigenbaum's sense of alarm. Anthony Battista, a leading staff member of the House Armed Services Committee, told Cooper in April 1983 that the Reagan administration in general and DARPA in particular appeared indifferent to "the Japanese threat." In Battista's view, the Japanese were "invading our microelectronics industry in Silicon Valley. . . . Look what they have done to us in the automotive industry," he enjoined Cooper.[35] "There is every indication . . . it is going to happen in computers." Far from questioning Cooper's request for $50 million in FY 1984 to start SC, Battista recommended that DARPA spend "a substantially higher" amount.[36]

Congressional alarm over Japan's computer development programs was apparent in other initiatives then taking shape. In 1982 and 1983, U.S. microelectronics firms launched a spate of cooperative research ventures designed to compete with the government-industry consortia then driving Japan's rapid development. The largest and most visible of the American ventures, the Microelectronics and Computer Technology Corporation (MCC), was formed in January 1983 and grew by 1985 to twenty-one members and a research budget approaching $70 million. Congress helped by passing the Joint Research and Development Act of 1984, limiting the antitrust liability of such ventures to one-third their normal level.[37] MCC was an attempt to realize some of the advantages of Japan's MITI without getting bogged down in the socialist mire of direct government intervention in the marketplace.[38] Because of its organizational similarities to Japan's program, MCC is often seen as the U.S. government response to Japan's Fifth Generation, a credit that others believe belongs to SC.

At the same time Congress was also approving yet another DARPA initiative driven in part by the Fifth Generation. SemaTech was a publicly supported, private consortium of microelectronics manufacturers, funded jointly by DARPA and industry. It was designed to reverse, or at least slow, the expansion of Japanese market share in chip—especially memory chip—manufacture. It was Japanese progress in microelectronics that led many in the United States to believe they could do the same in AI.[39]

And Congress had already approved the very high speed integrated circuit program (VHSIC), a joint industry-DoD program intended to accelerate the development of microelectronics for military applications. Once again, Japan was the stimulus. If chip manufacturing in the United States was driven to extinction as the television and consumer electronics industries had been, then the United States military could find itself dependent on Japan for weapons components.

Cooper, Kahn, and others, who had gone to Japan to see for themselves what kind of threat the Fifth Generation posed, came back with a very different view than the one that Feigenbaum had sold to Congress.[40] They thought the Japanese were far behind the United States in computer development and AI. What is more, they were sure that the Japanese were headed down the wrong path.[41] But if playing the Japan card would help to sell SC, then they would play it. "We trundled out the Japanese as the arch-enemies," Cooper later confessed, noting that in private conversations with congresspeople and senators he "used it . . . unabashedly." In fact, Cooper went so far as to assert that he went to Japan specifically to gather material for this argument.[42] The tactic worked. Congress formally approved the SCI in the Defense Appropriations Act of 1984. President Reagan signed it on the day it was passed, 8 December 1983.

In some ways, closure had been reached by interpretive flexibility. Strategic Computing was different things to different people. By packaging SC to sell, DARPA allowed the consumers to define what they were buying. Kahn, the true believer, always thought it would sell itself. Like the DARPA-IPTO community he admired and represented, Kahn foresaw strengthening of the technology base, realization of AI, and transfer of new capabilities to applications in industry and the military.

Cooper, in contrast, focused on industry. For him SC was a flagship program that would ornament his years at DARPA, enhance national security, and move the agency toward his model of industry-based, applications-focused projects. Conway saw it as her baby, whose success would be based on process. Her version of SC placed more emphasis on coordination, making developments both rise up Kahn's SC pyramid and move across Cooper's time line by careful management.

Richard DeLauer saw SC contributing to the Reagan agenda for the Cold War; in some ways, it, like SDI, was a technological fix to the Cold War. Some critics saw it as a thinly veiled program to create "killer robots." Others saw it as a front for SDI. Some believed it was an AI program, a response to the Japan's Fifth Generation.[43] In this sense it marked a renewal of support for this field after the cutbacks of the mid- to late-1970s. Some believed that supercomputing was the core technology.[44] It was ripe; it held promise for all fields, and it operated behind the public hoopla surrounding AI and applications. Even more specifically, some believed SC was a plan to achieve supercomputing through massive, scalable parallel processing.[45] Some believed it was an applications program, pure and simple. And some believed it was "a pot of money."[46] This last is a variation on the Kahn approach. As early as 1985, one participant at a conference on SC addressed the question of whether or not the program was successful. He said, "from my point of view, Strategic Computing is already a success. We've got the money."[47]

With the money came great risk and great opportunity. $1 billion was more than IPTO had spent in its first two decades, the golden era of interactive computing, graphics, ARPANET, and AI. To whom much is given, much is expected. What would be expected of DARPA ten years hence in return for this largesse? This is fundamentally a question of patronage. When the government functions as a patron of science and technology, what should it expect in return?[48]

In recent years scholars have explored the ways in which societies choose their technology, while paying less attention to the ways in which society patronizes technology.[49] Which projects are selected for research and development? What role is played by government? When does the government's role cross the line into national industrial policy? To what extent are favored projects self-fulfilling prophecies?[50]

The answers to such questions often turn on packaging. Programs win funding and achieve success based largely on expectations. The Apollo program, for example, was conceived as a gambit in the Cold War struggle for prestige, but it was sold as a first step in manned space exploration. It succeeded in the first role and disappointed in the second. Commercial nuclear power was packaged as offering "energy too cheap to meter." Americans turned against it when they discerned that it was more dangerous and more costly than advertised. The supersonic transport was rejected by Congress because it could not be packaged as economically or environmentally benign. All of these projects were technically successful: men went to the moon; commercial nuclear power generates seven percent of American electricity and even higher percentages for some other nations; supersonic transports fly for other countries. Yet whether one views them as socially successful turns on what one thought they were and what they were supposed to achieve.

So too with SC. The program won its funding because different groups viewed it differently. Even people within DARPA and observers entirely outside the decision-making process entertained disparate views of SC. Those views shaped the program as it unfolded and colored its history when it was done. Whether it succeeded or failed depended on what people thought it was supposed to be and do.

The Divorce of the Dynamic Duo

Just as there was no consensus about the nature of SC, so too was there no agreement about how to run it. Until October 1983 all attention had been invested in designing and selling the plan. That process had revolved around the personalities and agendas of Robert Kahn, Robert Cooper, and finally Lynn Conway. The search for a way to turn $1 billion into the results they sought would also swirl about them.

Kahn was the godfather of SC, and he had the first claim on organizing the program. His instinct was to run it fundamentally the same way he ran others. He carried SC around in his head, and he planned to manage it out of his hip pocket. One coworker has said that "the usual Kahn approach is that he gives you the brush and the fence and paints one board."[51] Kahn is a man of vision, but he is stronger on conjuring vision than articulating it. He can see in his own mind the enclosure that he wants to fence in, but he does not share such visions readily or well. His reticence, for example, had plagued the drafting of the SCI plan. For this reason, he seldom lays out a grand design and assigns tasks to his subordinates. Rather he meets individually with each subordinate, gives him a brush, and paints for him that one board. If the subordinate hesitates, Kahn does not insist. He simply paints another board. He leads by sweet persuasion and the power of his own convictions. People follow him because they have faith in his vision, not because they fully understand where it is going.

In awarding contracts for SC, for example, Kahn shared with his colleagues a matrix that gave some guidance on how he thought about types of systems and levels of support SC should fund. His architecture grid, for example, might have a vertical axis for types of systems and a horizontal axis for levels of support. The systems might be categorized as signal processing, symbolic processing, and hybrid-general purpose. The levels of support might be exploratory ($50–100 thousand), bread board (a few $100 thousands), and implementation ($\geq$$1 million). Kahn would invite projects that filled in these squares, implying that funding should be invested in different fields and different levels of development.[52]

As a management tool, matrices like these had some utility. They helped subordinates conceptualize the kinds of systems that Kahn envisioned and anticipate the level of funding that he thought might be appropriate. The latter anticipation, however, was presumptive. Kahn gave no indication of how funds were to be distributed among the three levels of support, only an implication that he expected a distribution among the levels. In other words, he did not want all small projects nor all big projects. He did not explain why systems should be divided into three categories instead of four, nor why there should be three levels of funding instead of two. He simply painted this board and enjoined others to do likewise.

With such devices, Kahn revealed how he envisioned the program unfolding. He did not, however, lay out for his colleagues a clear roadmap of how to get from here to there. In Kahn's mind, there could be no such roadmap. Research was by definition an exploration of the unknown. Where you might want to go tomorrow would be determined in part by what you discovered today. The leader, in this case the director of IPTO, had to stay informed and stay in front. Running such a program meant handing out new brushes day after day. As Cooper recalls it:

He [Kahn] managed things in detail and he had the structure of the arrangements, the goals and objectives, the tasks that needed to be done, down at the minute level in his brain. He didn't ever write things down; he knew what needed to be done and he knew who he had to give direction to and how he had to interact with them and review what they were doing and so on. And he went off and he pretty much did that. There was not as much process in the execution as there was in the creation of the plan. The plan was for public consumption, not for private management. We tried to make it [the SC prospectus] into an ARPA management document, but with Bob's style, that really was not possible, as much as probably the program needed in order to be completely successful.[53]

Conway had a different style. She was, in Cooper's words, "a process person, an organization process kind of person. . . . She was just what the program needed." Indeed Cooper thought she and Kahn had the makings of a "dynamic duo."[54] As Kahn himself stated it, she was "great at exposition, . . . at getting people stoked up."[55] Her style was to articulate a plan fully and then mobilize people to make it happen. She lacked the vision of a Carver Mead or a Robert Kahn to conceptualize something entirely new, but she had tremendous organizational skills at making a given vision materialize. It was that gift that made her attractive to Kahn and Cooper and brought her to DARPA. She was to run SC for Kahn, perhaps to succeed him when he left.

She saw SC as a process. For her, the specific goals and technologies were not as important as the mechanism by which the disparate parts of the undertaking would be coordinated. The SC plan had merely sketched out "a landscape of possibilities."[56] Now it was time to plot a course across the landscape. How were the programs within IPTO, to say nothing of the other offices in DARPA, going to pull together toward the same goals? Who defined the goals? Who determined if all the parts fit? How was the computer community to be recruited? Would the researchers want to do what DARPA thought was needed? How could they be educated to buy into DARPA's vision? How could university research be transferred to industry? In short, how could the disparate components of this sprawling program connect with one another?

DARPA was going to play the role of composer and conductor. It would invent a piece of music based on the tastes of the audience, the abilities of the available musicians, and the state of the art in composition. Then it would recruit the appropriate mix of musicians to play the score. It would get them all to play at the same time from the same sheet. And finally it would infect them with the sheer excitement of the undertaking. None of the musicians would have heard this score before. Most of them would not have played with each other before. Kahn had composed the score; Conway was going to conduct. That was what she was good at. That was what she thought she had been hired to do.

There was a fundamental disagreement here. Kahn may have been happy to have Conway conduct, but he was the musical director. He envisioned SC as being one of the programs that he ran from IPTO. It involved other DARPA offices, but the bulk of the funding was in IPTO and coordination of the program should flow from his office.[57] Conway saw SC as an agencywide program. When it was approved, she assumed the title of Assistant Director for Strategic Computing, meaning that she was assistant to Kahn. But the title could also be construed as assistant to the Director of DARPA, and apparently Conway interpreted it thus.

Robert Cooper might have clarified the relationship between Kahn and Conway, but he did not. If anything he muddied it. By dealing directly with Conway when Kahn was in the hospital in August 1983, Cooper fed the impression that Conway worked for him. Never entirely satisfied with Kahn's conception of SC, Cooper used Conway to keep Kahn from running away with the program. She rewrote the plan in a form he liked, and she promised to serve as a brake on Kahn's natural inclinations to take the program in the direction he had originally envisioned. Kahn remembers that Cooper used the term "creative tension" to describe the working relationship between Kahn and Conway.[58] Tension there surely was, but what it created was not good. In essence, Kahn and Conway found themselves locked in a struggle for control of SC. While they contended for control through the first half of 1984, the program went without unified leadership.

By July of 1984 Cooper finally had to step in. First, he began referring to the "Strategic Computing Program," dropping "Initiative" from the title. This signaled two changes. First, he wanted to de-emphasize that this program conveyed new money; the selling phase was over. "Initiative" waved a red flag in budget documents, provoking Congress to ask for new results. Second, Cooper wanted to regularize administration of the program, to move from an ad hoc, start-up activity to a routine program with standardized organization, procedures, and oversight.

With this end in mind, he made a series of organizational changes. He moved the Strategic Computing Program out of IPTO and into his own office. Kahn, while remaining director of IPTO also became the Director for Technical Development in the new entity. Conway, now moved out of IPTO, became chief scientist of DARPA and Assistant Director for Strategic Computing. Clearly she was assistant to the Director of DARPA, not to the Director of IPTO. On the face of it, Conway was now directing SC out of Cooper's office and Kahn, for purposes of SC, was restricted to the technical component. Part of this move was driven, no doubt, by Cooper's insistence that SC money be kept separate from IPTO money. But most of it appears to have been an attempt to ensure that Cooper and not Kahn would orchestrate SC.

Further evidence of this trend is found in the growing role played by Charles Buffalano, who appeared on the SC organization chart between Conway and Kahn. His title was Director of Applications, a topic dear to Cooper's heart. Buffalano was Cooper's right-hand man, a kind of chief of staff. Many executives like Cooper graft to themselves disciples like Buffalano as their careers take off. The executives need to attain a certain level of achievement and power before enjoying the luxury of taking followers from job to job. Cooper had reached that level, in part by moving from job to job. The ideal tenure is four to five years, the term that Cooper had set for himself at NASA and at DARPA. In that time, an executive has probably changed all that he or she is likely to change. It is best to move on before becoming type cast, and before, say the cynics, the bills come due for one's mistakes.

Buffalano had worked for Cooper at NASA Goddard, and though he did not follow Cooper directly to DARPA he joined him there in April 1984, just in time to help with the SCI problem. Buffalano was smart, calculating, tough, loyal, and efficient. DARPA people liked him, or at least respected him, even though his degrees were not of the right pedigree. He held a B.A. degree from Stevens Institute of Technology and a Ph.D. in Engineering and Applied Science from Yale.[59] His twenty-five years of experience in the aerospace field, including his interim service at BDM Corporation, the contractor that had provided a home for drafting the final SC plan, did nothing to enhance his reputation with the DARPA computer community. Given their respective roles and styles, it appears inevitable that he and Kahn would get crosswise with each other. Kahn refers to him as Cooper's "henchman," who began every sentence with "Bob Cooper wants . . . ."[60] There was about him an air of smug self-confidence that was offended by Kahn's stature within the agency. It was he who characterized Kahn as a cardinal, whose ring the IPTO faithful knelt to kiss.[61]

Buffalano's particular hobbyhorse was teaching DARPA how to do "Big Science." Both he and Cooper cherished their experience at Goddard and believed that they had learned there something about how to support and manage large-scale research and development. "Bob Cooper would remind us damn near every week about Big Science," recalled Paul Losleben in 1994, adding with undisguised glee that Cooper held up as an exemplary program NASA's Hubble Space Telescope, which was just then suffering public ridicule for a faulty mirror.[62] Cooper was characteristically measured in these attempts to impart his wisdom at DARPA, though many still came to believe that he was trying to force a NASA model down their throats. Buffalano was intemperate in the same pursuit.

From the days of Apollo program and the reign of NASA Administrator James E. Webb, NASA had cultivated the image of being unparalleled in the management of large-scale technology.[63] If there was a quintessential NASA management style it was clear definition of goals, control through PERT-style time lines, routine management reviews, and management by exception. The last technique called upon managers at every level to report periodically on their progress and seek help only when necessary. Their superiors up the chain of command could assume that component projects were on track unless a subordinate reported falling behind.[64] Perhaps the greatest strength of the NASA style was in systems integration, which was at the heart of bringing literally millions of parts from thousands of contractors and subcontractors together in an essentially fail-safe spacecraft.[65] The technique may have been better suited to development than research, to applying known ideas than searching for new ones. Nevertheless Buffalano, and to a certain extent Cooper, saw it as a model for what DARPA was trying to do with SC: connect disparate players and disparate components in a complex technological system.[66]

On one occasion Buffalano took Kahn and Keith Uncapher, an IPTO contractor, to visit Goddard and get the message first hand. The group met with the Chief Scientist of Goddard, who agreed to exchange views. He asked his guests to begin by explaining how DARPA worked. When his visitors were done, he confessed that he did not know why they were there because they obviously had a better system than NASA's. The meeting broke up quickly, leaving Buffalano with little to say on the long ride back to DARPA. Tales of the episode entered the folklore of the IPTO faithful, who saw it as a complete repudiation of Buffalano's conceit.[67]

Whatever his record as a tutor of big science, Buffalano's role in SC was clearly to keep Kahn and Conway from each other's throats, to ensure that the "creative tension" between them did not turn into a destructive feud.[68] He failed. By November 1984 the organization chart had been shuffled once again, producing an entirely new entity, the Engineering Applications Office (EAO) (see figure 3.1). It would be directed by Clinton Kelly, who had joined DARPA in 1981 as assistant director of the Defense Sciences Office and manager of the Autonomous Land Vehicle Program. The creation of EAO may be seen as a bureaucratic move to give Kelly and other managers of SC applications programs an institutional home conducive to cooperating with the services. Or it may be viewed as a vote on Cooper's part that he could not trust Kahn to give applications the attention and resources they deserved. It may also be seen as a bureaucratic mechanism to separate Kahn and Conway, whose relationship had continued to deteriorate. While retaining her titles as Chief Scientist of DARPA and Assistant [DARPA] Director for Strategic Computing, Conway moved into EAO with the title of Assistant Director for Applications.

It was as if the Kahn-Conway feud, which mirrored the applications and technology-base feud, had resulted in a divorce. The rift further eroded Kahn's control of the new program. Instead of lowering institutional barriers within DARPA and connecting participants from various

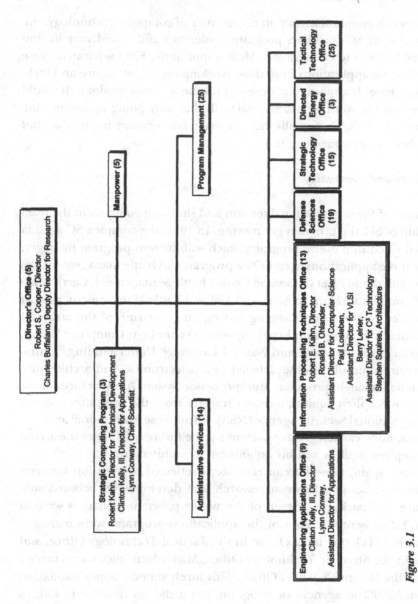

Figure 3.1
DARPA organizational chart, November 1984. *Source:* DARPA telephone directory.

offices in a common assault on the frontier of computer technology, the first year of SC fractured program leadership and raised new institutional barriers to cooperation. Most importantly, EAO separated those working on applications from those working on infrastructure and technology base. If applications were to pull the technology along, it would have to lift it over the institutional walls that were going up around bureaucratic fiefs. These walls did not make cooperation impossible, but neither did they facilitate it.

Process and Personality

In spite of the organizational tension and shuffling going on in the leadership of SC, the program got moving. In 1984, the complex SC agenda was divided into twelve programs, each with its own program manager. The three applications became five programs, as battle management was subdivided into Fleet Command Center Battle Management, Carrier Battle Group Battle Management, and AirLand Battle Management, the last designed to serve the changing war-fighting doctrine of the army. AI spawned four programs: Expert Systems Technology, Computer Vision, Speech Understanding, and Natural Language Understanding.[69] Infrastructure spun off a subset labeled Microelectronics. And architecture occupied a single program, Multiprocessor System Architectures, managed by Stephen Squires, a recent transfer from the computer division of the National Security Agency (NSA). Of all these institutional arrangements, none carried greater portent for the future of SC than the arrival of Stephen Squires and his assignment to architecture.

Once again, these bureaucratic steps reflected the tension between Kahn and Cooper, between research and development, between substance and packaging. Seven of the twelve program managers were in IPTO, five were not. Four of the applications programs were managed outside IPTO, one in EAO, two in the Tactical Technology Office, and one in the Strategic Technology Office. Microelectronics was managed from the Defense Sciences Office. This surely suited Cooper's ambition to make SC an agencywide program, but it did no violence to Kahn's ambitions either, for it kept the technology base largely within IPTO.

Each program manager prepared a plan that laid out how the program was to proceed, how it would fit into SC, and how it would transfer its results to other segments of the pyramid. As the plans emerged they were briefed to Cooper in monthly meetings attended by Kahn, Conway, Buffalano, and the other office directors and program managers involved in

SC. This process provided a kind of coordination for the program and ensured that all the principals knew what everyone else was doing.[70]

In the first year alone, this process produced thirty-nine research contracts and two for technical and management support.[71] The number of contracts let in different areas provides one measure of the ripeness of the fields and the confidence of the program managers that they knew where they wanted to go. Infrastructure and architecture, for example, let six contracts each and microelectronics let five. In contrast, expert system technology let only one, for an ongoing project at General Electric on reasoning under uncertainty. A similar pattern appeared in applications. The Fleet Command Center Battle Management Program let five contracts, while Pilot's Associate and AirLand Battle Management let only two and one contracts respectively for preliminary definition studies.

Of the thirty-nine research contracts, sixteen went to industry, fifteen to universities, and eight to government or private laboratories. The flow of dollars mirrored this distribution, but not exactly. Industry received 48.4 percent of the $49.1 million SC budget for Fiscal Year 1984, universities 43 percent, and the laboratories 8.6 percent. This pattern reflected the higher cost of the development work done by industry compared to the research work done in universities and laboratories. By program areas, the bottom of the SC pyramid received the lion's share of the early funding, with 29.8 percent going to microelectronics (most of that for GaAs pilot lines) and 23.4 percent going to infrastructure. Architecture received 19.1 percent, applications 14.7 percent, and AI 13 percent. This simply meant that promising activities were already up and running in areas at the bottom the pyramid. Applications and AI needed more preliminary investigation and program definition before they could absorb significant amounts of funding. Down the road, applications would eat up an increasing portion of the SC budget, as costly development work succeeded early planning.

These individual contracts varied from preliminary feasibility studies in the case of the pilot's associate to well-defined, mature projects such as the Butterfly Multiprocessor, which DARPA had already been supporting at Bolt, Beranek & Newman, Robert Kahn's former employer. As one of the program management support contractors, BDM Corporation, the company that had assisted in producing the SC plan, now helped to reduce the plan to practice. The task was daunting. In expert systems, for example, DARPA had so little institutional experience in how to proceed that it sent BDM out to interview ten manufacturers of existing experts systems and find out how long it had taken to produce their products. BDM supplemented this information with a literature search on the topic, designed to give some sense of what might be expected in the area of expert systems and where that contribution could be placed on an overall time line of the SC program. Other segments of the program entailed the use of different techniques. ARPA sponsored conferences and meetings to bring together experts in different areas. For some fields, advisory panels were created; pilot's associate, for example, put together a panel including F-4, F-15, and F-16 pilots with engineering, computer science, and flight test experience.

Every aspect of the program was projected as far as 1989 and beyond, and predictions were made about how contributions from one field would feed into others. The Butterfly Multiprocessor, for example, was recommended to run simulations of battle management even though the machine was not scheduled for evaluation until the third quarter of Fiscal Year 1986. In other words, they were betting on the come. Success would depend on integration of components that were yet to be developed. Developments like this were labeled Critical Path Technologies, meaning that they had to work on schedule to play an assigned role in one or more of the applications. Program managers of those applications were left to wonder if they dared rely on these developments or if they should explore off-the-shelf alternatives.[72]

Thus, even while SC got under way in a flurry of activity, some fundamental questions about its basic structure and philosophy remained unresolved. In the course of 1985, the problems were exacerbated by the departure of the founding leadership of the program—Kahn, Cooper, and Conway. SC, as a separate program in Cooper's office, disappeared by February. Remarkably, it did not reappear until June 1986. For over a year it appears to have gone underground, haunting the halls of DARPA without a home and without a head. Conway, the erstwhile "Assistant Director for Strategic Computing," left the agency in March 1985 to accept a position as assistant dean of engineering at the University of Michigan. Her DARPA colleagues sent her off with a framed drawing depicting her, incongruously, as Dorothy, headed toward Oz, while they flew off in a balloon (see figure 3.2).

It would have been far more appropriate for Conway to fly off in the balloon to her new home in the Midwest, while the DARPA stalwarts stayed on their yellow brick road to machine intelligence in the land of Oz. Lynn Conway's yellow brick road in Washington had hardly led to the fulfillment of her wishes. Instead it had been littered with enmity, disappointment, and bitterness. As one former colleague put it, she had

Figure 3.2
"Going Away" gift picture for Lynn Conway. *Source:* Lynn Conway.

been "hit by a buzz saw." She had played a critical role in getting SC launched, but she had not been able to bring her real organizational skills to bear on getting it going.

Following Conway's departure, a new figure in the SC story stepped forward to take her place. Craig Fields, a future Director of DARPA and perhaps the longest-serving professional in ARPA-DARPA history, had transferred from the Defense Sciences Office to become Chief Scientist of IPTO in June 1983. He had participated with Charles Buffalano in drafting the SC plan and he got on well with Kahn. More importantly, Kahn got on with him, which not everyone at DARPA did. Fields was by all accounts one of the most intelligent people ever to serve at DARPA, a quick study who knew a little bit about everything that was going on. But he never put his imprint directly on any program, for he moved from area to area, office to office, seemingly in search of the main chance. Some DARPA veterans saw him as being more concerned with his own advancement than the goals of the agency, winning him the sobriquet "Darth Vader" from one former colleague.[73] Fields moved again in November 1984 to become Deputy Director of EAO, working for Clint Kelly, who had asked for him.[74]

Robert Cooper left DARPA on 5 July 1985. True to his word when he accepted Richard DeLauer's invitation, Cooper had stayed at DARPA exactly four years, longer than any other director before or since. As his fourth anniversary approached he warned his superiors at the DoD that he would be leaving, but they found no replacement for him. So his deputy, Charles Buffalano, took over as Acting Director. Buffalano stayed only until September and then left to join Cooper at his new start-up company, Atlantic Aerospace Electronics Corporation. James Tegnalia joined the agency at that juncture to serve as acting director.

Three months after Cooper's departure and six months after Conway's, Robert Kahn left DARPA to form the Corporation for National Research Initiatives (CNRI). This not-for-profit organization functions something like a private DARPA, identifying and promoting national initiatives that hold out promise for the future of the country. From this base, Kahn would continue much of the work he began at DARPA. He had been planning this move for some time, and in fact had stayed at DARPA after Cooper's arrival only because he saw an opportunity to make a difference by working with him to launch SC.[75]

Within six months, then, the triumvirate that had conceived, designed, and sold SC left DARPA. So too did some of the program managers who worked in the trenches during these early years. Navy

Commander Ronald Ohlander, for example, had served as Executive Secretary of the Defense Science Board panel on military applications of computer developments and had been the lead program manager in computer science within IPTO. He retired from the Navy just before Kahn left DARPA and accepted a position with Keith Uncapher's Information Sciences Institute at the University of Southern California. Paul Losleben, a civilian program manager running VLSI, left at the end of 1985 to accept a research position at Stanford University.

The departure of this early leadership did not, however, stall the program. All had given direction and guidance to the program managers who were on the front lines. Projects were defined and researchers were identified. Funding flowed out the door and into the laboratories. Even some coordination was achieved, both among DARPA offices and among researchers in the field.

None of this, however, could substitute for a coordinated plan. If DARPA was attempting to move a whole field forward, how would it ensure that the component technologies were making the necessary progress? How would it attack reverse salients, stalled pockets on the advancing research front that threatened to retard progress on their flanks?[76] If applications were to pull the technology base, how would the program manager of the autonomous land vehicle, for example, find out what could be expected from the architecture program or from artificial intelligence? How did program managers in those areas know what the autonomous land vehicle needed? Would they shape their agenda to fit the application, to fit all three of the applications, or would they exploit targets of opportunity and trust that general progress must somehow convert down the road into a form applicable to ALV?

Kahn had a picture in his own mind of how this would happen, and he managed IPTO and his part of SC accordingly. Conway also had a notion of what the process would look like, and she managed the projects she took to EAO accordingly. Even Cooper had a view of how SC might come to look more like the management of big science he had come to know at NASA; he and Buffalano pushed DARPA in that direction from their respective positions. But none of them reduced their conceptualizations to a plan and none left behind a technique that could be taken up and deployed by others.

Most importantly, they left no management structure that would ensure coordination. Each area of SC had developed its own plan, including time lines that showed how their developments would be transferred up the SC pyramid (see figure 3.3). But all of those transitions were

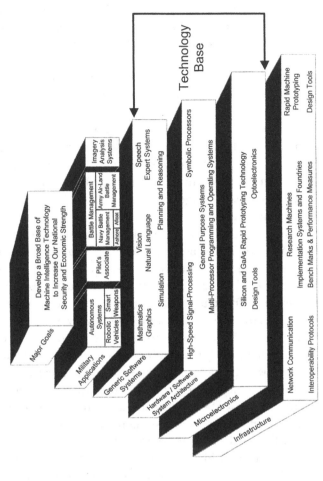

Figure 3.3
The fully articulated SC pyramid. Adapted from a viewgraph in Robert Kiggans's files.

estimates of future research results. All knew that research was unpredictable. As the program unfolded, who would ensure that the programs were coordinated? What would happen in the fall of 1986 if the Butterfly Multiprocessor was not ready to do simulations for battle management? How would the autonomous land vehicle see if the vision program fell behind schedule? Where would the next generation of multiprocessors come from if VLSI was slow to produce chips on demand? Would the architecture program deliver machines with enough power and memory to run the new generation of expert systems? All the programs were launched with these questions in mind, and with tentative answers laid out in milestones across their time lines. But nowhere within DARPA was there an institutional mechanism to manage coordination. A score had been written, but there was no conductor.

In time, a conductor would step forward from the orchestra. Steve Squires, the young math whiz brought in to run the SC architecture program, parlayed success in that realm into the directorship of IPTO, or rather of IPTO's successor, the Information Systems Technology Office (ISTO). Along the way he developed a scheme of program management equal to the goals of SC. Eventually his technique spread within DARPA and became part of the SC legacy.

The Plight of the Program Manager

Even before the SCI was formally approved by Congress, money began to flow. In anticipation of congressional approval, Cooper diverted $6.13 million in FY 1983 funds to SC.[77] Its first full year of funding, 1984, saw $49.1 million flow into the program, followed by $62.9 million in 1985. Plan or no plan, program managers had to get that money into the hands of researchers.

The initial steps were straightforward. Kahn and Conway explained SC to office directors and program managers from the DARPA technical offices other than IPTO—Defense Sciences, Directed Energy, Strategic Technology, and Tactical Technology. IPTO was the largest of the technical offices and the one likely to have programs appropriate to SC. But there were candidates in the other offices as well. Defense Sciences, for example, had programs on electronic sciences and materials science, both of which might have interests in microelectronics. Its systems sciences programs might contribute to SC applications. Similarly, the offices of Tactical Technology and Strategic Technology had naval warfare programs that might tie into naval battle management.

Participation in SC could be a two-way street. Program managers in IPTO or other offices could request SC funding for projects in their areas of interest that also held out promise of achieving the goals of SC. By the same token, funds already invested in those undertakings could be seen as leveraging SC support. And as always, program managers were encouraged to look for cosponsors outside DARPA. This last source of funding was to prove particularly important in SC's applications projects as the program evolved.

Of course, SC funding was new money. It was available to program mangers if they wanted it and if the projects they wanted to fund contributed to the goals of SC. But they need not take it. Barry Leiner, for example, chose not to seek SC money for networking. He was just then funding the research that would ultimately contribute to the creation of the Internet, a larger, more generalized version of the pioneering network that Robert Kahn and ARPA had done so much to initiate. Networking was clearly infrastructure of the kind envisioned in the SC plan. Indeed, it was in some ways the paradigmatic infrastructure. It connected this technology in the most fundamental way. But Leiner chose not to seek SC funding for his program. He felt that existing funding was adequate to his needs, and doing what he intended to do in any case would also serve the needs of SC.

Leiner had another reason for remaining outside SC. He worried about the size of the program. DARPA's success in the 1960s and 1970s had been based in large measure on working quietly outside the public glare. Small potatoes by Washington standards, its budget attracted neither attention nor jealousy. DARPA was free to tout its successes and bury its failures. But a billion dollars was real money, even in Washington. Leiner feared that funding on that scale would attract both scrutiny and opposition. He needed neither. His work was adequately funded on its own merits. When Kahn invited him to the first SC planning meeting, he politely declined.

Most program managers, however, saw SC as a windfall, a new pot of money from which they could draw resources to expand and accelerate the research already under way in their areas. Program management at DARPA rewarded entrepreneurial initiative. Out of the fixed pie that was the DARPA budget each year, PMs competed for their individual slice by arguing that their projects were more promising than the ones proposed by their colleagues. A new pot of money such as that plopped down by the SCI meant that for a while, at least, this was no longer a zero-sum game. During the finite time that this window was open, there was more money available than anyone had yet figured out how to spend. The SC plan entailed starting up some research projects that were not under way; funding for these would go to the best proposal, but some projects awaited definition. The enterprising PM could win funding for a field if he or she could represent research within it as serving the goals of SC. To the imaginative, it was an exciting time to be working on information processing at DARPA.

The excitement of getting the program going, however, was dampened not just by management conflicts at the top, but by contextual issues that intruded on SC coincidentally. First, the simultaneous creation of the SDI had important repercussions for DARPA. As a principal supporter of research on ballistic missile defense, DARPA might well have been the institutional home of SDI. But Cooper demurred. Believing that SDI would not work, he feared it might unbalance the agency, drawing funds away from other, more worthy projects. Additionally, because it was a large, politically charged program, it threatened to raise DARPA's cherished low profile. This was Barry Leiner's concern about SC, raised to an agencywide level. Rather than take those risks, Cooper transferred $300 million in ballistic-missile-defense research to SDIO. Thus, in the first year of the SCI, the overall DARPA budget actually fell.[78] The 1985 SCI budget dropped from $95 million to $72 million.[79]

Cooper believed himself well rid of SDI, but he paid a price. The DARPA Directed Energy Office was gutted in 1984, disappearing from the DARPA budget altogether in 1985. What is more, GaAs, one of the microelectronics technologies that had been highlighted as ripe for development in the various SCI plans, went to SDIO as well.[80] While GaAs was potentially faster than silicon as a chip material, its greatest attraction was increased resistance to electromagnetic pulse, a key concern of space-based weaponry.

A second blow to the start-up of SC came from Capitol Hill. The Democrat-controlled House and Senate reluctantly had given President Reagan the defense buildup he wanted in the early 1980s. At the same time, however, they renewed with increased vigor the perennial campaign against waste and fraud in defense procurement.[81] The Competition in Contracting Act of 1984 required government agencies to "obtain full and open competition" through the use of sealed bids or competitive proposals.[82] Agencies were empowered to circumvent these requirements for cause, but Cooper appreciated the political risks of abusing the privilege. Even before the act became law, Cooper directed in

January 1984 that ARPA contracts for research and product development be "competed" whenever possible.[83]

To further complicate the problems of launching the SCI, Congress had moved in the summer of 1983 to ensure that small business had the opportunity to bid on government contracts. Henceforth agencies would have to publish in the *Commerce Business Daily* (*CBD*) notice of procurements over $10,000 and wait at least thirty days before beginning negotiations.[84] Paired with the Competition in Contracting Act, this legislation transformed the old IPTO-ARPA style of funding research.

In the salad days of IPTO during the 1960s and 1970s, program managers solicited proposals in the research community, often by direct contact with the researchers. An office director such as J. C. R. Licklider, or a program manager such as Larry Roberts, for example, might call up a researcher to ask if he or she would be interested in doing a project for ARPA. If there was a meeting of the minds, the researcher would submit a proposal that was essentially approved before it came through the door. Program managers achieved results by knowing what was going on in the research community and who could deliver. They also accepted unsolicited proposals coming over the transom if they were good and suited ARPA's purposes. And they put some contracts up for competition, as they did the ARPANET contract that BBN won with Robert Kahn's help. The style, however, was still less formal than at agencies such as NASA or even elsewhere within the DoD.[85]

Cooper surely appreciated the advantages of the old informality—the speed, the efficiency, the absence of red tape. In some cases he agreed to waive the procurement restrictions in the new legislation. But in addition to sensing the congressional mood, he also felt that SC was a different kind of enterprise. This was big science of the kind he knew at NASA. It entailed far more work with industry, a constituency less suited than universities to the casual informality of the Licklider days. And it would soon be distributing $100 million a year, too much to give out to the old-boy network.

The result, said *Science* reporter Mitch Waldrop, was months of confusion and heated tempers.[86] Program managers now had to use an instrument called "sources sought." The PMs would define research or development projects with as much specificity as possible, keeping in mind the capabilities of the community and the need to attract as many potential performers as possible. Their description would be published in the *Commerce Business Daily* (*CBD*), noting that sources were sought who could do this research. Those who were able and interested were invited to submit a description of their qualifications. Based on these qualifications, some researchers would then be invited to submit proposals.

As proposals came in, program managers established their own procedures for reviewing them and selecting awardees. Some PMs used informal and ad hoc networks within ARPA—fellow PMs from related programs and other professionals with appropriate expertise. Others moved to more formal arrangements, setting up review committees composed of DARPA personnel and outside experts as well.

Particularly crippling to DARPA was the requirement in the new law limiting contact between procurement officers and potential contractors. Such contact was the taproot of the PM's success. It was by staying in touch with the community, by letting the community know what was needed at DARPA, by finding out what the community was capable of doing that the PM was able to map out a reasonable research agenda in his field and sign up the best workers to make it happen. In the new dispensation, such networking could be viewed as collusion. PMs in the future would simply have to be much more careful, much more circumspect, much more correct when dealing with researchers who were potential contractors. This did not make their jobs impossible, but it did make them harder and less enjoyable.

Program managers adjusted to these new circumstances in different ways. Steve Squires, for example, developed a new technique for soliciting proposals, a cross between an RFP and a fishing expedition. Out of his innovation would evolve the Broad Area Announcement (BAA), the instrument that came to be used throughout DARPA. To avoid the appearance or the fact of collusion, Squires relied upon DARPA contracting agents to act as a buffer between himself and potential contractors. He could tell the agents what he was seeking in a contract and allow them to negotiate with proposers without compromising his own ability to judge the formal submissions.[87]

Paul Losleben, the PM for infrastructure, remembers the early years differently. A DARPA veteran who had been around longer than Squires, he found the new restrictions frustrating. The VLSI program, which was already under way when SC started and the new regulations came into play, proved more difficult to fund than it had been heretofore. "I . . . had twice the budget that I had had before," says Losleben, "and I was accomplishing less because I had lost the freedom to build the program, to interact, to define direction, to work with the community."[88]

Ronald Ohlander, a senior IPTO PM, recalls yet a different experience. In the first two years of SC, from the formal approval of the program in October 1983 until his departure from DARPA in 1985, his budget went from $14 million to $55 million. His biggest problem was recruiting staff fast enough to keep up with the expansion of activity. He could cope with the new contracting requirements, but he needed more people to keep track of the increased volume of funding and the proliferation of paperwork.

Just as PMs responded differently to the new conditions, so too did they get different results. The "sources sought" for the autonomous land vehicle attracted only three proposals, all from industry. The call from architecture attracted ninety-two proposals, fifty-eight from industry and thirty-four from universities. Out of these, Steve Squires found one promising new departure, a start-up company called Thinking Machines; the five other contracts that Squires let in 1984 went to projects already under way and well known to DARPA before SC began, two of them at Carnegie Mellon and one at MIT. The RFP in expert systems produced seventy-two proposals, sixty from industry and twelve from universities. The one contract let in this field in 1984 was not drawn from these submissions.

Obviously the troops in the trenches experienced the launching of SC very differently. Some of these differences flowed from their personalities and styles. Some flowed from the nature of the programs they were running. These varied from established, ripe, and productive areas of research to new and unknown fields that required far more cultivation. Some were determined by where people were on the SC pyramid, from the secure and dynamic base of infrastructure up through the new and uncertain realm of applications.

This diversity is important because the organization and operation of SC were going to be invented primarily on the front lines of the program. The early leadership of SC had defined its goals and secured the funding that would fuel the program. They had not articulated, however, a plan or even a philosophy of how to make SC work. Furthermore, they had invested SC with two fundamental flaws that would haunt the program throughout its history. The plan did not make clear whether applications were a mechanism for technology pull or whether they were simply window dressing to rationalize a program of building the technology base. Second, the plan did not explain how simultaneous and uncertain research programs were going to deliver predictable products when these were needed higher up in the pyramid hierarchy. The notion sounded lovely and the time lines looked impressive. It would be up to the PMs to figure out how to make it happen.

Part II

4

Invisible Infrastructure: MOSIS

The Transformation of Microelectronics

As SC got under way, the artificiality of the SC pyramid was quickly revealed. The layers of the pyramid had been helpful in conceptualizing a hierarchy of technologies, and the broad categories of applications, technology base, and infrastructure held up over time. But the division between infrastructure and microelectronics broke down in the first year of SC's existence. A review of what was initially planned in infrastructure and microelectronics reveals how the transformation took place.

In the SC plan, infrastructure had six components. First, *Networks* were the easiest to grasp; they were the ARPANET and its successor, the Internet. More generally, they were any communication networks, including local area networks, that connected computer users and allowed for the electronic sharing of resources and information. Second, *research machines*, like networks, were straightforward and easy to understand. DARPA would help provide the latest technology to researchers, so that they could run their experiments on the most up-to-date equipment. This not only educated the researchers and made their findings current, it also helped to debug the latest machines and expand the range of problems to which they were applied. Third, *interoperability protocols* were those communications standards that allowed work in one realm to be translated into work in another. The TCP/IP that Kahn developed with Vinton Cerf is one example of an interoperability protocol; another is the "system interoperability kits," envisioned in 1983, which were standardized tool kits that designers could use to imbed and test components in existing systems. This is rather like providing mechanics with metric tools to ensure that the components they built fit into the metric machine for which they were designed.

The other three components of infrastructure were interrelated among themselves and furthermore directly tied to microelectronics. *Design tools* were analogous in some sense to machine tools, that is, tools that made other tools. In this case they were hardware and software capable of designing new hardware and software, which in turn could be tested on emulation machines before they were actually manufactured. *Implementation systems and foundries* would quickly turn new architecture designs into integrated circuit chips (ICs). And *rapid machine prototyping* would help researchers turn new chips and designs into complete machines.

Infrastructure, then, provided government-funded equipment and services. These would be made available to researchers who had a demonstrated ability to use them effectively. Costs would be born entirely by the government or shared with the researchers. The payoff would be better research, faster movement of ideas through the system, and products at the end that were compatible with existing technology and readily integrated.

The next level up the pyramid, microelectronics, was supposed to operate somewhat differently. It was the first layer of the technology base. Instead of providing wholesale services and equipment, microelectronics would target specific research projects that advanced the goals of SC. It was primarily a research field, as opposed to infrastructure, which was primarily a service field. But two developments, unanticipated when SC was first conceived, transformed microelectronics into an infrastructure program.

The reorganization was prompted by the changed status of two technologies, both of which had loomed large in early versions of the SC plan. The Josephson Junction (JJ), a switching device capable of high speeds and low power consumption, was attracting wide interest in the late 1970s and early 1980s. IBM, for example, had invested heavily in the technology beginning in the early 1970s. When the Microelectronics and Computer Technology Corporation (MCC) was organizing in late 1982 and early 1983, it included the JJ as a prominent part of its research agenda.[1] By the middle of 1983, however, enthusiasm for the JJ was waning. Problems of manufacturing and maintaining the switch outweighed its speed and power advantages. Most troubling was the requirement to operate it at temperatures near absolute zero. IBM cut back dramatically on its research program and the JJ disappeared from the final version of the SC plan.

The decline of the JJ elevated gallium arsenide (GaAs). Both in emphasis within the SC plan and in the commitment of dollars, GaAs took on new prominence in SC microelectronics. The GaAs foundry that DARPA was committed to cofunding with Rockwell/Honeywell was targeted for millions of SC dollars.[2] Indeed, GaAs was the primary microelectronics focus of SC. When Robert Cooper transferred ballistic missile defense technologies out of DARPA to the new Strategic Defense Initiative Office, he included GaAs. This relieved DARPA of a major commitment of its SC funding, but it also eliminated a second major pillar of the microelectronics tier of the SC pyramid.

The role of microelectronics in SC shrank dramatically. In the first year, the field had only four other contracts, three in optical technology and one in GaAs input-output devices. By the end of fiscal year 1986, that number was up to sixteen projects in nineteen contracts, but still it accounted for only 4 percent of SC spending.[3] Microelectronics was then collaborating in the operation of a second GaAs pilot line and four other GaAs projects; most of the remaining work was in optical technology. But these programs were run out of the Defense Sciences Office (DSO), not the Information Processing Techniques Office (IPTO) nor its successor after 1986, the Information Systems Technology Office (ISTO).

Microelectronics was never as well integrated and as important as had been envisioned in the original SCI plan. In fact, it disappeared from the SC pyramid, withdrawn from the technology base and assimilated within infrastructure. Its transformation in the SC plan reveals two of the many hazards besetting the IPTO visionaries. The waning of the JJ points up the unpredictability of a rapidly evolving research field, just the point that Kahn had been at pains to impress upon Cooper. The transfer of GaAs to SDI highlights the volatility of plans in the charged political environment of Washington. Robert Cooper had wanted GaAs but not SDI. Unloading one, he lost the other.

From VLSI to MOSIS

The microelectronics that turned out to be truly central to SC was done mostly within the infrastructure segment of the program, specifically in very large scale integration (VLSI) and in the metal oxide semiconductor implementation system (MOSIS).[4] To understand the pivotal role they played, it is necessary to revisit the area in which Lynn Conway had contributed before she came to DARPA. Her contribution, in turn, can be understood only in the context of the evolution of microelectronics.

Through the 1960s and 1970s, advances in computer performance were driven primarily by improvements in ICs—also known as microchips or just chips. At the most basic level, a computer consists of a collection of switches, on/off devices. These devices use binary mathematics and Boolean logic to convert the presence or absence of current flow into complex mathematical computations or sophisticated logical operations.[5] Microchips are semiconductors on which the switching devices, the transistors, are formed by doping alternating layers of the semiconductor material. The material doped with electron-donating impurities such as phosphorous, arsenic, or antimony, is called n-type (negative); that doped with electron-accepting impurities, such as boron or gallium, is called p-type (positive).

In the 1980s the alternately doped layers were organized on microchips in two basic patterns. The junction transistor accomplished switching through the behavior of ions collecting at the p-n junction between layers. N-type impurities provided an excess of electrons, while p-type provided holes into which those electrons might move. When the two materials were joined, some electrons migrated across the junction to the p region and some holes migrated to the n region. Together they formed a "depletion" region a few microns thick. This region resisted the flow of current. An external voltage applied across this junction could thicken the region, creating a diode; or, if reversed, the voltage could thin the depletion region and allow current to flow. Thus, application of the voltage, called biasing, could switch current flow on and off (see figure 4.1).

The junction transistors used in computers were called bipolar transistors. They came in the form of sandwiches, either n-p-n or p-n-p (see figure 4.2). If the center region, such as the p region in the n-p-n transistor shown in figure 4.2a, was forward biased by the application of a positive voltage at the gate (G), then electrons would pass through depletion region 1 and current would flow through the circuit from the source (S) through the gate. This voltage at the gate, however, would not break down depletion region 2; this would continue to act as a diode, that is, as a barrier to current flow across the entire device. If the center region was sufficiently thin, however, as in figure 4.2b, then the electrons crossing depletion region 1 would reach such concentrations in region p that they would overcome the barrier of depletion region 2 and flow to the drain (D). This produced current flow across the entire device. Such devices, then, depending on their geometry and their biasing, could serve either as insulators or as conductors.

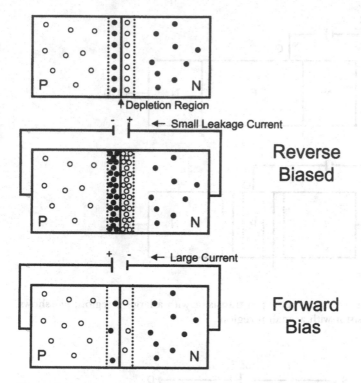

Figure 4.1
Junction transistor.

Field effect transistors (FETs) or metal-oxide semiconductor (MOS) transistors, the second major type, worked differently. They still relied on the characteristics of alternately doped layers of semiconductor material, but they achieved current flow in a different way. Figure 4.3 shows a cross section of a MOS transistor. In a substrate of p-doped silicon, two n-doped regions have been embedded. Above them is a layer of silicon dioxide (SiO_2) insulator, and above that is a metal gate. At the bottom of the device is another metal region called the bulk. When a positive voltage V_{GB} is applied across the gate and the bulk, an electric field is set up across the oxide insulator. It draws electrons in the substrate toward the gate, setting up a thin channel. When a voltage (V_{DS}) is established between the n regions, a current flows between the source (S) and the drain (D).

In the late 1970s and early 1980s, when MOSIS was getting under way and Robert Kahn was thinking about the needs of the computer research

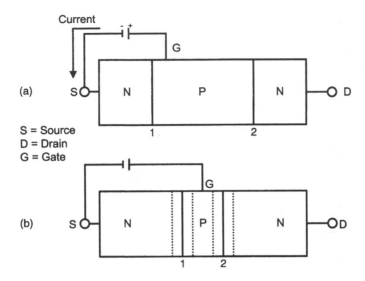

S = Source
D = Drain
G = Gate

Figure 4.2
NPN transistor. (*a*) shows an n-p-n transistor with a large p region; (*b*) shows an n-p-n transistor with a small p region.

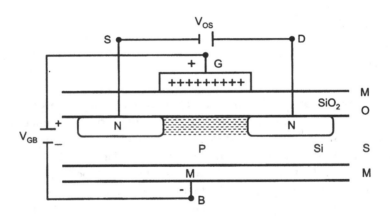

Figure 4.3
Schematic of a field effect (FET) or metal-oxide semiconductor (MOS) transistor.

community in the United States, bipolar or junction transistors had performance characteristics much different from FET or MOS transistors. Bipolar transistors are inherently faster than FET transistors, but they also require more power and therefore generate more heat. In the early 1980s, they were the transistor of choice for processing chips, while FET transistors were preferred for memory. Therefore, research on the design and manufacture of ever smaller and more powerful chips tended to concentrate on FET technology, for the comparatively low voltage of these devices meant less mutual interference as they were crammed ever closer together (see figure 4.4). Advances in design of FET chips often found their way into new designs for bipolar chips.

Figure 4.4
Field effect or metal-oxide semiconductor (MOS) chip.

Manufacturing technique was the primary constraint on the production of higher density FET chips. The techniques for manufacturing these chips, which were to be central to SC's MOSIS program, evolved at a rapid pace through the 1970s and 1980s. The fundamental process, however, remained the same (see figure 4.5). An integrated circuit design was transferred to a glass or polymer plate called a mask. Each layer of the chip had its own mask. Beneath the mask was placed a wafer of three layers: for example, pure silicon, silicon dioxide, and photoresist, a material that hardens when exposed to ultraviolet light. After exposure through the mask, areas blocked by the design lines on the mask remained soft. These were removed chemically, leaving the circuit pattern of the mask etched on the wafer. The exposed silicon was doped and the process was repeated for each layer of the final chip. As many as twelve layers could be built up this way. Then a film of conducting metal was evaporated over the entire wafer and etched away to form the connecting leads to the circuit. Finally, an insulating layer was placed over the entire wafer and pierced to provide external contacts. After testing, the wafer was cut into individual chips, which were packaged and retested.[6]

As manufacturing techniques improved, a hierarchy of miniaturization emerged[7]:

IC Generation	Size, bits	Period
SI: small-scale integration	1–10^2	1960s
MSI: medium-scale integration	10^2–10^3	1970s
LSI: large-scale integration	10^3–10^5	early 1980s
VLSI: very large-scale integration	10^5–10^6	mid-1980s
ULSI: ultra-large-scale integration	10^6–10^9	1990s
GSI: giga-scale integration	$>10^9$	after 2000

Achievement of this level of miniaturization appeared physically possible. The obstacles were in design and manufacturing technique.

In the 1960s and early 1970s the United States, where the microchip was invented, dominated advances in this technology, both in memory chips and in processing chips. Beginning in the 1970s, however, Japan began to challenge the United States in chip development. Its first target was the manufacture of dynamic random access memory (DRAM) chips. Starting from 0 percent of world market share in 1974, when the United States had 100 percent share, Japan surpassed the United States in 1981 on its way to 80 percent market share in 1986.[8]

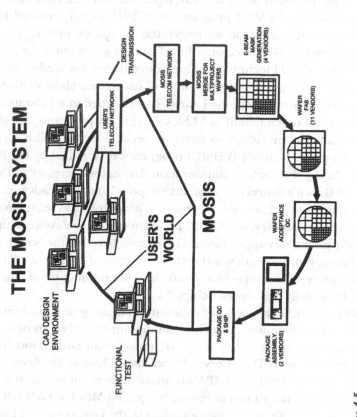

Figure 4.5
The MOSIS system. Adapted from "Review briefing," 12 May 1987, Toole collection.

Japan achieved this coup by combining superior manufacturing technology with aggressive marketing practices. While this overthrow of American dominion was under way, Japan launched an even more ambitious challenge. Its VLSI program, from 1976 to 1979, moved beyond manufacturing technique to target the design of new chips. Japan invested over $300 million, one-third from the government, in six different phases of VLSI technology. These ranged from the traditional realm of manufacturing to the new territory of basic research in VLSI design.[9]

To remain competitive, the United States needed to accelerate its own research and development in VLSI. Of the many programs already pursuing that goal in 1980, two warrant special attention. The very high speed integrated circuit (VHSIC) program was launched by the DoD in 1979, during the Carter administration. Driven by Secretary of Defense Harold Brown's concern that "smart weapons" and other advanced military equipment did not embody the most powerful chips being produced by American industry, the VHSIC program sought to develop chips specifically for military applications. The demon here was the Soviet Union, not Japan. But the result was the same: infusion of hundreds of millions of dollars over the course of a decade into the production of large-scale (not necessarily high-speed) chips.[10]

DARPA's VLSI program had different roots and goals. It originated in the DARPA community, primarily among university researchers. Three advocates had an especially powerful impact. Ivan Sutherland (the former director of IPTO), Carver Mead, and Thomas Everhart wrote a RAND study in 1976 at DARPA's request. Sutherland was on the RAND staff when the report was prepared but joined Mead at CalTech by the time it was published; Everhart was chair of the Department of Electrical Engineering and Computer Science at the University of California, Berkeley. Extensive visits with chip manufacturers convinced the authors that "U.S. industry generally appears to persist in incremental development,"[11] a conclusion similar to that reached by Robert Cooper when he toured the industry in 1982. In contrast, they recommended to ARPA that "the integrated circuit revolution [had] only half run its course"; another four orders of magnitude improvement in chip manufacture was possible in the foreseeable future, approaching 0.1-micron feature sizes and device densities of 10^8 per chip, that is, ULSI. This was far more ambitious than the subsequent VHSIC program, which had an interim goal of 1.25 microns on the way to submicron chips.

Under this impetus, Robert Kahn proposed in 1977 the ARPA VLSI program, targeting submicron design, semiconductor fabrication, and computer architecture and design. The approved program was divided between IPTO and the Defense Materials Office (later the Defense Science Office), which ran semiconductor fabrication. DARPA accounts do not mention the Japanese threat,[12] but a sense of national urgency pervades the RAND report and surely weighed in the decision to approve Kahn's proposal.

Up to the late 1970s advances in both design and manufacture had been monopolized by large firms, such as IBM, Motorola, and Intel.[13] A plant to manufacture chips could cost hundreds of millions of dollars. Companies that made such investments used them to develop new manufacturing techniques and to test their own designs. The primary purpose of their investment was to improve their own competitiveness. They might make their facilities available to produce a chip designed by someone else, but only at a high price. A single wafer could carry a price tag of tens of thousands of dollars. Independent researchers could not afford such fees, and businesses with new designs would be loath to place them in the hands of competitors.

This is where Carver Mead and Lynn Conway entered the picture. In the 1970s, Mead had wanted to teach his students at CalTech how to design VLSI chips. But there was no standard to teach to. Each manufacturer had its own protocols establishing such variables as standard space between features. To choose one set of industry standards—say Intel's—would be to prepare his students to work only at that company. So Mead developed a generalized set of protocols out of the various standards. From these students could learn all the principles and still have a base from which they could translate into the protocols of a specific company.

Furthermore, Mead's design standards were scalable. All dimensions in his system were proportional to a standard unit λ. His protocols told designers how close together features could be or how far apart they had to be in increments of λ. If, for example, the holes through the insulating layer were 1λ across, then they had to be centered over a segment of the metal layer that was at least 2λ across, and adjacent strips of the metal layer had to be separated by at least 3λ (see figure 4.6). A design following these rules could then be scaled.

It might, for example, be implemented with a feature size of 3μ (3 microns, or 3 millionths of a meter), meaning that the devices and the connections among them are of that scale. The same design could be "bloated" to 5μ or "shrunk" to 1μ without compromising the physical integrity of the design. Furthermore, scaling promised not just smaller

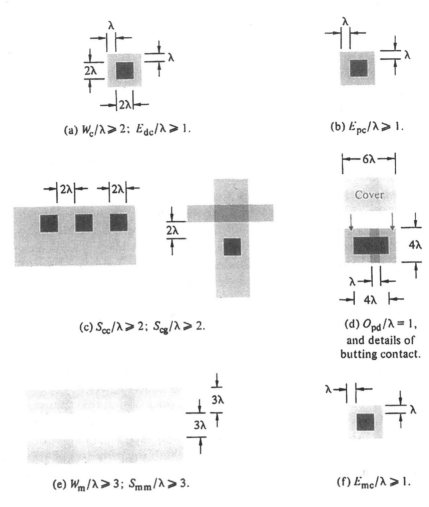

(a) $W_c/\lambda \geqslant 2$; $E_{dc}/\lambda \geqslant 1$.

(b) $E_{pc}/\lambda \geqslant 1$.

(c) $S_{cc}/\lambda \geqslant 2$; $S_{cg}/\lambda \geqslant 2$.

(d) $O_{pd}/\lambda = 1$,
and details of
butting contact.

(e) $W_m/\lambda \geqslant 3$; $S_{mm}/\lambda \geqslant 3$.

(f) $E_{mc}/\lambda \geqslant 1$.

Figure 4.6
λ rules. *Source:* Mead and Conway.

devices but power savings as well. For example, an instruction processor
in 2μ CMOS on a 5 mm square running at 20 MHz clock rate, if scaled
to 0.2μ would take up only 1/100 the chip area, draw only 1/100 the
power from a 0.5 volt source, and operate at 200 MHz. With some limitations, it would experience the advantages of the cube law of scaling of
switching energy.[14]

The implications of this were enormous. Not only would designs remain useful as manufacturing technology evolved to allow chips of

smaller feature size, but also designers could work out standard routines for series of devices to perform complex functions. For example, all the electronic steps necessary to arrange a list alphabetically could be built into a single alphabetizer design. Because that design would be scalable, any designer could thereafter take it off the shelf and insert it to do that function in the chip he or she was working on. It would always scale to the task at hand.

To help him package and sell his design standards, Carver Mead turned to Lynn Conway, head of the LSI System Design Area at Xerox PARC (Palo Alto Research Center). Together they wrote *Introduction to VLSI Systems*,[15] which soon became the bible among VLSI designers. Prepublication chapters of their text were used to train instructors and teach courses at CalTech, Berkeley, Carnegie Mellon, and MIT. Mead understood that teaching VLSI design would be truly effective only if the students could implement their designs, that is, produce and test a chip. By pooling many designs on a single wafer, the costs could be divided into reasonable chunks. What was needed was a "silicon broker," as Mead called it, an agent who would gather the designs, negotiate a chip run with a manufacturer, translate the Mead-Conway protocols into the manufacturer's protocols, and test and distribute the resulting chips.[16] Conway convinced Xerox PARC to serve this role for Mead's students at CalTech.[17]

In 1978 Conway herself went to MIT to teach a VLSI design course at the high temple of American computer research. To do that successfully, she needed one other resource. She asked Robert Kahn if she could use the ARPANET to transmit her students' designs electronically back to Xerox PARC. He agreed and Conway pioneered the network transmission of VLSI designs for remote implementation. With that, all the components were in place for what would become MOSIS.

Xerox PARC continued its role as the "silicon broker" after Conway's return from MIT. Back in California, Conway organized the first truly interinstitutional, multichip projects. In the fall semester of 1979, 124 designers at 11 universities contributed 82 integrated designs to a wafer implemented by Xerox PARC. The following semester, in the spring of 1980, 250 designers from 15 universities and R&D organizations submitted 171 projects for implementation. Chips were fabricated by Hewlett Packard in June and delivered in July.

With these demonstrations in hand, Conway convinced Kahn that DARPA should take up the mantle of "silicon broker," which Xerox PARC felt it could no longer sustain. Kahn could already see the potential good, not only in training students and facilitating design innovation but also in providing this service to DARPA researchers. Though the program had no direct utility for the military services, nor any tangible application of the kind that Cooper would later insist upon, Kahn went ahead and approved it. To replace Xerox PARC as the implementing agent, the "silicon broker," Kahn turned to the Information Sciences Institute (ISI) at the University of Southern California.

ISI was the creation of Keith Uncapher. While conducting computer research at the RAND Corporation in the early 1970s, Uncapher had come to realize that RAND had neither the mission nor the organizational structure to support the work he wanted to do.[18] RAND did paper studies such as the Sutherland-Mead-Everhart report; Uncapher wanted to direct research and work on the technology itself. He organized a group of select coworkers and sought DARPA support to move his team to a new site. With a commitment of DARPA interest, Uncapher found a home for his team at the University of Southern California. ISI became an unfunded research arm of the university; USC would provide administrative services for an overhead fee, but ISI would pay its own way through grants and contracts. With the ARPA connection clearly in mind, Uncapher chose quarters in Marina del Rey, California, eleven miles from the USC campus but only two miles from Los Angeles International Airport.

In the ensuing years, ISI prospered, mostly on DARPA funds, performing service for the agency, conducting a wide range of research, and creating several technology transfer programs, including those at the headquarters of the Commander in Chief of the Pacific Fleet, the headquarters of the Strategic Air Command, and the Air Force Logistic Command. It was ideally positioned, by geography and organization, to take up VLSI implementation. Contracts were let in 1979 and 1980 under two separate ARPA orders to begin work.[19] Danny Cohen, a USC researcher who had worked with Mead at CalTech, took the lead in establishing the effort at ISI. By August 1980, the project had adopted the name MOSIS and completed its first experimental, multichip fabrication. It was formally instituted in January 1981.

Within DARPA, Kahn assigned responsibility for the project to Duane Adams, DARPA's first VLSI program manager.[20] As the program got going, Adams in turn recruited Paul Losleben from the National Security Agency (NSA). Losleben had advocated vendor-independent design rules at NSA, one of the earliest users of MOSIS, and he relished the opportunity to run a program based on those principles. With Losleben in place, the program was well under way—more than two years before SC came into being. Its later absorption into SC provides a window on both programs.

The Valued Added

As MOSIS evolved, five goals came to dominate the "silicon broker."[21] First, the Mead-Conway design rules had to be continually evaluated and updated to keep pace with VLSI developments. This meant, for example, that scalability had to be verified in practice, not just in theory, and adjustments made to ensure that features designed at the 5μ scale could function at the submicron level. Constant adaptations also were necessary to ensure that the standard MOSIS protocols were translated into the forms used by the commercial foundries that the program employed.

Second, designs had to be processed economically. The greatest cost in the manufacturing process was the foundry run. MOSIS director George Lewicki reported in 1988 that 74.3 percent of MOSIS expenses went to fabrication costs compared with 13.1 percent for salaries and 11.4 percent for ISI overhead.[22] ISI had to gather chip designs via the ARPANET, sort them by type, marshall all those of a specific type on a wafer (taking up as much space as possible), and schedule a run with a commercial foundry that offered high quality, good price, and appropriate schedule. The actual cost of the service was divided among the researchers, or rather among their research grants and contracts. Many of these came from DARPA.

ISI reported in 1988 that a production run, averaging between ten and twenty wafers, cost between $50,000 and $60,000; but single 7.0 × 7.0-mm chips of 2μ CMOS design on one of those wafers could be purchased for $258.[23] In 1987 ISI reported to DARPA that it had processed 1,382 projects that year for 36 different customers. The average cost was $3,635 for a part measuring 43.26 mm^2.[24] The higher cost per part resulted from on-demand runs and chips with smaller feature sizes.[25] DARPA's direct role in MOSIS, in addition to supporting research on implementation, was to underwrite the process, ensuring that ISI could pay for a run even if there were not enough chips to fill up a wafer.

Third, MOSIS was responsible for quality control. It did not evaluate designs per se, but it did screen them for proper use of the design protocols. When runs were complete, ISI tested the chips for defects, sliced them from the wafer, and packaged and shipped them to the researchers. But because these chips had tens of thousands of devices, it was impossible to test them all. Therefore, ISI also estimated from experience how many copies of each chip were necessary to achieve 90 percent reliability that one chip would work satisfactorily. For example, if experience with a certain kind of design and a certain foundry and maskmaker revealed that 40 percent of the chips produced would be defective, MOSIS would place three of that design on the wafer. There was a 94 percent probability that one of the three would work.

Fourth, MOSIS strove to lower turnaround time to the designer. Turnaround time was defined by MOSIS as the gap between e-mail receipt of an acceptable design to shipping of the finished, tested, packaged chip. Although it was possible to get turnaround times as low as three weeks under certain circumstances, MOSIS generally operated closer to two months. Part of the reason was economy; it simply made more sense to gather sufficient numbers of similar designs to fill a wafer. Indeed scheduling was so complex as to require difficult trade-offs between time and money. Eventually, MOSIS came to publish a schedule of runs to be made, so that researchers could plan on when they might get a chip back and what deadline for submission they would have to meet. By committing itself in advance to a particular run, MOSIS risked not filling a wafer and thus incurring increased costs. But that risk diminished as experience built up over time, and the risk in any event appeared equal to the advantages of having predictable productions schedules.

Fifth, MOSIS promised security and confidentiality. The program serviced researchers who were often working at the cutting edge of VLSI design. Proprietary rights were a serious concern. Designers might be loath to put their best ideas in the hands of a manufacturer who was a potential competitor for commercial exploitation. But they came to trust ISI and its agents to guarantee the confidentiality of their designs. Only ISI had the electronic version of their designs; versions on masks or even on the chip itself were prohibitively difficult to decode.

Because of these advantages, MOSIS flourished in the 1980s and into the 1990s. The SCI components of MOSIS and machine acquisition were grafted onto one of the two contracts that ISI held with DARPA. The contracts had changed over the years, as had the DARPA orders authorizing them; the SCI addition, for example, was just one of sixty-four amendments to ARPA order 5009 made between 6 February 1984 and 2 September 1987.[26] A twelve-month period in 1987–1988 saw $9 million flow to the MOSIS program, $6.8 million for fabrication of chips and $2.2 million for salaries and overhead at ISI.[27] By May 1987, DARPA had committed $89,480,357 to ARPA order 4848, for work by ISI, about half of the funding apparently coming from SC.[28]

The number of MOSIS projects, that is, wafers, rose steadily in the early 1980s, from 258 in 1981 to 1,790 in 1985. After sharp declines in 1986 and 1987 and stagnation in 1988, due to DARPA budget problems and a worldwide drop in chip demand, output rose again in 1989 to an all-time high of 1880.[29]

In this period, MOSIS expanded the range of implementation technologies it offered. In 1981, for example, it offered only nMOS implementation, the model technology presented in *Introduction to VLSI Systems*.[30] This produced an n-p-n transistor of the type described in figure 4.2 above.[31] These devices have inherently higher speed than pMOS devices, which reverse the role of the n and p regions. They offer speeds and efficiencies approaching those of the bipolar transistors that had previously been the mainstay of processing chips.

In 1982 MOSIS began offering complementary MOS service (CMOS). This technology combines both nMOS and pMOS gates in a way that reduces power loss with no significant loss in speed.[32] The availability of CMOS implementation through MOSIS helped to accelerate the development of this improved technology. With the introduction of this technology, FET-MOS transistors became competitive with bipolar transistors. Increasingly, then, improvements in these transistors served both memory and processing applications. By the end of the 1980s, all MOSIS chip implementations were CMOS. In the early 1990s, combination bipolar-CMOS chips appeared, giving designers an even wider range of choices in the tradeoff between speed and power consumption.

From feature sizes of 4μ and 5μ in 1981, MOSIS worked down to 1.6μ in 1986 and 1.2μ in 1988.[33] As MOSIS moved toward submicron (ULSI) implementation at the close of the Strategic Computing Program, it could anticipate what historian Edward Constant has called a presumptive anomaly, that is, a point at which the laws of nature appear to set an impenetrable barrier to further development along this technological trajectory.[34] At some point, then projected to be around .15μ, feature size becomes so small relative to the electrical charges in the components that electrons simply jump the microscopic distances between components and the integrity of the circuit collapses.[35]

It was not just this absolute barrier, however, that retarded movement to smaller feature sizes. At virtually every step of the fabrication process, technical problems mount as feature size declines. Masks had to be exposed with ever greater precision, with electron beams and x-rays replacing photolithography for etching photoresist material. And clean rooms had to be cleared of ever smaller particles. As a general rule, integrated circuits can tolerate particles up to 10 percent of their feature size; larger than that the particles threaten to short the circuits.[36] Thus, a 5μ device can tolerate particles of .5μ, but a submicron device requires that particles in the fabrication area be smaller than .1μ.

MOSIS bore an unusual relationship to industry in lowering these technical barriers. On the one hand, MOSIS was not itself directly involved in manufacture. Rather it contracted with mask makers and foundries to produce chips for its clients.[37] Those foundries were pushing the state of the art for commercial reasons; that is, they engaged in their own research and development to remain competitive. Companies such as Intel and IBM wanted their manufacturing techniques to be the best available so that their chips would have the best qualities at the lowest prices. They might then make that manufacturing capability available to MOSIS, but they were understandably anxious to have proprietary information protected. For its part, MOSIS wanted to see as much progress as possible in chip manufacture and it wanted to stay abreast of what progress there was so that it could make those developments available to its clients.

Therefore, MOSIS pursued two routes simultaneously. First, it cultivated manufacturers to have access to the best foundries at the best price. By respecting proprietary information and by negotiating the translation of its own design standards to the standards of the different manufacturers, it convinced those foundries to share information about their manufacturing capabilities. Second, MOSIS conducted its own research program on problems such as a functional test language (SIEVE) and new packaging technologies for finished chips.[38] It was not just a silicon broker, but an honest broker. The program worked in part because it earned the confidence of both its own clients and the manufacturers.

Service expanded accordingly. Design rules, for example, increased in number. MOSIS introduced its own rules in 1983, an adaptation of the Mead-Conway rules. The following year, MOSIS added NSA design rules to its repertoire, a reflection of the coordination provided by program manager Paul Losleben.[39] In 1985 MOSIS accepted fabricator design rules for CMOS, meaning that it could accept designs executed to the protocols of specific manufacturers. All of these steps ensured the widest possible latitude to designers and thus stimulated research diversity.

Networking services multiplied as well. MOSIS became available on line in 1981, accepting submissions over ARPANET and Telenet, the commercial network pioneered by Kahn's former employer Bolt, Beranek, and Newman (BBN). It added CSNet in 1983 and MILNET in 1984, opening itself to the network joining operational military bases. As the Internet absorbed ARPANET in the 1990s, MOSIS converted to that wider resource.

As communications sources expanded, so too did the MOSIS user community. Universities were the first MOSIS users supported by DARPA, along with some industrial contractors such as BBN and some government and quasi-government laboratories such as NASA's Jet Propulsion Laboratory. In 1981 approved DoD contractors were added, along with institutions approved by the National Science Foundation (NSF). DoD and DARPA contractors got the service for free, while NSF researchers paid for their runs out of their research grants. Commercial firms gained access to the service in 1984, though they had to pay the full cost of their implementations. In 1983 DARPA waived the charges to NSF researchers, bringing them into the same category as DoD and DARPA researchers.[40]

Simple economics dictated free service to government researchers. The government was going to pay for the service in the end. Either it could pay ISI directly and make the service free to its contractors, or it could include money in its contracts to allow the researchers to pay ISI. If it chose the latter course, however, then both the contractor and ISI would charge the government overhead. If it paid ISI directly, it would pay only the cost of the implementation service plus ISI's overhead and fee. Rather than pay institutional overhead twice, it simply funded ISI directly.

As it expanded its client base and its communications services, MOSIS also expanded the resources it made available to the community. As it processed designs, for example, it collected nonproprietary ideas in an on-line library. In 1984 it provided access to its collection of common circuits, input-output pads, and other designs. The following year it expanded the accessible library to include standard logic and computational functions and memories. By 1989 it was offering the DoD standard cell library incorporated into five commercial design tool sets. The following year it was brokering design kits for computer-aided engineering (CAE) and computer-aided design (CAD).

The overall record of success and improvement could not entirely escape setbacks and disappointments. Perhaps the greatest was the

difficulty experienced in reducing feature size. Plans to introduce bulk
1.25µ service in 1983 suffered through five years of frustrations and re-
versals before finally coming on-line in 1988. Furthermore, efforts to
further reduce turnaround times ran up against the economic realities
of scheduling sufficient chips to fill a wafer. A 4µ nMOS device in 1984
cost about $233 and took 9.9 weeks to turn around. A 2µ CMOS device in
1989 cost about $258 and took about 8.1 weeks. Those six years witnessed
marginal improvements in cost and service, but a feature shrinkage of
50 percent.[41] In general, MOSIS was comparable in quality with the best
commercial practice of the time.[42]

Public Good, Political Liability

Even before the creation of the Strategic Computing Program, MOSIS
was having a significant impact on computer research—especially architecture research—in the United States.[43] In computer-aided design and
graphic tools, MOSIS supported the work of James Clark at Stanford
University on the Geometry Engine. This path-breaking development
was later commercialized in Silicon Graphics, Inc., one of the most successful pioneering firms in the field of television and cinema graphics.
Through the 1980s this emerging industry rivaled the DoD in the promotion of computer graphics technology.

MOSIS also serviced early experiments with reduced instruction set
computer (RISC) architecture. RISC is a design technique intended to
simplify the architecture embodied in traditional machines known as
complex instruction set computers (CISC). Pioneered by Seymour Cray
while he was at Control Data Corporation in the 1960s and at IBM's
Watson Research Center in the 1970s, the technique was put in its modern guise by David Patterson at the University of California, Berkeley,
and John Hennessy at Stanford.[44] In essence it entails a simplified architecture that depends upon compilers to adapt to the problem at hand
instead of hard-wiring a more complex architecture that is capable of
automatically handling all possible cases. MOSIS was employed to develop RISC I and RISC II architectures, which became the basis of the
SPARC computers of SUN Microsystems, essentially the first and still one
of the most successful lines of work stations. MOSIS also supported the
development of MIPS and MIPS-X, follow-on projects by John Hennessy
that were commercialized in MIPSCO.

On an entirely different front, MOSIS was employed in designing parallel processing architectures.[45] Even before SCI was under way, DARPA

was supporting H. T. Kung's research at Carnegie Mellon University on WARP, a systolic array architecture. Systolic arrays are banks of parallel processing elements that lend themselves to repetitive computations of the kind often encountered in science and mathematics. They are well suited, for example, to signal and image processing, matrix arithmetic, polynomial arithmetic, and some nonnumerical applications such as searching, sorting, and pattern matching.[46] As iWARP, Kung's work found its way into INTEL chips of the 1990s. SC also supported BBN's Butterfly parallel processor, Charles Seitz's Hypercube and Cosmic Cube at CalTech, Columbia's Non-Von, and the CalTech Tree Machine. It supported an entire newcomer as well, Danny Hillis's Connection Machine, coming out of MIT.[47] All of these projects used MOSIS services to move their design ideas into experimental chips.

The list of MOSIS contributions goes well beyond the small selection presented here. More than 12,000 projects were implemented during the 1980s. Many of these were student designs, no doubt as useful for the mistakes and misconceptions they reified as the breakthroughs. This too was part of the goal of SC. Besides these, many runs surely implemented promising, even good, ideas that fell by the wayside because other people had still better ideas or at least more practical ones. But MOSIS also had enough certified, bona fide winners to justify its existence on their merits alone. There is little doubt that the program paid for itself.[48]

None of this, however, could entirely protect MOSIS from three sorts of criticisms. The first came from industry. To chip producers, such as former DARPA Director George Heilmeier, then an executive at Texas Instruments, MOSIS looked like a competitor. Firms such as Texas Instruments, INTEL, and IBM had built up their own research laboratories and foundries at great expense. Was it not unfair that a private researcher at a university could get free access to comparable resources? If the researcher sought just the expansion of knowledge, that was one thing, but many of these university professors patented their inventions, formed companies to produce them, and showed up in the marketplace as competitors to the very firms that were at first locked out of MOSIS.

It did little good to open up MOSIS to industry in 1984. Industry had to pay full price for the services, while university and government researchers got them free. Worse still, this tack opened up MOSIS to a second and entirely different line of criticism, that it was a government subsidy of industry. Even though commercial firms had to pay for the service, they were paying simply marginal costs, that is, the cost of implementing their design. The real cost of MOSIS was in the infrastructural investment to set up, develop, and improve the services that MOSIS made available to all its clients. After 1984 industry could exploit that investment without repaying the true cost. Any such use by industry had the potential to keep the U.S. computer industry ahead of the rest of the world—one of the underlying purposes of SC—but the potential for private gain at government expense never entirely disappeared.

A third criticism of MOSIS was that it was a never-ending program. Like SC, most DARPA programs had a predicted life expectancy, say five to ten years. DARPA did not fund projects indefinitely. It was in the business of funding "high-risk/high-payoff research at the cutting edge." As MOSIS became routine and settled into predictable patterns of client service, it hardly appeared to qualify for DARPA support. It always had a research component in product and service development that met the DARPA standard, but its most fundamental activity was a routine service, hardly the stuff of the cutting edge. It might be high payoff, but it could hardly be construed as high risk. Justifying it to DARPA directors often proved a thorny problem.

Bringing MOSIS under the SC umbrella offered some protection from this last criticism. For Kahn, VLSI was exemplary infrastructure, an essential component of his grand scheme to advance the entire field of computer research by bringing along its interrelated parts in tandem. Furthermore, in the SC plan, he had articulated a relationship between this kind of infrastructure and the more concrete goals that Cooper had forced upon the program. Even though Kahn did not believe that the SC applications were appropriate, he did believe that advances on every level of the SC pyramid would support "a broad base of machine intelligence technology to increase our national security and economic strength."[49] It was clear in his mind how MOSIS would support advances in VLSI architecture that would make possible more sophisticated applications of artificial intelligence to achieve these broad national goals. The SC plan had laid out just that case. When MOSIS moved into the SC program, it moved under the protective cover of that argument—now blessed by DoD, the Reagan administration, and Congress.

The budget crisis of 1986, however, brought all DARPA programs under renewed scrutiny. James Tegnalia, who had been acting director of DARPA after Cooper's departure in the summer of 1985 and then served as deputy director until 1987, had never been sold on MOSIS.[50] He requested a study of commercialization potential. Paul Losleben, the former program manager for VLSI and MOSIS, conducted the study for

Penran Corporation of Menlo Park, CA, a company that was a potential locus of commercialization.

Not surprisingly, Losleben believed that MOSIS was "possibly one of the most valuable projects ever undertaken by DARPA." Still, he doubted that it could be successfully commercialized.[51] The obstacle was not demand. He believed that the DoD and commercial customers could provide enough business to support a commercial venture. The problem was risk. Venture capital gravitated toward products, not services. Moreover, this particular service was intensely execution dependent; only sustained high levels of efficiency such as those achieved by ISI could guarantee a continuing customer base. Finally, any such business would face the possibility of ruinous competition from semiconductor manufacturers, which had the resources to undersell if they chose.

The only prospect that Losleben saw for commercialization was a venture arising out of a consortium of potential customers. But even that prospect was not good, because commercial customers were still too few. In 1986 university research and teaching, funded by both DARPA and NSF, accounted for 89 percent of MOSIS use. DoD funded 9.5 percent, while commercial orders amounted to only 1.5 percent of the total. There simply did not appear to be much demand outside the government. A private firm, SYNMOS, had attempted to offer the service in the early 1980s when MOSIS was just getting started; by 1984 SYNMOS was out of business.

Thus DARPA faced a dilemma. MOSIS was by all accounts among its most successful and important ventures, a key component of infrastructure and a cornerstone of the SC pyramid. It was not, however, primarily a research project in the traditional ARPA-DARPA style. It was too good to terminate, but not good enough to spin off.

The only help for it was to emphasize the strengths of the program and gloss the philosophical problems lurking in the shadows. The strength of the program was different to different constituents. To researchers, the program provided service. For industry, this service trained tomorrow's employees, it brought new ideas into the American market, and it avoided competition with industry. To DARPA executives, MOSIS was the most cost-effective way to support a certain kind of important research and teaching. And to Congress, MOSIS developed new techniques for chip manufacture and simultaneously supported advances in VLSI architecture. To ensure that all interested parties saw their particular face of MOSIS in the most favorable light, the program hired a topflight professional photographer to take high-quality color pictures of

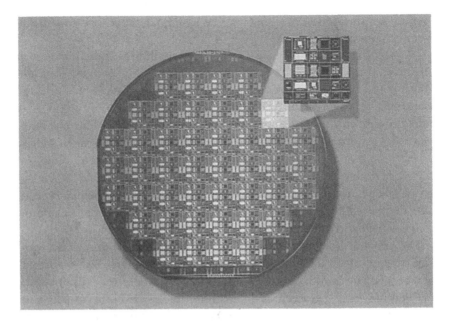

Figure 4.7
A wafer with multiple chips.

the chips implemented in the program. The results were renderings of technology that approach high art, color photographs that captured the coincidence of form and function (see figure 4.7).

Computer scientist Frederick Hayes-Roth has said that infrastructure works best when it is invisible.[52] The same might be said for ARPA-DARPA. By being a high-risk, high-gain agency, it necessarily sponsors a large number of failures. It is better that the public and the Congress not explore with any regularity its rejection bin. Similarly, infrastructure is of necessity mundane and ancillary. It does not entail the development of exciting new ideas; rather it facilitates their development. It cannot claim to be responsible for any such achievements, only to have helped. Thus it does not stand up well to close scrutiny or to logical argument about its consonance with ARPA-DARPA philosophy. It is far better to direct attention to the marvelous photographs of the designs it implements and let those speak for the enterprise.

Other Infrastructure

As MOSIS was absorbed within SCI and expanded its services, other elements of infrastructure were set in place as well. When SCI began, the most pressing infrastructure concern among IPTO contractors was lack of computing power.[53] Most research groups then were sharing time on large, mainframe machines. The machines of choice for AI research, for example, were the DEC 10 and DEC 20, computers limited by their lack of addressing capacity for virtual memory. This ill suited them to run LISP, the language of choice among AI researchers.

LISP, which stands for LISt Processing, was the most popular of a series of computer languages that used linked lists to represent data. First developed in the 1950s and 1960s by John McCarthy and his colleagues in Project MAC at MIT, LISP had real power to represent symbolic concepts in a readily manipulable machine language. It was useful, in short, for problems of AI.[54] Instead of requiring that the programmer direct every step the computer had to make, LISP and other high-level languages allowed the programmer to write out complex instructions in a readily understandable form, depending upon a compiler to translate those instructions into the signals understood by the computer's circuitry.

Frustrated by the inability of machines such as the DEC 10 and DEC 20 to process LISP effectively, researchers in the late 1970s began to experiment with programming environments to run LISP on different architectures. They also turned their attention to machines specifically designed to exploit LISP. The MIT LISP Machine Project, funded by DARPA, developed both a LISP work station and a new LISP environment, Zeta LISP, which integrated object programming with traditional LISP. The success of this program bred two rival start-up companies, Symbolics and LISP Machine Inc. (LMI). These competed with Xerox Corporation and Texas Instruments (TI) to market machines custom designed for LISP processing.[55]

All four companies faced the same problems. Servicing a comparatively small market, LISP machines could not be produced in sufficient quantities to generate economies of scale; thus, they sold for $50,000 to $100,000. Second, they were not compatible with non-LISP machines, so they could be linked or networked only to each other. Third, only AI researchers were well prepared to program them, a further limit on their market potential.[56]

As SC was getting under way, then, AI researchers lacked adequate computing equipment and could not afford the specialized tools just coming on the market. Into this breach stepped the Strategic Computing Machine Acquisition Program. Directed by Ronald Ohlander out of the rapidly growing Computer Science section of IPTO, the machine acquisition program allocated more than $8 million by the end of 1985, purchasing more than 100 LISP machines for delivery to key researchers as "government furnished equipment" (GFE). This meant that the government paid for and retained title to the equipment but deposited it indefinitely in the custody of the researcher. In practice, such equipment was seldom recovered by the government; it simply stayed with the user as its value depreciated. It was then sold as surplus or simply discarded.

The program had great strengths. It realized significant economies by negotiating bulk purchases from vendors.[57] It placed in the hands of some researchers expensive equipment ideally suited to the work they were doing for SC. It gave them experience in programming in LISP and sped up their research. It trained younger researchers in what looked like the next generation of software and hardware. And it helped to ensure that research in one segment of SC would be compatible with research elsewhere.[58]

Besides providing machines to the research community, SC also promoted the development of a standardized LISP language that would help researchers to collaborate and share results. Even before SC got under way, Robert Englemore, a DARPA-IPTO program manager from 1979 to 1981,[59] had launched an initiative to get the AI research community to agree on a common language. At the time, one popular variant, Interlisp, had already been taken over by Xerox and made proprietary. Four or five other versions were under development in laboratories across the country. Englemore empaneled a group of five researchers to hammer out a common LISP language. Their work continued past the departure of Englemore in 1981 and the beginning of SC in 1983.

The SC Machine Acquisition Program thus began with several principles clearly established. It would use SC resources to purchase sufficient LISP machines to meet the needs of the research community. It would use its bulk buying power to win significant discounts from the competing manufacturers. It would allow researchers to decide which machines best suited their needs. And it would push the entire SC AI research community toward a common version of LISP. Though the focus was on AI research, the overall goal of the program was to make available to researchers "any needed class of architecture from any vendor at a significant discount."[60]

In August 1983 ISI proposed to serve as the machine acquisition center for SC.[61] In its first installment, it offered to purchase thirty Symbolics 3600s at about 35 percent off the regular price. ISI would charge DARPA no overhead or fee for the purchase. DARPA would pay only for the cost of the machines and the time and material expended by ISI personnel in acquiring, testing, and distributing the equipment. Having received no other proposal for this service, DARPA issued ISI a contract to purchase machines on its behalf, though it initially gave no authorization to acquire the Symbolics 3600s.

After exploring alternatives, ISI sent to Ohlander on 13 March 1984 a sole-source justification for purchasing the Symbolics 3600s. ISI had tested an LMI machine and found it to be a wire-wrapped prototype running at one-half clock speed with an inoperable memory cache. Xerox's Dorado machine was not yet stable; the company declined to provide a prototype for testing. TI's machine was still far from production. Furthermore, the Symbolics 3600 ran a version of Zeta LISP that was suitable for conversion to DARPA common LISP language. On 20 March 1984 DARPA authorized ISI to purchase 30 Symbolics 3600s at a cost of $2,759,325. It noted that future purchases would depend on a reevaluation of competing machines.

As part of its acquisition service, ISI operated the Symbolics machines at its California offices for thirty to sixty days. In this first block of machines it discovered a major problem in the display units, which Symbolics corrected before the machines were delivered to the end users. This kind of testing saved time, money, and aggravation, and it facilitated the research for which the machines were intended.

As the first round of acquisitions proceeded, Ronald Ohlander accelerated the specification of a common LISP language begun by Robert Englemore. The five-person committee, including members from Carnegie Mellon University, Symbolics, and Stanford University, completed a draft in early 1984 and circulated it in the research community for comment. In the middle of that year, Ohlander convened a meeting of interested parties from industry, academia, and government at the Naval Postgraduate School in Monterey, California. The participants agreed to accept the new protocol, Common LISP, as a public-domain, de facto language standard. Vendors could still achieve competitive advantage by developing proprietary programming environments that embedded the language, but researchers would nonetheless enjoy a lingua franca that would ease collaboration. The participants at the Monterey meeting formed a nonprofit governing group to oversee development and dissemination of Common LISP. For its part, DARPA required that Common LISP be offered on every machine purchased by Strategic Computing.

After the Monterey meeting, a second round of machine acquisitions began. It soon became apparent that Symbolics had developed a troubling competitive advantage. The company was now offering second-generation machines in two models, the 3670 and the smaller 3640. Over thirty requests for these machines had accumulated since the first round of purchases. LMI now had machines ready to ship, but it refused to discount them substantially. Xerox and TI machines were now available, but without a track record they attracted no requests from researchers. Even some work-station manufacturers wanted to enter this market but held back for fear that Symbolics had it locked up.

Ron Ohlander took two steps to solve the problem. First, he authorized the purchase of 20 Symbolics 3540s at a cost of $1,092,000. Second, he sought to restore competition by making clear to researchers all their options. At his direction, ISI advised all eligible researchers that the machine acquisition program would purchase for them any combination of commercial equipment that (1) met their research needs and (2) fell within the category of equipment that SCI approved for their project. The letter undertook to explain what options in performance and cost were available from the various vendors. At the same time, it tried to avoid revealing the deep discounts that the vendors had offered to DARPA. Thus, one researcher focusing on graphics might choose equipment well suited to that application, while another who needed symbolic processing power might choose another combination. Each would be free to mix and match components from various vendors to get the best package for his or her project within the constraints imposed by SC.

The letter, distributed in March 1985, appears to have worked. Some researchers began to choose TI Explorers, which had adequate capabilities for many purposes at less cost than Symbolics machines. Most found LMI machines costlier than those of Symbolics with no significant performance advantage. The performance of Xerox machines was generally considered low. Neither LMI nor Xerox sold many platforms to the program in spite of the best efforts of DARPA to make then available to the community.

These principles of researcher choice and maximum economy in machine acquisition had another effect unanticipated in the March 1985 letter. By early 1986 researchers were choosing general-purpose work stations, such as the Sun Sparc, over LISP machines. The specialized architectures and low-volume production of the latter machines kept their prices near $50,000 to $100,000. Even with SC discounts, these machines were still three to four times as expensive as the new work stations entering the marketplace. Made possible by dramatic commercial advances in architecture and in chip design and manufacture, the work stations were powerful enough to run LISP programs and yet versatile as well. The computing power now available to individual researchers was unimagined by most when SC was first conceived. Capabilities then available only through time-sharing on mainframes could now be delivered by an affordable work station.

This development had a swift and profound impact on DARPA's machine acquisition program and on the LISP machine manufacturers. The start-up companies that pioneered LISP machines experienced stagnating sales in the mid-1980s. LMI held its ground and refused to offer DARPA steep discounts on its expensive machines and thus provided only one to the machine acquisition program; it became the first to go under, in 1987.[62] Symbolics, Xerox, and TI offered DARPA significant discounts on their LISP machines, and respectively sold eighty-three, five, and ten to the program.[63] But these sales were not enough. Symbolics struggled on before filing for bankruptcy in 1993; Xerox and TI abandoned LISP machines years earlier.[64]

By early 1986 machine acquisitions had gone over from LISP machines almost entirely to the new work stations. Twenty-three Suns were purchased in February of that year; by June a total of forty-five were on order. At that juncture the budget crisis of 1986 was being felt and the machine acquisition program contracted in fiscal years 1987 and 1988. When it revived in 1989, the primary focus had shifted to the acquisition of parallel processing machines developed in the SC architectures program.

From 1984 to 1992 the machine acquisition program spent more than $15 million on equipment for its contractors. It won from vendors discounts of 35 to 45 percent, and it provided additional services of testing and support that sped the process of getting equipment into the researchers' hands. In addition, it negotiated favorable maintenance agreements with vendors; Symbolics, for example, provided software maintenance on its machines for $76,176, compared with the book value of $495,000.[65] ISI charged DARPA about 2 percent of the cost of the machines to run the program.[66]

Machine acquisition was a popular SC program, because it put new equipment in the hands of researchers. Whether or not it was a success depends on one's point of view. From one perspective it appears that IPTO misread the field and supplied its researchers with an obsolescent technology. Program manager Ron Ohlander protests that ARPA simply bought the machines that people asked for. If their choice of machines was wrong, it was the fault of the researchers, not DARPA. That defense has some merit, but it casts doubt on DARPA-IPTO's ability to predict even the short-term future let alone the next ten years of computer development. It also belies the conceit that DARPA-IPTO was a tough taskmaster, bending researchers to the needs of the government. Rather, the experience suggests that the "DARPA community" in computer research was inbred and self-referential, as likely to reinforce each other's mistakes as to correct themselves by intramural criticism. In this case, the AI researchers proved to be out of touch with hardware developments outside their realm, and DARPA-IPTO followed them blindly down a dead end.

In spite of the misstep on LISP machines and the failure to draw networking into the SC orbit, the infrastructure program proved to be a virtually unalloyed good. It was a Kahn fundamental-research program in contrast to a Cooper applications program. Giving researchers access to better tools ensured that all research, both good and bad, would be done better. It could not ensure that any particular application would come to pass, but Cooper nevertheless supported it because it increased the chances of success for all applications.

As for the research trajectory, infrastructure shaped both direction and velocity. Wherever architecture, software, and applications were going, they were likely to be accelerated by the infrastructure. The direction of the trajectory, however, was constrained, not liberated, by the infrastructure program. For a while LISP machines tied researchers to specific implementation of a flexible programming language. Even more constraining, the adoption of the Mead-Conway standards limited microelectronics designers to one version of solid-state electronic devices, those with λ proportions.[67] While subsequent research or different applications might have demonstrated the superiority of other geometries, the λ rules made it unlikely that alternative devices would be implemented. Industry and the research community had invested too much in the λ standards to reverse course.

This phenomenon is well recognized in the economics literature.[68] It arises whenever standardization settles upon technological practice.[69] Historians of technology who investigate what they call "social construction" refer to this phenomenon as "closure." Once it is reached, the path of a given technological trajectory is usually fixed for the indefinite future. Once VHS format was selected over Beta, for example, the future of magnetic videotaping was settled for decades to come.

But because the Mead-Conway rules were embedded so deeply in computer infrastructure, MOSIS did not have that kind of impact. The program may have closed off some semiconductor geometries, but it did not determine where Kahn's technology base would go. And because

MOSIS facilitated research throughout the technology base, it proved to be about as close to a sure thing as a DARPA program manager might fund. Virtually all researchers were granted access to the service, so all had reason to praise it. New designs that fell by the wayside could hardly be blamed on MOSIS. Those that succeeded could be offered as evidence of the program's value. This was high gain without high risk.

After VLSI and machine acquisition, the dimension of SC infrastructure that is perhaps most remarkable is the one that did not materialize. Networking, the quintessential infrastructure, the great connector, did not at first become part of SC. In spite of Kahn's background in networking, and the prominence of networking in the SC plan, program manager Barry Leiner elected to keep his program within IPTO but not bring it under the SC umbrella.

Leiner's reasons, presented in chapter three, warrant revisiting in the context of the infrastructure program.[70] First, he already had adequate funding for what he wanted to accomplish. The ARPANET had demonstrated the importance of networking; the concept of an Internet to connect networks practically sold itself. He faced many technical problems in making this larger and more versatile network a reality, but funding was not one of them.

Second, Leiner feared that the influx of SC funding would draw public and congressional attention to the program. DARPA works best out of public view, free of scrutiny. "High-risk/high-payoff" also means high failure rate; it goes with the risk. Leiner preferred to work without the oversight and criticism that he felt SC would surely attract.

Third, Leiner realized many of the benefits of SC without actually joining the program. The increased level of research activity supported by SC raised the level of Internet use. And the work stations purchased with SC funds sparked increased demand for local area networks. Vendors installed Ethernet cards in their architectures, using the same TCP-IP protocols used on the Internet. All of this stimulated and facilitated use of the Internet and gave impetus to Leiner's program.

Finally, Leiner could be assured access to other funds, in part because SC relieved budget pressures on all research areas in information processing. By 1986, for example, 53 percent of DARPA VLSI work had been funded by SC, and 47 percent by "base," that is, the regular IPTO budget.[71] The new money supported not only new projects but old ones as well. All ships did rise on this tide, whether or not they joined the SC yacht club. Infrastructure, in particular, achieved Kahn's goal of supporting the entire SC pyramid and spreading its benefits upward through the technology base.

5

Over the Wall: The SC Architectures Program

The Connection Machine

Danny Hillis had an idea for a parallel processor. The conventional architectures that dominated the computer industry in the early 1980s, he pointed out, were based on technology that had been in use decades before in the early days of electronic computing. In this model, the computer essentially consisted of a single, central processing unit (CPU) and a memory that stored both the program instructions and data to be operated on. The instructions and data were placed in memory via some input/output (I/O) device, and from there they were delivered to the processor for execution. The results of the computation were then deposited back in the memory for either further use by the processor or return to the user through the I/O unit. All this was done in proper sequence, one instruction or data item at a time.

This "von Neumann-style" of computing made sense in the days when the speeds of the various components were relatively slow compared to the speed of transmission between CPU and memory, and the memory and processor units employed different technology and materials. But by the 1980s, the processors and memory units were faster, and the rate of computation was less likely to be slowed by their switching speed than by the time it took to shuttle the data back and forth between them. This problem was known as the "von Neumann bottleneck." Furthermore, by the 1980s processor and memory components were made of the same material, silicon, and Hillis saw no reason to keep them as physically defined and distinct units.[1]

In 1981 Hillis, then a graduate student at the Massachusetts Institute of Technology, proposed to build a computer he called "the Connection Machine."[2] This was to be of a type that would later be called a "massively parallel processor" (MPP). Instead of a small number of relatively powerful processors, as found in many parallel processing designs, the Connection Machine would have a large number of relatively simple processing components with their own associated memory units, which were correspondingly small. Thus the memory was not physically concentrated as in a conventional machine, but was "distributed" throughout the machine. The close association between processors and memory would avoid the von Neumann bottleneck.

To permit these processor-memory "cells" to communicate with each other and exchange data as needed, they would be wired together in some grid-like pattern and would function as nodes on a switching network, much like the nodes on the ARPANET but on a vastly smaller scale. A cell could send out a "message" that would be passed along from cell to cell by the shortest available route until it reached its destination, which could be any other cell in the entire machine. A problem could therefore be divided up among the various cells, each of which was responsible for only a small part of the whole computation. The pattern of communication among the cells, instead of being fixed as in some parallel systems, could be reconfigured as needed to suit the problem.[3]

The idea of parallel processing was certainly not original with Hillis. As early as 1840, Charles Babbage, one of the early pioneers of computing machinery, conceived a mechanical computer, the Analytical Engine, to perform mathematical operations in parallel. A century later, one of the first electronic computers, the ENIAC, was constructed to operate in this way, though the design was later changed to the von Neumann model. Since the 1950s most computers have used the principle of parallelism in some form or other. In performing calculations, for example, processors generally operate on all the various bits representing a number at once instead of one at a time—bit-parallel, it is called, as opposed to bit-serial.

Separate processors and data paths called "I/O channels" allowed data to be input into and output from the machine concurrently with the processing of other data. Computer manufacturers found ways to speed up certain operations, for example, by adding "vector processors" that could multiply entire arrays ("vectors") of numbers concurrently, an important function for scientific calculations. By the 1980s rapidly evolving microprocessors stimulated researchers to propose different configurations to connect processors, memory, and communications networks. A few private companies were producing and selling parallel machines commercially.[4]

Nor was Hillis's idea for a distributed-memory machine using many simple processors entirely original either. Such machines had been proposed for years and several were under construction or on the market. Charles Seitz and Geoffrey Fox, at the California Institute of Technology, had devised the highly influential hypercube design, which would later be marketed by a start-up company called nCube. The Goodyear MPP that Robert Cooper had authorized for NASA was then under construction, and the Connection Machine itself bore a strong resemblance to a British machine then currently being sold, the ICL DAP.[5]

What was new and revolutionary about Hillis's proposal was not the basic plan, but its scale. Instead of hundreds or thousands of processors (the DAP had 4,096), the Connection Machine would have tens and hundreds of thousands, even (Hillis predicted optimistically) "a few million."[6] Such a fantastic machine could not be built at once, but the Connection Machine was designed to be "scalable," that is, capable of great expansion without significant reengineering. Hillis initially proposed to build a prototype with over 128,000 processors, which, if successful, could then be "scaled up" to its full size of a million merely by adding more racks of identical circuit boards.[7]

More breathtaking still was Hillis's vision, his motivation for proposing the design. This is what set Hillis apart from most designers of parallel architectures. He wanted to make a "thinking machine," one so sophisticated, he said, "that it could be proud of me."[8] Although fascinated with computer hardware—as an undergraduate he made a computer out of Tinker Toys that played tic-tac-toe[9]—he was not a computer engineer, but an AI researcher. Indeed, Hillis had originally come to MIT in 1978 as a neurophysiology student to study the human brain, but he was quickly drawn to Professor Marvin Minsky, the brilliant, idiosyncratic director of the AI Lab at MIT.[10]

By the late 1970s, Minsky and other AI researchers were trying to determine how knowledge is represented and used by humans and, therefore, how it could be used by machines. Knowledge ultimately consists of facts—thousands and millions of them—as well as abstract concepts. These facts and concepts could be represented in a computer as symbols that could be manipulated the way numbers are; the LISP language had been developed for just this purpose. Doing this in a meaningful (or at least useful) way was the crucial problem, because an unstructured mass of individual, unrelated facts is as useless to a computer as it is to a person, like a book with the words in random order. Minsky proposed the theory of frames in a famous paper in 1974, suggesting a means of structuring knowledge for any given situation or problem. Roger Schank of Yale proposed a similar concept, called "scripts," for the problem of understanding language. Other researchers explored the idea of "semantic networks" by which "chunks" of knowledge were connected into meaningful ideas and concepts.[11]

Traditional computer architectures were not suited to such work. They were designed to perform arithmetical operations on numbers, sequentially. Data were read into the processor a chunk at a time, and operations were performed on it one at a time. The data had to be returned to memory and a new chunk fetched before another operation could be performed. In response to a growing demand among the scientific and engineering communities for the ability to perform operations on arrays of numbers, representing two-dimensional grids, supercomputer manufacturers such as Cray Research were adding vector (i.e., array) processors, semiparallel devices that could operate on all of the elements of the array concurrently. Yet such specialized vector processors did not meet the needs of the AI researchers, who needed machines with a lot of memory and the ability to conduct searches and pattern matches quickly and efficiently. This was the need that stimulated development of LISP machines.[12]

These machines were still sequential computers. They did not take advantage of the emerging parallel technology and apply it to the management and manipulation of large knowledge bases and semantic nets. It was for this very purpose that Hillis proposed the Connection Machine. He designed it not as a stand-alone machine nor as a general-purpose computer, but as an extension of the LISP machine for the sole purpose of performing AI applications. Indeed, he originally considered the machine not so much as a collection of processors, each with its own memory, but as a collection of individual memory units that happened to have their own processors. This was, said Hillis, "putting the processor where the data is, in the memory. In this scheme, the memory becomes the processor." The user would input the program into the LISP machine (known as the "host" or the "front-end"), which, as usual, would assign the data to memory addresses. In this case, however, the memory addresses were distributed throughout the Connection Machine, with their attached processors. Thus the user did not have to worry about "decomposing" the problem and "mapping" it onto the machine, the bane of many parallel programmers; this was all done automatically. The machine itself opened communication paths and routed messages according to the relationship of the various data elements in the semantic network, so that the pattern of communication and computation exactly matched the problem itself. Each chunk of data stored in memory had its own processor. When the LISP machine broadcast the program's instructions to the Connection Machine, the same instruction went to every cell. Each processor executed the operation on its own chunk of data. This was much more efficient than lining up the data in memory and operating on them serially, one chunk at a time.[13]

The machine would have other applications besides semantic networks. Hillis expected the machine to be useful for any AI applications involving the storing, sorting, and connecting of large masses of data, such as language comprehension or the processing and understanding of visual images. For example, an image, to a computer, is a huge array of numbers, each representing the brightness value of one of the thousands of pixels (picture elements) into which the image is divided. On a Connection Machine, each pixel could be assigned to its own processor-memory cell. Thus the machine could almost literally hold a physical representation of the image in its memory. Processing the pixels could be done quickly and simultaneously. For example, the image could be enhanced by averaging the value stored in each cell with its immediate neighbors, in effect smoothing rough edges and improving the clarity of the image. Again, conventional computers with a single processor were unsuited for such operations. The Connection Machine, Hillis would later say, "is a tool to match a method."[14]

During 1980 and 1981 Hillis and his colleagues simulated pieces of this design on a LISP machine, especially the communications network that would connect the processors. In January of 1981 they implemented a message-routing chip of their own design through DARPA's MOSIS.[15] In the fall of 1981 Hillis secured DARPA funding to develop a prototype. The project went well. By the following summer the basic system design had been completed and new chip designs had been sent to MOSIS for fabrication. Though Hillis had promised a machine of over 128,000 memory-processor cells by the end of 1983, DARPA appears to have expected one of a million cells. Meanwhile, programmers developed experimental application software for the simulated machine, including a program to support VLSI design.[16]

By 1983 Hillis had apparently caught the commercial fever that was sweeping the AI community. The success of the Lisp Machine had already emptied the MIT AI Lab of many long-time programmers who had gone off to join competing start-up companies, such as Symbolics and LISP Machines Inc., to manufacture and sell the devices. Evidently it was Minsky who conceived the idea for a new company that would act as a "halfway house between academia and industry," marketing the

Connection Machine in the same way as the LISP machines were being marketed.[17] Even though the prototype had not yet been built, the Connection Machine idea clearly had tremendous promise, and Minsky and Hillis began looking beyond the prototype to the commercial manufacture and sale of the machine. Hillis joined forces with Cheryl Handler, the head of an economic consulting firm who held a doctorate in urban planning from MIT. In June 1983, with Minsky's encouragement, they formed a company called Thinking Machines Corporation, (TMC) reflecting Hillis's ultimate goal. The new company set up shop in a house in Waltham, Massachusetts.[18]

Having a company with a snappy name and a clever idea was not the same as having a product in hand. Chip manufacture was expensive, especially early in the life of a new chip, when the low usable yield of the first few batches meant high costs. With the funding Hillis was receiving from DARPA through the AI Lab—$250,000 in 1983[19]—he could not build a prototype with 65K cells, let alone 128K, or a million. He needed more government assistance. Fortunately for Hillis and Handler, many in DARPA shared his vision for a thinking machine, or at least one that could perform intelligent tasks that could be of great service to the DoD. They also shared the same belief in the necessity of government assistance to realize such new ideas. SC came along at the right time for Thinking Machines—and the Connection Machine came along at just the right time for SC. The Connection Machine did for architecture what SC was trying to do for artificial intelligence; it connected the best technology to produce a system that was more than the sum of its parts.[20]

The Wall

From the perspective of 1983, the architectures program was probably the single most critical component of SC. Even those who considered SC to be an AI program believed that truly intelligent systems could never be achieved without massive amounts of computing power, far more than were available in the early 1980s. Kahn had estimated that computing power would have to increase by a factor of a thousand for intelligent software to run in real time. DARPA did not expect this sort of performance from the current manufacturers of computers, or from the machines then being designed and built. Conventional computing had come a long way in the previous fifteen years or so, since the Control Data Corporation (CDC) had come out with what is considered the first true "supercomputer," the CDC 6600, in 1964, followed by the 7600 in 1969.

Four years later, Seymour Cray, the designer of these pioneer machines, left CDC to found his own company, Cray Research. Just three years after that, in 1976, he produced the fastest computer in the world, the Cray-1, which introduced specialized vector processing, and performed at a peak rate of 160 gigaops (billion operations per second). In 1982 CDC delivered the Cyber 205, similar to the Cray machine but faster. Cray topped it with the Cray X-MP, which essentially linked up to four Cray-1s to operate as a single unit.[21]

Yet these machines were not suited to the goals of SC. They were serial, von Neumann-style machines that performed one operation at a time. They made some modest use of parallelism, such as pipelined vector processing, but they were essentially designed for mathematical operations ("number-crunching"), not symbolic operations. CRAY-1s and CYBER 205s were of little use to AI researchers.

Furthermore, it appeared evident to Kahn and others at DARPA and elsewhere that supercomputers like the Cray series could not continue their breakneck advances in speed. The gains of the 1970s were based largely upon developments in fundamental component design, especially in VLSI. Designers were cranking up the clock speed, that is, the speed at which the electronic signals coursed through the machine, reducing the number and length of the wires. In the days when computers used slow and expensive vacuum-tube processors and mercury-delay-line or drum memories, the delay—and the expense—caused by the wires that connected them were insignificant. By the 1970s, however, communication delays were a major factor in computer design, and builders such as Cray were taking extraordinary measures to minimize them.[22] Communication speed had replaced switching speed as the desideratum of fast computers.[23]

From DARPA's perspective, the supercomputer manufacturers were engaged in incremental improvement of a design introduced in the 1940s. The laws of physics and economics appeared to place an upper limit on this trajectory. The projected ceiling in chip design and manufacture was already in sight, somewhere around ULSI, with feature sizes of 0.1 or 0.15 μ. The machines in which such chips were embedded also had foreseeable physical limits, another of Edward Constant's "presumptive anomalies," calling for a paradigm shift, a technological revolution.[24]

The ceiling for supercomputer architecture was created by heat. By increasing clock speeds, the size of chips, and the number of devices, supercomputer manufacturers were raising the electrical current and thus the heat produced in their machines. This was especially true of bipolar transistors, which were still preferred by supercomputer manufacturers over the newer MOSFETs because of their advantages in speed. Increasingly the machines depended on liquid coolant to dissipate this heat. Freon ran through the CRAY-1 in special channels in the circuit boards, while the Cray-2 was literally filled with the coolant. Other designers turned to even more exotic packaging technologies to dissipate the heat.[25] Design and manufacturing costs rose while gains in performance slowed. Many believed that there would come a time when such efforts would fail. There was a wall, they said, a fundamental, asymptotic limit beyond which performance could not be improved.

Yet Kahn still hoped for—and needed—a thousand-fold increase in computer performance to achieve a "super intelligent" machine. As he saw it, there were three options. The first was to seek gains in chip performance, the goal of the VLSI program. The Josephson Junction and gallium arsenide raised some hopes before being overtaken by events. Optoelectronics and wafer-scale integration also held some promise and won modest IPTO support.

Kahn's second option was software improvement. Contemporary commercial software was still geared toward the traditional von Neumann-style machines and the numerical problems to which they were generally applied. Imaginative, well-developed software environments, carefully tuned to the nature of the problems that an intelligent machine would have to solve, could add much to the speed of that machine.[26]

Yet software development begged the hardware question. It is difficult (and potentially pointless) to develop software for machines that do not and perhaps cannot exist. The SC plan flirted with this dilemma by an act of faith, projecting that developments in one layer of the SC pyramid would appear in time to connect with projected developments higher up the ladder. The most promising development sequence for architecture, however, was to design new, more powerful machines and then create software to run them. Designing machines to suit the software, as had been done with LISP machines, risked repetition of that sorry history.

Kahn's third and most promising option, therefore, was machine architecture, especially parallelism. In this view, Kahn was on firm ground. The potential of parallelism was widely recognized in the computer community in the early 1980s, especially within the ARPA community. As one assessment noted in 1981, "parallelism has become accepted as a necessary vehicle (desirable or not) toward greater computational power."[27] The definition of the Mead-Conway design rules for VLSI, the extraordinary drop in the cost of VLSI fabrication, and the establishment of MOSIS all presented exceptional opportunities for further advancement.

In 1983, however, Kahn did not know what form the machines would take, or even exactly how they would be applied. He doubted that a single architecture would support all of the capabilities he wanted, at least not for the foreseeable future. Rather, he anticipated that development efforts would focus on special-purpose machines effective in different realms of AI. For example, certain architectures would be better at signal processing, essential for vision or speech understanding, while others would be better for searching large knowledge bases required by AI programs.

In his own version of the SC plan, Kahn suggested that five different types of machines would be needed: (1) signal processors, to handle sensor inputs; (2) symbolic processors, to do the actual logical calculations of the AI programs; (3) database machines, to manage the large knowledge-bases; (4) simulation and control computers, primarily for robotics applications and simulations; and (5) graphics machines, to promote a more efficient and user-friendly graphical interface. These various machines would be designed as modules that could be mixed and matched as needed for a particular application. Thus, the superintelligent computer would not be "a single box of microelectronics which fills all needs," Kahn predicted, but "a family of computing capability built up out of all of the combinations of the basic modules and their communications systems." Someday, perhaps, they could be integrated into a single machine, but that would be in the as-yet unforeseeable future.[28]

There was no shortage of architectural ideas for DARPA to choose from. Some, such as systolic arrays, were relatively straightforward and well-understood, and they could be prototyped right away. Others, such as the tagged-token data-flow technology developed at MIT, were more complex and theoretical in nature; they should be tested only in simulation until more was known about them. The intellectual ferment surrounding parallel computing in the early 1980s made it a highly unsettled, even chaotic, field. As one observer noted, "the modern age von Neumann has yet to be identified. Instead of having one person pointing the way we ought to go, we have a thousand people pointing every which way."[29]

With the exception of vector processing, the theory and programming concepts behind parallel computing were as unclear as the hardware, and its practical value had yet to be proven. The technological risk would necessarily be high. Nonetheless, to those who studied the field and looked closely at current trends ("the curves"), the solution to the computing problem was so obvious, so peculiarly self-evident, that it hardly admitted of debate. Parallel processing was the technology of the future. One such believer was the man Kahn asked to organize and manage the SC Architectures Program, Stephen Squires.

Stephen Squires, Human Agent

Kahn insists that the story of SC is best understood with the people relegated to the background. It is, in his mind, a story driven by the logic of the technology. Highlighting his role, and that of Robert Cooper and Lynn Conway, obscures the real narrative thread with personalities and politics. Human agency, for him, plays a supporting role. Not even the intervention of Stephen Squires warrants prominence in this view, for it detracts attention from massive parallel processing and its contributions to SC. But SC is unimaginable without Squires, whose impact on the program was exceeded only by Kahn's. Squires's career with SC illustrates the many ways in which individuals can still shape large technological developments, for better or for worse.

Squires had worked for years at the National Security Agency (NSA), having been recruited directly out of college. NSA had long been one of the leading users of supercomputers in the federal government, because of its interest in code breaking and the analysis of intelligence data gathered from a wide variety of sources. The value of computers for cryptanalysis had been recognized as early as World War II. It was the motivation behind the development of one of the world's first electronic computers, Britain's Colossus, at Bletchley Park. Colossus, a closely guarded secret as late as the 1970s, played a central role in the cracking of the German Enigma codes that gave the Allies the priceless advantage of reading Germany's operational radio traffic.

Valuable as computers were for intelligence analysis in the 1940s, they were all the more so in the 1970s and 1980s. Increasingly, intelligence agencies were being swamped with data from a large and growing number of sources, satellite and radar imagery, for example, as well as radio intercepts and print communications. Computers helped analysts sift through this data, a task to which several of the later SC applications efforts were directed. NSA had played a large role in high-performance computer development in the 1950s. Though it influenced computer research less in the next two decades, it supported the supercomputing industry heavily, purchasing the latest machines and experimenting with new technology.[30]

At NSA Squires wrote systems software. He developed an appreciation for hard computational problems and for the difficulties in achieving high performance. He gained extensive experience with serial machines, especially Crays, and concluded that they would soon reach their performance limits. He believed the agency was on the wrong track technologically. There was much interest at NSA in Josephson Junctions, for example, but Squires, who worked in the research group investigating them, had little faith that they would prove viable. The agency wanted to increase Cray performance by an order of magnitude; Squires insisted that the only way to do that was to link ten separate Crays with a switch, or to develop a new parallel machine.[31]

During the late 1970s, Squires became convinced of the importance of parallelism. He attempted to interest the agency in a program in scalable parallel computing, but with little success. He also became convinced that software was an increasingly critical obstacle to the advancement of the field. Typically during the 1970s, users wrote their own application and even system software. The first Crays delivered to NSA came with no software at all, not even an operating system.[32] To Squires it appeared that writing system software from scratch for a single machine was a losing proposition. Not only did it absorb the user's precious resources, but it hindered the development of computing as a whole, by creating a multiplicity of incompatible, "stovepiped" operating systems and applications.[33]

Instead, Squires believed, one had to advance the entire system—software, hardware, components, and so forth—on many levels all at once. To overthrow the von Neumann paradigm and bring about the widespread acceptance and use of parallel processing would require not just building a machine here and writing a paper there, but many changes on many levels, all occurring in such a synchronized fashion that the process would become self-reinforcing, self-sustaining, and ultimately self-promoting. The writers of applications programs to be executed in parallel needed new languages, compilers, and tools; the compiler and language developers needed new system software and software concepts; the systems software writers needed workable parallel hardware; the hardware designers needed new chip designs; and the chip designers needed the means to design, prototype, and test their chips. All these levels were interdependent and had to move forward together.[34] All had to be connected.

Squires believed that major change occurs at a fundamental, systemic level. Such change could be manipulated and guided by a relatively few people, as long as they understood the process and how to manage it. This involved identifying and influencing key "leverage points," at which a relatively small amount of effort could bring about a disproportionately large gain, in the way that a small investment can reap large profits. The field of computer science could be changed by manipulating a few key leverage points—providing infrastructural services like MOSIS and nurturing the development of certain key technologies such as VLSI and multiprocessing—always encouraging, without dictating, the course of technological change. Squires thought in terms of research trajectories, path dependence, and agency. The pace and direction of research were manipulable; successful outcomes depended on setting the proper course. He would come to rival Kahn in his vision for SC and his ability to think of computer development as a whole when making incremental decisions about the pace and direction of SC projects.[35]

Squires moved to DARPA in August 1983. With SC in the offing, he had sought an interview with Kahn and convinced the IPTO director to take him on. Hired to work in system software, Squires quickly shifted to architecture. His first task was to revise and refine that portion of the SC plan.[36] DARPA had little experience managing a major computer development program of this nature. Prior to SC, IPTO had sponsored only one such program, ILLIAC IV. Cooper asked Kahn to evaluate that program and to draw lessons from it for SC. ILLIAC had been a controversial program, begun in 1966 with the goal of producing the world's most powerful computer. The machine, consisting of four quadrants of sixty-four processors each, was expected to achieve a gigaflops (a billion floating point operations per second). Yet a number of the experimental components did not work, and the program was plagued by cost overruns.

By the time construction ended in 1972, the original price of $8 million had ballooned to $31 million. Software costs ultimately doubled that total. Although completed in 1972, the machine had so many flaws and bugs, and so little software, that it did not become fully operational until 1975. Despite all the expense, however, only one of the four quadrants was constructed, and the machine delivered only a twentieth of the performance expected of it—about 50 megaflops instead of a gigaflops. The full machine was never completed, and no production model was

ever manufactured and sold commercially. The project was widely regarded as a failure in the computing community.[37]

Yet Kahn reported back favorably on the ILLIAC experience. The machine was a concept demonstration to see if such a powerful parallel
processor could work. It did. Indeed, after 1975 ILLIAC was the fastest
machine in the world, at least for certain types of problems, such as
calculations of fluid flow. It was decommissioned in 1981 only because
it was no longer economical, given changes then taking place in the
industry. Furthermore, the machine pioneered a number of new technologies that spun off to become industry standards.

It was the first machine to use the new emitter-coupled logic (ECL)
in its processors in place of the lower, obsolescent transistor-transistor
logic (TTL), and the first to employ all-semiconductor memory in place
of the old magnetic core memory. The circuit boards were the first to
be designed with the aid of a computer. The storage technology was no
less novel, consisting of sixty-four disks that could be read from or written
to concurrently, permitting very high speed input/output.[38] Finally,
Kahn considered the project a model of an effective university-industry
partnership. The University of Illinois produced the ideas, and Burroughs Corporation built the machine. Universities cannot fully implement or commercialize their new ideas, nor should they; but in
collaboration with industry, which did not often invest in basic research,
they could get those ideas out into the marketplace, where they could
support military as well as civilian needs. This university-industry collaboration was one of the goals of SC shared by both Kahn and Cooper.

When Squires studied ILLIAC, he came to a different conclusion than
Kahn: ILLIAC was a model of what not to do. On the technological level,
Squires believed, ILLIAC pushed too many new and untried technologies at once. It was, he later said, "pure frontal assault." Not only was the
overall architecture a new departure (with all of the attendant software
problems), but so were virtually all of the component technologies, including the processors, memory, storage, and circuit boards. Indeed,
nearly everything inside the machine was an innovation. Problems with
any one of the components could have serious ramifications to the project. This is exactly what happened.

The designers planned to use custom-made medium-scale integrated
(MSI) circuits, with twenty of the new ECL gates on each chip. Texas
Instruments, which was fabricating the chips, fell a year behind on the
work, forcing the builders to resort to scaled-down chips from Hewlett-
Packard with only seven gates each. These smaller chips upset the entire

design, functioning poorly and causing subtle short-circuits that were difficult to detect. The new chips, though smaller, still took up more room in the machine than expected, forcing Burroughs to switch to semiconductor memory chips in place of the planned thin-film memory. The circuit boards contained literally thousands of manufacturing flaws, most of which did not show up until after the machine was in operation. The communications network connecting all the processors could not deliver all of the signals on time, and the clock speed had to be halved before the machine would function satisfactorily.[39]

Squires concluded that DARPA should reduce the risk involved by following what he called a "gray code" approach.[40] DARPA should limit new technologies on each project, preferably to one. That way, problems could easily be traced, isolated, and corrected, and the chances of a catastrophic failure of the system would be minimized if not eliminated. Advance of the field would not occur in one fell swoop by "frontal assault," but through the selected development and installation of key technologies, all well-timed and well-coordinated. Advance would occur over a series of generations. Once a technology or concept had proven reliable, a new component or design would be implemented and tested. The program would be cautious, deliberate, and scientific.

Furthermore, emphasis should be given equally to software. As Kahn noted, DARPA's entire ILLIAC effort had been focused on hardware, on getting the machine built. Only afterward was thought given to the applications that would make the machine usable; the software effort doubled the cost of the machine and added to the delay in rendering it operational. In the SC program, software development would neither precede not follow hardware; rather it would be an integral part of any architectural prototyping effort.

The larger lesson of ILLIAC was the danger of focusing too heavily on a single idea or technological approach. No one can predict the best approach, especially in such a new technology as parallel processing. To pour resources into one project is a heavy financial gamble. More importantly—this was the key point for Squires—focusing on a single project strayed from DARPA's mission. ILLIAC was a "point solution," a single machine. Though Squires did not make the comparison, the project may be seen to resemble NASA's quest to reach the moon. With its objective so well defined, NASA necessarily focused all of its efforts narrowly to accomplishing it. It achieved only what it set out to do, nothing more and nothing less. ILLIAC obviously did more than that, producing valuable spinoffs, but as a by-product, not a conscious goal of the project.

The real goal of the SC architectures program, and indeed of SC, was to build not a machine, but a technology base. Though he and Kahn interpreted ILLIAC differently, Squires nonetheless extrapolated its lessons to arrive at the same conclusion that Kahn had argued from the outset: build the technology base. Ironically, Kahn saw the ILLIAC as a useful application that had exerted a healthy "technology pull" on the field, just as Cooper had insisted that the Pilot's Associate, the Autonomous Land Vehicle, and Battle Management would pull computer technology along in SC. Squires took up Kahn's original position, making the technology base an end in itself, just as he would eventually take up Kahn's mantle as the don of SC.

For Squires the enabling technologies that made an advanced computer possible were far more important than the machine itself. A sound technology base generating mature enabling technologies would produce high-performance machines and impressive applications almost of its own accord, whereas a high-performance computer project without an adequate technology base was at best a point solution and at worst a futile effort. DARPA should focus on developing that technology base, promoting good new ideas and getting them out of the labs and into the marketplace; industry could take it from there. The specific projects within the technology base should be as generic as possible, so as to have the broadest application and impact and the greatest likelihood of bringing about a real advance in the state of the art.

Furthermore, Squires decided that to move a system one must move all parts of it in proper coordination. Developing isolated technologies without reference to each other or to the broader objectives would only squander DARPA's precious leverage. Spin-offs should be consciously sought, not merely as a by-product of each effort, but as an essential— and carefully controlled—part of the process. Finally, Squires believed (with Kahn) that DARPA should not invest in a single technology, but should follow several trajectories at any given time, watching carefully to see which offered the most promise.[41] Assuming that the technology might be path dependent, they would hedge their bet by pursuing several paths at once. The VLSI program supported many designs; architecture would support many concepts. If one succeeded, it would more than compensate for all the failures. Furthermore, Kahn believed that there might be multiple solutions to the architecture problem.

Armed with these ideas, Squires helped prepare the final SC plan. The architectures portion of the plan largely followed the lines laid previously by Kahn. Like Kahn, Squires projected that high-performance machines would consist of integrated modules, each performing specialized functions. He reduced the number of basic architectures to three. Two of these, signal processors and symbolic processors, had been on Kahn's list. But Squires also added multipurpose parallel computers as a goal of the program. These would be more general-purpose machines than the others, capable of performing a wider array of tasks although with less power and speed. Eventually all three types, developed as modules, would be integrated into "a composite system capable of addressing a significant problem domain." For example, a system for the control of an autonomous vehicle could include signal processors for the low-level vision systems, symbolic processors for the high-level analysis of the visual images and the reasoning and general control systems, and a multifunction machine to control robotic manipulators.[42]

In the methodology and timing of the program, Squires's plan was relatively cautious. It projected three phases, each of approximately three years. The first phase would identify and refine promising architectural ideas, and it would develop the microelectronics components that would be used. During the second phase, full prototypes of the most promising architectures would be completed and tested. During the third phase, the various modules would be integrated into the "composite systems" that could support the planned applications. The plan anticipated that prototypes of the signal processing and multifunction machines could be completed by the middle of 1986 and the symbolic processors by 1990. The composite machines were to be ready by 1993, when the SC program would presumably conclude. Unlike the applications and machine intelligence portions of the plan, however, the architectures section was prudently vague on performance milestones. Indeed, the only specific milestones were that the signal processors should achieve 100 megaflops by 1987, a gigaflops by 1989, and a teraflops by 1992.[43]

Several other features in the architectures plan deserve mention. Working prototypes or simulations should be benchmarked, that is, tested for speed and efficiency by running standard test programs and operations. This would both measure their performance and establish a standard to compare the relative merits of various architectures. Serial and parallel machines already made or under development, including the Cray-1, Cyber 205, and the Navy's new S-1 parallel computer, would also be benchmarked for comparison. Second, the plan emphasized software. Existing languages did not serve parallel programming. New languages as well as compilers, programming tools, and fundamental programming concepts, were required. Work on the software would proceed concurrently with the development of the hardware prototypes. Third, designs had to be scalable. Prototypes would be small-scale machines; if successful, they would be expanded by increasing the number and sophistication of the processors.

Squires clearly shared Kahn's view of the nature of research. He aimed to build the research base but refused to decide in advance what trajectory might reach that goal. Instead, he opted for contingency, launching several promising trajectories and reinforcing those that worked. Cooper's applications appeared in Squires' plan not as "point solutions" but rather as capabilities that would exist within the realm of the technology base. Squires thereby found a language that recognized Cooper's applications while remaining true to the contingency of Kahn's research process.

Defining the First Generation

Kahn had realized from the outset the importance of starting the architectures programs as early as possible to get computing power into the hands of the researchers quickly. This priority was reinforced by Cooper's insistence that the applications programs begin concurrently with the technology base; the applications would need enhanced computing power to achieve their early milestones. By the time Kahn produced the spring 1983 version of the SC plan, several institutions had proposals ready. As with VLSI, these projects entailed expansion of work currently sponsored by DARPA. Kahn moved quickly on these proposals, requesting funding as early as June 1983. In most cases, however, the agency withheld funding authorization, or delayed the writing of contracts, until the final plan was completed and congressional funding was assured.

The first project authorized, in August 1983, was BBN's Butterfly computer. DARPA funding of this machine had begun as far back as 1977. Designed as a communications switch, the Butterfly was significantly different from the massively parallel model represented by the Connection Machine. The Butterfly was what was called a large- or coarse-grain, shared-memory machine. The grain size referred to the processor. Unlike the simple processors found in abundance in the Connection

Machine, the Butterfly used a small number of very powerful processors, microprocessors in fact, the excellent, commercially-manufactured Motorola 68000, which would soon achieve fame as the basis of the Macintosh personal computer.

These processors were capable of running entire programs independently of each other, unlike the cells of the Connection Machine. And unlike the Connection Machine, in which each processor had its own reserved memory unit, in the Butterfly all the processors shared all the memory, access being gained by a "Butterfly switch," which could connect any processor to any memory unit.[44] Like the Connection Machine (and all of the other early SC parallel systems), the Butterfly used a conventional computer, in this case a VAX, as a front-end for the user to input the programs and data.

The 1983 Butterfly had only ten processors. BBN proposed to expand this number to 128, to test the effects and feasibility of scaling up a machine by adding processors. The completed prototype could also be used to benchmark parallel architectures, that is, to establish standards for comparison of cost and performance. Work began in mid-October 1983, before the SC plan was published, and the first 128-node machine was installed in February 1985. In the fall of 1984, as this work was proceeding, DARPA expanded BBN's effort, ordering the construction of ten 16-node machines for eventual insertion into SC applications. The Butterfly was considered the only relatively mature machine DARPA had available to run the expert systems to be used by the applications.

Not long afterward the program was moved to the Engineering Applications Office (EAO), which specialized in the transition of technology to applications. By this time the machine was in a mature state, with its own parallel operating system (Chrysalis) and application software. The first machines to be transferred went to the University of Rochester and the University of California-Berkeley in August 1984. By March 1986 nineteen Butterfly computers had been installed at SC research and applications sites, including two 128-node and two 64-node machines.[45]

Carnegie Mellon University (CMU) was another early entrant in the SC architecture program. It proposed several projects in the spring of 1983 as part of its "supercomputer program," and it, too, began work in October. One effort was called the Software Workbench. This was originally intended to create a hardware and software environment and tools to simulate multiprocessors, thereby exploring and testing parallel designs without the expense of implementing them in hardware. Another project was the Production System Machine, a special-purpose processor attached to a general-purpose host. Its sole function would be to run a type of expert system known as production systems, also called rule-based systems because they consisted entirely of "if . . . , then . . ." rules. At this early stage the project was still under design.[46]

Further along was Carnegie Mellon's systolic array machine called Warp. A systolic array processor is a pipelined machine, meaning that the processors are connected sequentially and the data passes from processor to processor in a steady flow, like the flow of blood in a body (hence the name "systolic"). A better analogy is an automobile assembly line. Like cars moving down the line, each chunk of data passes from processor to processor in sequence, with each processor performing a different operation on it. For most of the computation, most of the processors are at work concurrently on different data chunks, hence the parallelism and the speed-up. CMU had demonstrated the feasibility of the systolic approach with its programmable systolic chip (PSC), and it now wanted to build a prototype array.

The original purpose of the machine was for processing signals (such as visual or radar images), but DARPA and the CMU researchers wanted to see what else could be done with it. A good deal, as it turned out. Just as work on the Warp was commencing, a major advance occurred in commercial chip technology in the form of the Weitek floating-point chip. This development allowed the construction of more powerful systolic cells that could perform more functions than simply signal processing. "Running as an attached processor [to a general purpose host]," the researchers later noted, "Warp forms the high-performance heart of an integrated, general-purpose system"—a possibility that made the project doubly attractive to DARPA.[47]

From MIT's Laboratory for Computer Science came a proposal for a Dataflow Emulation Facility. This project would simulate and study the data-flow architecture, the operation of which is driven by the data. Processors execute a computation only when two operands (themselves products of previous separate computations) arrive at the same node. Computation could proceed at a rapid pace as long as the data kept flowing steadily. It is a complex and difficult approach, and MIT had been studying it for years. An effort to construct the machine using custom-made VLSI was begun in 1983 as part of IPTO's VLSI program, but evidently no machine was ever built. Instead, with SC funds, MIT assembled an emulator consisting of a number of LISP machines networked together, each machine representing one node of the system. Kahn had tried to begin this program prior to SC, but the agent, the

Office of Naval Research, refused to award a contract until after the announcement of SC. MIT began work in January 1984.[48]

Columbia University proposed to continue its work on two machines, DADO and Non-Von, started in 1981. These were both "tree-structured" architectures. A central processing unit was connected to two others, each of which in turn were connected to two others, and so on; the network of interconnections spread out like the branches of a tree. DADO, a coarse-grained machine, ran production systems (like CMU's machine), while the Non-Von, a finer-grained, massively parallel system, managed and manipulated large databases, especially knowledge bases. Work had proceeded smoothly on both of them in the pre-SC period, and in July 1983 Columbia sought money to develop them further and build prototypes. After delaying until the announcement of SC, DARPA funded a two-year effort beginning in March 1984 to build a 1023-element DADO2. But the front office held up the Non-Von, which Larry Lynn, the Deputy Director of DARPA, called "a questionable expense" for SC. The project was finally approved in the summer of 1984, evidently after Squires mandated changes in the program. Work began that December.[49]

One project that greatly interested the DARPA management was Texas Instruments' proposal to develop a compact LISP machine. This miniature version of a full-scale LISP machine (of which TI was attempting to market its own model), would implement a LISP processor on a single chip. The machine would be fabricated with high-density CMOS devices and would have the large memory (2 megabytes) expected of a LISP machine, but would use far less power and dissipate correspondingly less heat. The entire machine would fit on nine small printed circuit cards. This had exciting possibilities. A miniature LISP machine could be embedded in the computers of various military systems, such as Pilot's Associate, to run expert systems and other AI programs. In December 1983 DARPA approved $6 million for a twenty-seven-month program to develop the hardware. TI agreed to fund the software development, thus providing a matching $6 million in cost-sharing.[50]

There were other projects as well: multiprocessor architectures being designed at Princeton, Georgia Tech, and the University of California at Berkeley; an Ada compiler system at FCS, Inc.; the work on Common LISP at ISI. It is important to note, however, that although all of these projects (with the exception of the Butterfly and the compact LISP machine) were incorporated into the architectures program, Squires himself did not initiate them. For the most part he had little to do with them during the first year or two of the program. Squires devoted his own attention and efforts in 1984 and 1985 to identifying and starting up a new set of projects at various levels of maturity to round out the architectures program.

The first step for Squires was to develop a more complete understanding of the state of the technology and the various participants in the field. After the SC plan was published in October 1983, he spent the rest of the fall traveling around the country, visiting sites, meeting researchers and manufacturers, and holding workshops.[51] Squires then prepared a "Call for Qualified Sources," which was published on 3 February 1984 in the *Commerce Business Daily* in accordance with federal acquisition regulations. This notice asked prospective contractors to describe their qualifications and explain what they would do in the areas of signal, symbolic, or multifunction processing if they were selected for the program.[52] DARPA received ninety-two responses, two-thirds from industry and the rest from universities.[53] Squires circulated the proposals for evaluation by about a hundred colleagues in the government. He then convened a week-long review, attended by DARPA personnel and about twenty interested government officials.

On the basis of this review, Squires claims to have settled upon a multipart management strategy for the program.[54] The first part was the "risk reduction model," limiting DARPA's technological risk via the conservative "gray code" approach. The second, the "ensemble model," dictated investment in a variety of technologies and technical approaches, as opposed to betting heavily on only one. The third, the "technology system model," called for investment in all of the various aspects and levels of the field—component design, packaging, systems architecture, software, and so on—in a carefully coordinated manner.[55]

The ensemble model and the technology system model were implicit in Kahn's vision of SC. They were not, however, made explicit until Squires articulated them. Nor is it clear when Squires himself gained full intellectual control over the models. It appears likely that they represent the culmination of his thinking and experience in managing SC, rather than concepts fully developed when he called for sources in 1984. The technology system model, for example, resonates with historian Thomas P. Hughes's model of technological systems, laid out most fully in his 1985 book *Networks of Power*, which Squires claims to have read.[56]

Whenever Squires may have settled upon his management scheme, however, it nonetheless represents one of the hallmarks of the SC program. Kahn, Squires, and others at DARPA were attempting to see computer development as a complete field. From that lofty perspective they hoped to discern how the entire field might be moved forward as a technological system by making key or strategic investments in ripe subsystems. Squires's research strategy illustrates even better than the SC plan itself that such an all-encompassing objective required a combination of sweeping vision and disparate technical competencies. This bold attempt may have been SC's greatest achievement; the inability to see the field whole may have been its greatest failure. In either case, Squires's intervention was shaping SC.

The two final keys to Squires's strategy were the maturity model and the technology transition model. According to the maturity model, all architectural development programs passed through various stages of maturity from initial conception to commercial production. Squires defined four such stages: (1) new concepts that were still being worked out on a theoretical level; (2) simulations and small-scale prototypes demonstrating the feasibility of fundamental concepts; (3) full working prototypes that showed not only if the architecture was useful and beneficial, but if it could be manufactured practically; and (4) fully mature machines that were ready for commercial production.[57]

DARPA would fund the development of technology at all of these stages, but the level and nature of DARPA's support for any given project, and the deliverable results expected from it, would depend on the maturity of the technology. If a project at any given stage had made sufficient progress, it would be permitted to pass on to the next stage, with a corresponding increase in support. Projects that did not progress significantly in a reasonable time, usually two years, would be terminated. Even these dead ends might produce some worthwhile ideas or technology that could be transferred to another project. All contractors were required to allow the technology produced with SC funding to be used by anyone.[58]

While DARPA would purchase some stage-four machines at cost for research purposes—the Butterfly was in this category—Squires was much more interested in the first three categories, which obviously required the most support to bring to fruition. Perhaps the most critical step was stage three, getting a simulation or a small-scale, workable prototype developed into a working production model. This forged the real link between the lab and the market, the step that Cooper and Kahn had found missing before SC.

In Squires's strategy, developing a production model of the machine in hardware was only one part of stage three; clearing the way for its introduction into the marketplace was another. This is what Squires called the technology transition model. In his view, transition could be facilitated in several ways. First, Squires usually required universities to team or subcontract with industrial partners to build the production prototypes. With the signal processors, he also insisted that prototypes be not just demonstration machines, but usable computers that could be put into the hands of other SC researchers to work on real-world problems.[59] Squires further required that stage-three contractors produce system software and some application software, perhaps focusing on a problem area specified by DARPA. He did not want to repeat the experience at NSA with the early CRAYs. Researchers, and especially commercial users, would have much less interest in the machines if they had to write their own software from scratch. Finally, Squires specified that simulations and prototypes at all levels of maturity should be usable by graduate students to fulfill the SC goal of building the technology base by promoting academic training in advanced computer science.[60]

Squires intended to advance the entire field by carefully managed stages, with the initial projects being merely the first of several waves. The first wave was intended primarily to identify the promising forms of parallel computing, provide experience in the construction and use of parallel systems, and establish the industrial base for the continued advance of the field. The projects were to use no exotic, experimental component technology (such as gallium arsenide chips or complex chip packaging) and were to be easy to build. They would not be stand-alone machines, but would be accessed through a conventional host computer, as the Connection Machine was accessed through a LISP machine. The first generation prototypes did not have to perform brilliantly or match the capabilities of the conventional, state-of-the-art supercomputers; they only had to function well enough to demonstrate that parallel computing worked. Meanwhile, stage-one and stage-two projects were intended to explore more sophisticated ideas and technology that would ultimately lead to the next wave of parallel systems.[61]

The first solicitation of February 1984 was for informational purposes only; it was not expected to produce any contract awards. It did, however, lead to one: Thinking Machines. Squires recalls being particularly drawn to the big black notebook that TMC sent in response to the notice in the *Commerce Business Daily*. He took it home, sat up all night with it, and "completely consumed the thing." The following week he and his colleagues at DARPA discussed the proposal. DARPA personnel and TMC representatives exchanged visits, discussing the technical merits of the Connection Machine and TMC's business plan. It appeared that Hillis and Handler had carefully thought out their venture from both angles.

Squires was attracted to the fact that the proposed technology was conservative, and that TMC already had a workable VLSI design ready for the processor chips, a major source of risk in such a project. DARPA drew up a detailed technical development plan, specifying certain changes and milestones. For example, Squires specified that instead of the simple, flat-mesh communications network, TMC would have to add an additional set of connections that would permit processors to communicate with distant destinations more quickly. Having such "global" communications would enable the machine to tackle a wider variety of problems than just the AI problems for which Hillis had designed it.[62] Otherwise, the design was to remain conservative.

The machine was to use only the simplest, 1-bit processors instead of more powerful 4-, 8-, or 16-bit devices. Also, the machine was to have no floating-point capability. Most number-crunching computers, including conventional supercomputers, performed calculations using floating-point arithmetic, in which all numbers were converted to scientific notation prior to calculation. For example, the number 2,400 would be converted to the form 2.4×10^3. This sped up numerical operations considerably but required extra hardware or software. The early SC architectures projects were not permitted this capability because it would add expense and distract from the demonstration of basic parallel processing concepts. Furthermore, floating-point capability was not essential to AI applications, which dealt with symbolic as opposed to numeric processing. An additional, unstated reason was that DARPA did not want to antagonize the numeric supercomputer manufacturers such as Cray and ETA by making them think that DARPA was sponsoring the development of competing number-crunchers.[63]

Squires was not the only one at DARPA intrigued by the Connection Machine. Both Cooper and Kahn discussed the machine at length with him.[64] Early in 1984, with unusual speed, funding was approved for $3 million dollars over two years for a "small" prototype of 16K processors, with an option for another $1.65 million for a scaled-up version with 64K processors. Thinking Machines used this award as validation to raise another $16 million in private capital. Work on the prototype began that spring.[65]

Meanwhile, Squires began the selection process for the rest of the program. About forty of the responses to the February solicitation looked promising. Squires prepared a lengthy notice soliciting more detailed proposals in a form that would later be called a "broad agency announcement" (BAA). The solicitation invited interested institutions to propose projects in symbolic, signal, or multifunction processing. The developers were responsible for producing software, programming languages, and design and analysis tools as well as the basic hardware. They were to propose complete systems, not simply hardware machines.

Following the maturity model, the solicitation specified that the offerors select one of three types of projects: to produce models and small-scale simulations (i.e., stage 1), large-scale simulations and small-scale prototypes (i.e., stage 2), and medium- and large-scale prototypes (i.e., stage 3). The goal of the program, the notice specified, was "to support a new generation of machine intelligence technology" using "a scientific approach of constructing models, designing experiments using simulations and prototypes, and measuring the results in well defined environments." The agent for the architectures program, the Naval Electronics System Command, or NAVELEX (which soon changed its name to the Space and Naval Warfare Systems Command, or SPAWAR), sent the BAA to each of the selected respondents.[66]

Following standard federal procurement regulations, the resulting proposals were divided into separate categories and evaluated by a team of DARPA and NAVELEX officials based on their technical merit, cost, scalability, and potential for transition. By March of 1985, DARPA had selected six contractors, all in industry, to produce stage-three prototypes. ESL, Inc., would work on a modular signal processing architecture called MOSAIC. AT&T, in collaboration with Columbia University, would develop a digital signal processor (DSP) to be attached to the DADO. General Electric would work on a prototype of the "Cross Omega Connection Machine" for real-time AI-based systems. BBN would attempt an advanced, fine-grained, massively parallel variation of the Butterfly called Monarch, which would retain the basic butterfly-type switch but would employ 8,000 RISC-like 1-bit chips in place of the Butterfly's powerful microprocessors. Encore set out to develop a massively parallel system called UltraMax that would use 16,000 processors to perform signal, symbolic, and multifunction processing. The system would build upon a current Encore commercial product by linking together eight Multimax machines to create what the company called "a multi-multiprocessor."[67]

Finally, IBM, in collaboration with New York University, would develop a shared-memory machine called the RP3, consisting of 512 processors connected by a communications system invented at NYU called the omega network. At least four other companies—Fairchild, Hughes, System Development Corporation (SDC), and Harris Corporation—were selected to produce stage-two multiprocessor simulations and small-scale prototypes. Other organizations, including the Kestrel Institute, were selected to develop advanced parallel software tools. All these contractors began their efforts in June 1985, when the selection was formalized, though most were not finally put on contract until the following year.[68] Altogether, at least $4.5 million was allocated for these new projects in FY 85.[69] Overall, the architectures program allocated about $85 million dollars for the fiscal years 1985 through 1987.[70]

Applying the Gray Code

Thus the architectures program fell into place during 1985 and 1986. As of mid-1986, it included twenty-three projects: eight stage-one prototypes, eight stage-two simulations-prototypes, four programs in software and development tools, and three projects for benchmarking and performance metrics.[71]

The various projects all represented different architectural concepts. DARPA sought to try as many as possible to gain experience with them and to determine the best approaches. Thus, the Warp explored the uses of systolic arrays, for example, while the Connection Machine explored the "data parallel model," in which identical operations are performed simultaneously on large, uniform data structures (such as a visual image). Other machines were shared-memory multiprocessors, meaning that many individual processors communicated with the same memory. They did so either through a "bus" (central communications channel), the way an interstate highway connects many cities, or through a multistage switch that functions like a rotary junction connecting several smaller roads. In either method, any message could go from any starting point to any destination. The Encore MultiMax was an example of a bus-based shared-memory machine, while the Monarch and the RP3 were examples of the switched network.

Still another approach was the multicomputer, in which each processor had its own memory and could perform all the functions of a computer. Columbia's tree machines, DADO and Non-Von, were such systems. Unlike the Connection Machine, in which the simple processors all executed the same instructions in lockstep, in a multicomputer each processor had sufficient power and memory to execute its own programs independently of the others. According to one popular taxonomy, multi-computers were classified as MIMD (Multiple Instruction stream/Multiple Data stream), whereas the Connection Machine was a SIMD (Single Instruction stream/Multiple Data stream) system. MIMD systems were much more flexible than SIMD systems, and they could tackle a broader and more complex set of problems. They were also correspondingly harder to program, however, because the programmer had to break down the problem into separate pieces and apportion them out to the various processors for execution.

The MOSAIC, by ESL, represented a heterogeneous multicomputing architecture. While the other systems interconnected identical processors—and were thus "homogeneous" systems—MOSAIC was a crossbar switch that would connect a variety of different processors, including both general-purpose processors (GPPs) like the VAX and special-purpose processors (SPPs) like the Warp. Theoretically, such a system would be able to take advantage of the special capabilities of the different processors for any given problem. The real challenge was to find means to permit the users to program the machines to exploit this capability easily.[72]

Still other projects were accelerators. These were typically special-purpose devices, fabricated on a single chip called an application-specific integrated circuit (ASIC), that improved the performance of another (usually general-purpose) machine in some particular task. AT&T's DSP was an accelerator for the DADO, while Berkeley's Logic Programming Accelerator was meant to improve the performance of work stations such as the Sun 3.

The progress made by the program exceeded the conservative and relatively limited expectations of 1983. The first generation prototypes did indeed prove the feasibility of multiprocessing—and not just as discreet, limited functional modules, but as general-purpose machines. They did more than demonstrate the concept; quite a number of the prototype projects made it to stage four: commercial production. AT&T's DADO/DSP effort led to the commercial ASPEN, and Encore's Ultramax spawned a follow-on project called the TeraMax.

Warp's success was particularly noteworthy. H. T. Kung's team at Carnegie Mellon completed a small, two-cell prototype in June 1985, and it subcontracted with two industrial partners, General Electric and Honeywell, to produce identical, ten-cell wire-wrapped prototypes. GE delivered its system in February 1986, while Honeywell delivered four months later. During the fall of 1985, Clint Kelly, then director of the Engineering Applications Office, requested seven of the as-yet unfinished machines for the SC applications programs. After soliciting bids from GE and Honeywell to produce a more refined commercial version of Warp mounted on printed circuit boards, CMU awarded a $2.6 million contract to GE. The first "PC Warp" was delivered to CMU in April 1987, incorporating changes specified by Carnegie Mellon and the applications contractors.

Meanwhile, in 1986, with DARPA's support, CMU formed a partnership with Intel to produce the next iteration of Warp. In this version, each cell (including processors, memory, and assorted other functional components) would be implemented on a single chip. The first two 64-cell integrated Warps, or iWarp, were delivered to the Naval Ocean Systems Center in San Diego in 1990. As the hardware was being developed, the researchers were busy developing system and application software, including libraries of algorithms for general image processing, image processing for robotic navigation, signal processing, and scientific computation.[73]

Yet by far the most spectacular success, the showcase of the architectures program (and even, to a large extent, for the whole new industry of parallel processing), was the Connection Machine. Right from the start, Hillis persuaded DARPA to install a LISP machine at the foundry, so that he could test the chips where they were made in California instead of having them shipped back east.[74] As fast as the tested chips came in to TMC, they were mounted on circuit boards, 512 to a board, until all 16,000 were installed. This first prototype was finished on 15 May, 1985, a month and a half ahead of schedule. DARPA officials went to the site to try out the machine—they "kicked the tires," Squires recalled—and immediately invoked the option for the scaled-up, 64K processor version.[75]

Hillis preached the Connection Machine to the research community with boundless enthusiasm. "[W]henever I go to a workshop like this," he said at one meeting in early 1985, "at the time I get to my talk I think that if everybody had a Connection Machine all their problems would be solved."[76] His Ph.D. thesis, which was published in 1985, was as much a call to parallelism as it was a technical description of the machine. It concluded with a chapter entitled, "New Computer Architectures and Their Relationship to Physics or, Why Computer Science Is No Good."[77]

More importantly, Hillis and Handler proved effective at drawing good people into the fledgling start-up, including talented AI researchers from MIT and elsewhere. The scientists worked to prepare languages, application software, and tools. They developed parallel-processing versions of LISP called *LISP and CMLisp and a version of the increasingly popular software language C called C*. They devised algorithms for vision and for VLSI design, real-world applications that demonstrated the machine's capabilities. The staff scientists made themselves available to provide expert assistance to users of the machine, an excellent public relations tool. In an observation reminiscent of IBM philosophy, Handler later said that "our customers were able to buy more than just the product: They were buying the company."[78]

By the end of 1985 the full 64K prototype of the Connection Machine was complete, again well ahead of schedule. By the following spring the CM1, as it was called, was being manufactured and offered for sale, jumping from stage three to stage four and then out into the marketplace in a matter of months. The chips used were the custom VLSI, each containing sixteen separate 1-bit processors and a device called a router that permitted chip-to-chip communications. Thirty-two such chips, each with four associated 4K memory chips, were mounted on a printed circuit board, called a "module"; sixteen modules were plugged into a backplane, and two backplanes were mounted on a rack, each of which formed a quadrant of the entire machine. Although the full machine included 64K processors, a quadrant of 16K processors could stand alone as a scaled-down model. Physically the machine was sleek and stylish, a black cube roughly five feet on a side with thousands of flashing, blinking red lights that would warn of faulty chips (see figure 5.1).

The real beauty of the machine, from DARPA's point of view, was that the technology was so simple as to be virtually risk-free. The VLSI chips, while custom made, were relatively basic and slow compared to the state of the art. They were built, Hillis would write, "by methods similar to those for making personal computers and pocket calculators."[79] No special packaging or cooling technology was required; the 1-bit processors themselves produced so little heat that the machine was entirely air-cooled. No site preparation was required either; the machine could be installed quickly and easily and hooked up to its host, either a LISP Machine or, for a UNIX programming environment, a VAX. The machine also required no revolutionary programming concepts. CMLisp and C* were easy to learn by anyone familiar with the conventional forms of these languages. Because the compiler took care of the memory assignments, the user worked within the LISP or UNIX environment of the host without having to worry about the structure of the Connection

Figure 5.1
The Connection Machine; CM-2 is the square structure to the left; on the right is the data vault. The footprint of CM-2 was 56″ on a side. *Source:* Thinking Machines Corporation.

Machine. It was therefore easy to program; it ran a number of standard LISP programs in particular with relatively little adaptation required.

By the summer of 1987, when TMC came out with an improved version of the Connection Machine called the CM2, it had sold a dozen CM1s, about half of them purchased by DARPA for its contractor community. Researchers found uses for the machines, not just in AI, but in science and engineering, solving fluid flow dynamics problems for example. One researcher at the University of Illinois, working in conjunction with a TMC scientist, modeled fluid flows over complex surfaces, an application to which the machine was particularly well suited. Using a technique called "virtual processors," each processor could mimic a number of processing units, each handling one particle of the flow, so that the behavior of millions of particles could be followed. TMC scientists also explored the use of the machine for large database searches, where many stored textual documents are searched simultaneously for a match with a given search string.

Squires was quick to take advantage of the opportunity offered by TMC. At the beginning of 1986 DARPA agreed to fund a two-year, $3.8-million-dollar effort to exploit the potential of the Connection Machine. The effort included four projects: to improve the software environment, to develop an education program for users through workshops and instructional materials, to adapt the machine for use as a network server, and to add a parallel mass storage system to improve the machine's capabilities in data-intensive problems. DARPA also optioned the purchase of additional machines of various sizes for the SC research and applications community.[80] Ultimately three versions of the Connection Machine were marketed: the CM1, the CM2, and the CM5, this last being delivered in 1992. Of the 150 machines of all three models that TMC sold by 1994, at least 24 were purchased by DARPA.

Few of the prototypes produced by the first-wave architectures projects had this kind of staying power. A number of projects were terminated at the end of their initial contracts. After intense debate at DARPA, Columbia's projects, DADO and Non-Von, for example, were canceled at stage two, the full-scale prototype, because Squires saw few fundamentally new and useful ideas in them, especially Non-Von.[81] GE's effort in Cross-Omega connectionist architectures was considered redundant with other related projects and was canceled in early 1988.[82] IBM's project was eventually cut off, not because of technological shortcomings in the RP3, but because of institutional conflict. The computer giant considered the prototype to be an experimental test bed for studying the application of massive parallelism. It jealously guarded its proprietary interest in the machine, refusing to make additional copies available for use by researchers. IBM eventually went on alone, developing the RP3X without DARPA support.[83]

Another disappointment was BBN's Monarch. This was the project to build a massively parallel, shared-memory machine. Ultimately, BBN hoped, it would hold 64K processors, approaching the level of parallelism of the Connection Machine. Yet the Monarch, with its shared-memory design and sophisticated RISC microprocessors, would be far more powerful. Unlike the Connection Machine, which was a SIMD system, Monarch would be MIMD—the different processors would each execute a separate stream of instructions. The system was expected to offer a well-developed software environment, running current applications developed for the Butterfly as well as new programs written in Common LISP.

Because of its complexity, however, the Monarch effort entailed far greater risk than the Connection Machine. It would use a number of custom-designed components, including the processors, switching chips that controlled the communications between the processors and memory, VLSI "memory controllers" that would have charge of the standard RAM chips in the memory modules, and a specially designed packaging subsystem, including two new custom-made printed circuit boards. All this would be incorporated into a novel design frame that would provide certain services every processor would require, such as their own system clocks and facilities to ensure that all the processors worked in synchronization with each other. The processors themselves were the riskiest components and caused DARPA the greatest concern. They would have custom-made VLSI chips using CMOS technology, eventually incorporating features as small as 1.25 microns, which MOSIS could not yet deliver.[84] The array of unproven technologies incorporated in this machine warrants comparison with ILLIAC and begs the question of why Squires would support it. The experience reinforces the impression that Squires' management scheme evolved over time and did not spring full blown into his mind after studying the ILLIAC history.

DARPA specified a cautious development schedule. The first major goal of the program was a medium-scale prototype with 1,024 processors. The processors would be designed and implemented initially with 3-micron features using the MOSIS scalable design rules; thus, subsequent generations of the chips could be implemented with 2 micron and eventually 1.25 micron features when the fabrication technology allowed. Meanwhile, the rest of the Monarch would be constructed as a "pre-prototype test bed," so that the interface and operation of the other custom-made components could be tested and made operational by the time the processors were ready for installation into the machine. This prototype would be tested using software being produced concurrently with the hardware. If the tests went well, DARPA would invoke its option to have two more such machines constructed, and, perhaps, have the machine scaled up to an 8K model. The completion of the 1K prototype was expected in twenty-five months, and the 8K model eighteen months later.[85]

Predictably, the Monarch project ran into trouble. By 1989 it was facing serious cost overruns. Part of the problem was that BBN had changed the design, apparently without DARPA authorization. Like all of the first-generation machines, Monarch was to have no floating-point capability, but BBN decided that many potential users of the machine wanted such capability for scientific applications. The company therefore added floating-point capability and increased the word-size of the processors, thereby increasing their power and their cost. Compounding this cost escalation, chip prices were rising throughout the industry in the late 1980s, driven in part by increasing wages for scarce VLSI designers. Finally, MOSIS was years late in delivering implementation below 3 microns, an example of the vulnerability of the SC plan to delays at the lower levels of the pyramid. These problems, plus others that BBN encountered with the VLSI-design tools provided to SC contractors, caused further delay and expense.[86]

DARPA also conflicted with BBN over data rights. Evidently BBN wanted much of the new technology to remain proprietary and patented some of the SC-sponsored developments. Agreeing that the initial cost estimate ($4.8 million) for Monarch was overly ambitious, DARPA provided another $385,000 to complete the first prototype, but it did not invoke the options for additional copies or for the scaled-up large prototype. BBN quietly ended the Monarch project in 1989.[87]

As the experience with Monarch shows, however, even dead-end projects (from DARPA's point of view) contributed to the technology base and the advancement of parallel computing. BBN used some of the technology developed for Monarch in its advanced commercial system, the TC2000.[88] Furthermore, such projects could and did contribute to other efforts in the architectures program. Such was the case with Non-Von. While Squires saw little in its overall architecture, he did like the multiple disk-heads it used for a scalable, parallel, mass-storage system, an area of particular interest to DARPA. TMC picked up on the idea (possibly on DARPA's suggestion), and, because all SC-funded developments were open for all contractors to use, incorporated it into the Connection Machine as the "data vault." This consisted of forty-two commercial Winchester disks with 5-to-20 gigabytes of total storage, which the machine could access in parallel through eight high-bandwidth channels at 320 megabits per second. With up to eight data vaults attached to it, the machine could achieve between 40 and 160 gigabytes of easily accessed storage, a remarkable capability.

The key to Squires's approach was to seek not specific machines but a set of technologies and technical capabilities that would permit the field to go forward step by step. The projects in Squires's architectures program were elements in a larger system. They were expected ultimately to produce modules that would physically plug into a larger architecture. The contracted efforts were of interest only if they offered a significant contribution to the overall effort. Only by carefully coordinating and promoting the best ideas and the best technology available could he help bring structure and order to what was then a chaotic field consisting of scores of individual, unrelated research programs with little interrelation.[89]

The advantages of Squires's strategy for the architectures program are most readily apparent when compared with Project THOTH, an attempt by Squires's old employer, NSA, to build a general-purpose supercomputer almost concurrently with the first generation of the SC systems. THOTH[90] was planned and executed with a more traditional approach. NSA had built its own high-end machines before and assumed it could do it again using the same methods. The goal of the project was to produce a "high-speed, high-capacity, easy-to-use, almost-general-purpose supercomputer" that would achieve 1,000 times the performance of a CRAY, plug into the agency's operational environment, run state of the art software, and be sold commercially. The program was planned to run in three phases. During the first phase, six competing contractors would study the various architectural possibilities and present high-level design options. From among these, the two or three best would be selected. During the second phase, the selected contractors would submit proposals for a detailed design from which one would be chosen. During the third phase, the selected contractor would build the machine.[91]

The project went wrong from the start. It took the agency over a year to prepare a statement of work, and another year to get some of the contracts let. As the program went on it suffered delays, partly due to problems with the contractors, some of whom dropped out, suffered internal reorganizations, or showed little commitment to the program in the first place. The delay brought rising costs. Meanwhile, the state of the art in technology was continually advancing, rendering the plans obsolescent. The agency found itself with a moving technological target. NSA finally killed the program in 1988 before phase three could begin.

It was clear to NSA afterward what was wrong with Project THOTH: everything. To begin with, the agency based its plans on what it wanted the machine to do, not on what could be done. It relied on Cooper's demand-pull instead of Kahn's technology-push. It did not take into account ongoing technological developments, and it assumed far too much technological risk, expecting to make satisfactory progress in both hardware and software. The technical failure of the program then derailed a follow-on effort that relied heavily on what had been done during THOTH. NSA also trusted the contractors to build the machine it wanted and burdened them with increasingly detailed requirements.

Furthermore, the program was conceived as a competitive process from start to finish, unlike the generally cooperative strategy of SC. Therefore, the agency was unable to work with the contractors to keep them on track as they were preparing their design proposals during phases one and two. It could only pose questions and hope that the contractors would take the hints.

Finally, and most importantly, THOTH itself was a point solution. As Larry Tarbell, who managed the project for over two years, observed, "The real killer was that when we were done, we would have had a one-of-a-kind machine that would be very difficult to program and that would not have met the goals of being a system for researchers to use."[92] The approach taken by the agency had worked reasonably well in the early days of computing, the days of ENIAC, Whirlwind, and STRETCH, but it could not work in the technological and political milieu of the 1980s. "We had built our own machines in the past," Tarbell noted. "We even built our own operating system and invented our own language. We thought that since we had done all that before, we could do it again. That really hid from us how much trouble THOTH would be to do."[93]

NSA tried to build a machine and failed. DARPA succeeded in building several viable machines, but that, to Squires, was almost beside the point. Whether a project went on to commercial fruition like Warp or the Connection Machine, or whether it languished in the lab like the Non-Von, each contributed to the building of the technology base, training designers and graduate students, giving users experience in multiprocessing technology, contributing ideas and technology, and showing which of them were fruitful and which were not. Most of all, the first stage of the SC architectures program proved unquestionably that parallelism was viable and did have promise in solving many real-world problems, including some of those posed by SC. According to Professor John Hennessy of Stanford University, a respected authority on microelectronics and computer architectures, computer designers before SC, in their quest for greater speed and performance, kept running into the wall posed by the fundamental physical limits to the improvement of conventional systems and components. Each architect would try, and each would join the pile at the base of the wall. It was Squires, he said, who showed the way, and, through parallelism, led the designers over the wall.

Hillis and Squires scaled the wall, but they failed to raze it. Hillis reached the parapets in the early 1990s and then fell back to join the pile at the base of the wall. Gross revenue at Thinking Machines peaked at $92 million in 1992. Profits peaked at $1 million in 1990. After losing

$20 million in 1993, Thinking Machines filed for bankruptcy protection on 17 August 1994.[94] Explanations of the failure varied widely:[95] a power struggle between Handler and Hillis; lack of business experience for both of them; Handler's extravagant corporate style, which spent $120 million in capital investment over eleven years and realized less than $2 million in profits; Hillis's pursuit of machine speed over market share; a softening market for supercomputers; competition in the market for massively parallel machines from such powerhouses as Cray Research and Intel; the end of the Cold War; over-dependence on government in general and DARPA in particular; and disregard for the commercial sector, which bought only about 10 percent of the 125 Connection Machines the company sold. As important as any other reason, perhaps, was the difficulty of programming a Connection Machine. "The road to programming massively parallel computers is littered with the bodies of graduate students," said one computer science graduate student.[96] Many Connection Machines appear to have sat idle in research laboratories, awaiting software that could connect them to real problems.

Perhaps the most intriguing explanation for the demise of Thinking Machines is that Hillis became obsessed with the goal of a teraflops machine. Instead of tailoring the TM5 to real applications in the marketplace, Hillis appears in the last days of Thinking Machines to have been configuring his machines to achieve maximum computing power. If so, he may well have been responding to pressure from Squires. As early as 1987, before the SC decade was half over, Squires drafted a plan to bend the SC trajectory. Instead of pursuing powerful machines capable of the symbolic processing required by artificial intelligence, Squires proposed to concentrate on supercomputers for scientific, that is, mathematical, computation. His new goal, indeed his fixation, became a teraflops machine, a computer capable of a trillion floating-point operations per second.

The ironies abound. Hillis invented the Connection Machine to do AI. Squires took up the SC architectures program to develop machines capable of AI. Squires and Hillis began their collaboration because it promised AI. But somewhere along the way, they changed their goal, redirecting the research trajectory they had launched together. The full implications of this shift will be explored in chapter 9. Suffice it to note here that the demise of Thinking Machines, the company, also marked the abandonment of thinking machines, the goal. The Connection Machine, the literal and figurative embodiment of the principle behind SC, came a cropper. Its components connected; its market did not.

6

Artificial Intelligence: The Search for the Generic Expert System

Artificial Intelligence (AI)

If parallel processing got over the wall, would it not open the gates for AI? Many people thought so in the early 1980s. AI had experienced alternating periods of optimism and skepticism since its foundational meeting at Dartmouth College in the summer of 1956. Early claims that machines were already thinking gave way to doubts in the late 1960s that they could move beyond toy problems. Enthusiastic work on speech recognition and machine translation programs was followed by withdrawal of government support. By the late 1970s, however, new optimism sprang from the success of knowledge-based expert systems. If this progress could be reinforced by the computing power of parallel processing, then the promise of AI might finally be realized. This renewal of enthusiasm for artificial intelligence infected SC. AI's power to excite flowed in part from its tendency to disappoint. The cycles of optimism and pessimism through which AI had passed in its first quarter century left its supporters longing for the breakthrough that the faithful were forever promising. This history of raised and dashed expectations helps to explain the role that SC assigned to AI.[1]

To understand that promise and how it connected with the other components of SC, it is helpful to briefly review the field's history. The first incarnation of AI, before it even took on that name, sprang from the fertile mind of British mathematician Alan Turing. Drawing on his work in cryptography in Great Britain during World War II, and his role in developing Colossus, the pioneering electronic computer designed to decipher German code, Turing provided not only the mathematical theory behind the problem-solving capability of a serial computer but also the test of when such a machine will have achieved the power to think like people.[2] Turing proposed that a person engage in a remote conversation with an instrument that could not be seen. If a five-minute interrogation of this instrument left the questioner unable to determine with 70 percent certainty if it was human or machine, then machine intelligence would have been achieved.

The Turing test became the Holy Grail of machine intelligence.[3] For many researchers, the quest took the form of replicating human thought processes. This path led Allan Newell and Herbert Simon to Logic Theorist, the computer program they developed at the Carnegie Institute of Technology in 1956. The program used recursive search techniques to solve mathematical problems, just as humans might solve them. From the experience, which also required the invention of a list processing language, IPL, Simon concluded that they had "invented a computer program capable of thinking," that is, capable of replicating human thought processes in the way envisioned by Turing.[4] Simon and Newell followed that achievement with General Problem Solver, a program that used means-ends analysis to solve a broad variety of problems, not just mathematical ones.

Logic Theorist made Simon and Newell the centers of attention at the Dartmouth conference of 1956. Organized by John McCarthy and Marvin Minsky, this meeting brought together the handful of researchers in the United States then working on some aspect of machine intelligence. It achieved no technical or conceptual breakthrough, but it adopted McCarthy's term "artificial intelligence" to describe the new field and it created a community of collaborators who would form the core of future growth.

Disparate disciplinary and conceptual forces, however, were giving the field a heterogeneous character. One dimension that took on increasing prominence was already under way, even before Turing wrote his seminal 1950 paper. Warren McCulloch and Walter Pitts had hypothesized as early as 1943 that a machine might achieve brain-like capabilities if it were modeled on human neurons, and if those neurons were seen to function as a network of on-off switches.[5] This was not inconsistent with the propositions of Turing or the inventions of Simon and Newell. Nevertheless, it introduced new threads of development that would recur through the history of AI and change the complexion of the problem. Thinking might not be linear, as in a Turing machine, proceeding one step at a time. Rather, multiple paths might be stimulated, with complex interactions among the various trajectories. Furthermore, the conceptualization of machine intelligence changed over time, as insights emerged in related fields such as human physiology and cognition.

Sometimes related disciplines shaped AI, and sometimes AI shaped those disciplines.

One way to categorize the relationship between AI and its cognate disciplines is to see AI as pursuing two different goals—thinking and acting—by two different methods: human modeling or rationality. The possibilities may be represented thus[6]:

	Human Modeling	Rationality
Thinking	Systems that think like humans.	Systems that think rationally.
Acting	Systems that act like humans.	Systems that act rationally.

Since the 1950s, researchers from differing disciplinary perspectives have construed the problem in each of these ways and pursued programs to achieve these goals. All of them are consistent in a way with Turing's general formulation, yet each of them has a slightly different research trajectory.[7]

Philosophy, for example, always undergirded the most fundamental thinking about AI, but in the 1970s it became a source of penetrating criticism as well. The best-known example is philosopher Hubert L. Dreyfus's scathing manifesto *What Computers Can't Do*. This popular book argued that AI fell short of its promises because it was attempting the impossible. "In the final analysis," wrote Dreyfus, "all intelligibility and all intelligent behavior must be traced back to our sense of what were *are*, which is . . . something we can never explicitly *know*."[8] But it was not just the premise of AI that offended Dreyfus; it was the practice. Likening AI researchers to the alchemists of the middle ages, Dreyfus claimed that "artificial intelligence [was] the least self-critical field on the scientific scene."[9] Such an accusation might have alarmed ARPA, the principal supporter of AI research through the 1960s and 1970s, but there is no evidence that it did.

Psychology contributed to AI by exploring the difference between thinking and acting. Behaviorists argued that stimulus and response offered the only reliable measure of cognition. Cognitive psychologists argued that the brain interposed a step between stimulus and response that might be likened to information processing; indeed developments in information processing helped to shape conceptualization of this field. Since the 1960s the information processing view has dominated psychology; during much of that time AI and cognitive science were seen by many as different sides of the same coin.[10]

Finally, an entirely new discipline arose in tandem with AI and contributed as well to the changing perceptions of that field. Linguistics as an independent discipline can be traced to Noam Chomsky's seminal observation in 1957 that language is not entirely learned; the human brain comes prewired for certain aspects of communication, such as syntactic structure.[11] This insight, and the evolving discipline that flowed from it, have provided fruitful models for the development of computer languages and for knowledge representation, that is, putting information in a form that the computer can understand.

The interaction with different disciplines, however, does not fully account for the up-and-down record of AI. The achievements and credibility of AI waxed and waned as the field shifted from one research trajectory to another. Its early period in the 1940s and 1950s, before it even took a name, was theoretical and inward looking. Its first heyday came in the late 1950s and early 1960s when the field adopted its name and its ethos. Herbert Simon set the tone in 1956, confidently predicting that "in a visible future" machines would have thinking capabilities "coextensive" with those of humans.[12]

It was during this period that Simon and Newell turned from "logic theorist" to "general problem solver." Other researchers were writing programs for checkers and chess, prompting Simon to predict in 1957 that a computer would be world chess champion in ten years.[13] John McCarthy and Marvin Minsky moved to MIT, where they would lay the foundation of Project MAC. Soon they were attracting the ARPA funding that buoyed AI through the 1960s and 1970s. In one year, 1958, McCarthy invented LISP,[14] time-sharing, and his own AI program. "Advice Taker," as he called it, operated in a general world environment without being reprogrammed for specific tasks such as checkers or chess. Those heady years in the late 1950s marked the first pinnacle of AI optimism.

The early 1960s witnessed AI's descent into the first valley. Some important successes continued to appear, but major failures cast doubt on the optimistic early projects. On the positive side, a series of bright young students mastered some "toy" problems in "microworlds," limited domains where all the variables could be programmed and controlled. Some of these programs solved simple problems in geometry and algebra; the most famous ones moved colored blocks around a table top by combining work on vision, natural language understanding, robotics, learning theory, and planning.[15]

The assumption behind many of these early demonstrations was that techniques developed in these limited domains could be scaled up to more complex problems. Learn how to move blocks around a table and you could soon move rail cars about a freight yard or airplanes in and out of an airport. The problems, however, did not scale. For reasons revealed by NP-completeness theory in the early 1970s, more memory and more processing power in computers were not sufficient to scale up the complexity of the problems. Complex problems required entirely different programs; some problems proved intractable no matter what machine and program were brought to bear.[16]

Nowhere did the limits of AI become more obvious than in language translation. In the early, optimistic years of linguistics, it appeared that machine translation was on the horizon. Dictionary-based translation of words need only be combined with reconciliation of the two languages' syntaxes, but such techniques ignored the importance of semantics. The most cited example is an apocryphal account of a machine translation from English to Russian. The sentence "the spirit is willing but the flesh is weak" is said to have come out "the vodka is strong but the meat is rotten."[17] It turned out that encyclopedic knowledge would be required to understand natural language.[18] Computer technology in the 1960s offered neither hardware nor software of that capacity. In 1966 all U.S. government support for machine translation ended.

By the late 1960s AI was moving in several different directions at once. Herbert Simon was turning to a new though related interest in cognitive psychology and Allen Newell was settling in with his colleagues at Carnegie Mellon University (CMU, successor to the Carnegie Institute of Technology) for the long haul of AI development. John McCarthy had left MIT in 1963 to organize an AI laboratory at Stanford, to which he soon added Edward Feigenbaum, a Simon student. Marvin Minsky stayed on at MIT, set up Project MAC, broke off to form a separate AI laboratory in 1969, and finally turned over day-to-day operations to Patrick Winston.[19] These three centers of AI research received the lion's share of ARPA funding in these years and trained many important new workers in the field.

One of the areas to which they turned their attention was speech understanding, the translation of spoken language into natural language. On the advice of his principal investigators (PIs) and a study group chaired by Carnegie Mellon's Allen Newell, IPTO Director Lawrence Roberts initiated in 1970 a five-year program in speech understanding research (SUR). The largest effort took place at CMU, which developed two programs, HARPY and Hearsay-II. In spite of significant advances, including development of blackboard architecture in the Hearsay-II program, no machine at the end of five years could understand human speech in anything like real time. IPTO Director George Heilmeier cancelled plans for follow-on work and AI descended once more into reduced activity and icy pessimism.[20] The late 1970s became another age of AI winter, or at least AI autumn.

But just when disappointing results in one branch of AI seem to have discredited the entire field, promising developments in another branch revived expectations and set off another scramble for the promised land. In the 1960s machine translation had restored some luster to a field tarnished by toy problems that would not scale. Speech recognition revived the faithful when machine translation crashed. In the late 1970s, expert systems sallied forth to take up the fallen banner of AI.

Expert Systems

The new standard bearer was Edward Feigenbaum, director of the Stanford AI Lab. As an engineering student studying with Herbert Simon in 1956, he had heard in class the professor's famous announcement that "over Christmas break Allen Newell and I invented a thinking machine."[21] That night Feigenbaum took home the operating manual for the IBM 701 and stayed up all night reading it. He was hooked. After taking his undergraduate and graduate degrees at Carnegie (still the Institute of Technology), Feigenbaum taught at the University of California at Berkeley before moving to Stanford at the invitation of John McCarthy.

In California he grew dissatisfied with the kind of work he had done at Carnegie, the development of thinking machines embodying theoretical models of human cognition.[22] Instead, he wanted to explore a specific problem and instantiate the way in which a specialist would work a solution. By chance he met Joshua Lederberg at a gathering of San Francisco Bay–area individuals interested in machine intelligence. Lederberg, the renowned geneticist and Nobel laureate, invited Feigenbaum to explore the problem he was then working, how NASA's Viking spacecraft might recognize life on Mars. Lederberg and his colleagues had developed an algorithm that described the entire matrix of possible molecules for a given chemical formula. Feigenbaum developed a scheme, through heuristics, to search the matrix.[23]

Feigenbaum put together a research team, including philosopher Bruce Buchanan, and began work. By interviewing Lederberg and Carl Djerassi, head of Stanford's Mass Spectrometry Laboratory, Feigenbaum's team learned the thought processes by which these experts made their analyses. They translated these processes into the language of a rule-based computer program, that is, one that proceeds through a logic tree by a series of Boolean "if . . . then" switches. The result was DENDRAL, widely regarded as the first expert system. Feigenbaum and his colleagues announced their results at the 1968 Machine Intelligence Workshop at the University of Edinburgh,[24] less than four years after beginning the collaboration with Lederberg.

The great insight of this first paper was what Feigenbaum called "the knowledge principle": success depended upon the amount and the quality of the expert knowledge that the program captured.[25] The method of processing the knowledge was also important, but the heart of the system was the knowledge itself. Feigenbaum emphasized this point repeatedly, both in subsequent reports on DENDRAL and in MYCIN, an expert system for medical diagnosis that Feigenbaum helped to develop in the 1970s. By 1975 Feigenbaum's success had precipitated what he calls "the first era" of expert systems,[26] the era of "knowledge-based systems (KBS)".[27]

Enthusiasm for Feigenbaum's achievements, and optimism that the results could be replicated and enhanced in other realms, swept through the AI community and infected researchers in related fields. In 1980 John McDermott of Carnegie Mellon delivered to Digital Equipment Corporation the first version of XCON, an expert system designed to help the computer manufacturer configure its machines to suit customer demand; by the middle of the decade the company estimated it was saving $40 million annually by use of XCON. The CBS Evening News reported in September 1983 that the expert system PROSPECTOR, instantiating the knowledge of nine geologists, had helped a company discover molybdenum deposits in Washington State's Mount Tolman. Comparable stories proliferated. Companies scrambled to buy expert systems. New companies sprang up to service them. Bright students were lured away from graduate school with salary offers of $30,000. Attendance at the meetings of the American Association for Artificial Intelligence swelled through the early 1980s.[28] And in the midst of it all Japan announced its Fifth Generation Project.

Small wonder, then, that Robert Kahn and the architects of SC believed in 1983 that AI was ripe for exploitation. It was finally moving out of the laboratory and into the real world, out of the realm of toy problems and into the realm of real problems, out of the sterile world of theory and into the practical world of applications. This was exactly the trend that Kahn and Cooper wanted to promote in all areas of information processing. AI would become an essential component of SC; expert systems would be the centerpiece. Other realms of AI might allow machines to "see" or "hear" or translate languages, but expert systems would allow machines to "think."

By the time SC got under way, the concept of an expert system was fairly well defined. It consisted of a knowledge base and a reasoning or inference engine. The knowledge base collected both the factual data and heuristic rules that experts used to solve the problem at hand. The inference engine would process input, applying the knowledge and rules to come to conclusions that would form the output of the system. In DENDRAL, for example, spectrographic data put into the program would be matched against stored knowledge of what spectral lines matched what molecules. Through logical analysis the program would arrive at a definition of what molecule must have produced the spectrograph.

While these basic principles took hold, many fine points of expert systems remained to be worked out. Knowledge acquisition was a long, tedious process and it had to be done anew for each field in which an expert system was to operate; efforts were under way to streamline and perhaps even automate this process. The best means of knowledge representation was still unresolved. Rule-based systems depended on knowledge in "if . . . then" form to actively lead input through a logic tree to a conclusion. Object-oriented systems instantiated their knowledge in units called objects—things or ideas—with characteristics against which input was measured. The knowledge base consisted of an elaborate set of objects, including rules and information about the relationships between objects.

Refinements were also being sought in the subtlety and sophistication of expert system "thinking." How would such a system reason under uncertainty, for example, as humans often have to do when data or rules are incomplete? Could rule-based systems combine both forward-chaining of logic, that is, reasoning from given conditions to an unknown conclusion, and backward-chaining, that is, reasoning from the desired conclusion back through the steps necessary to achieve it? Could expert systems communicate with one another, to update their own knowledge bases, and to infer and acquire new knowledge? Could expert systems exploit the blackboard architecture developed at Carnegie Mellon in the 1970s as part of the Hearsay-II speech system? The blackboard architecture was intended to permit software modules to function independently while communicating through, and accessing information stored in, a common database called a blackboard. The blackboard directed the activities of the various modules (called "knowledge sources") and acted as a go-between, supplying data to those knowledge sources that needed it and receiving the output from those that had it. Because the knowledge sources were not "tightly coupled"—that is, they did not have to work in close lockstep with each other—they could be constructed independently and plugged into the system. The only requirement was that they successfully interface with the blackboard.[29]

Most importantly for SC, could there be such a thing as a "generic expert system"? E. H. Shortliffe's expert system for medical diagnosis, MYCIN, had already been converted to EMYCIN, a so-called expert system "tool" or "shell." This is a program that has certain general characteristics of an expert system designed into it, qualities such as an inference engine with a fixed processing method already programmed, a knowledge-base structure already engineered to be, for example, rule-based or object-oriented, input-output devices that facilitate the acceptance and generation of certain forms of data, compilers that allow data to come in multiple languages, and architecture that is well suited to recursive operations.[30] With such a tool or shell in hand, a programmer need not address all those basic functions. Rather he or she can simply load the knowledge base, that is, the data and rules, for a particular application and begin processing.

Taking this development one step further, SC sought to develop "generic software systems that will be substantially independent of particular applications,"[31] that is, shells or tools of such general capability that a programmer could take one off the shelf, load up a knowledge base, and go to work. Robert Kahn introduced the term "generic" into the SC discussion of AI in his initial draft of the SCI program document in September 1982. He noted that the Japanese Fifth Generation Program recognized the potential of "generic applications that are AI based."[32] He even offered examples of generic applications, such as display management systems, natural language generation systems, and planning aids.[33] The next iteration of the SC plan specifically said that "we need increasing[ly] sophisticated 'empty' expert systems for military applications."[34]

Nowhere in these documents, however, did Kahn or other members of DARPA use the term "generic expert system," which is being used here.[35] Saul Amarel maintains that "generic expert system" is a contradiction in terms.[36] Expert systems, after all, mimic human experts in certain domains; neither people nor machines can be experts in everything. But in another sense, intended here, human beings are generic expert systems. Just as people can be trained to perform certain kinds of thought processes, such as art appreciation or deductive logic, so too can machines be loaded with inference engines of specific capabilities. And just as trained experts can then be given knowledge bases peculiar to the tasks they are about, so too can expert systems be loaded with such information. The term "generic expert system" here stands for the twin, related goals of developing "generic software applications" and seeking as well "machine intelligence technology . . . to mechanize the thinking and reasoning processes of human experts."[37]

That such a goal appeared within reach in the early 1980s is a measure of how far the field had already come. In the early 1970s, the MYCIN expert system had taken twenty person-years to produce just 475 rules.[38] The full potential of expert systems lay in programs with thousands, even tens and hundreds of thousands, of rules. To achieve such levels, production of the systems had to be dramatically streamlined. The commercial firms springing up in the early 1980s were building custom systems one client at a time. DARPA would try to raise the field above that level, up to the generic or universal application.

Thus was shaped the SC agenda for AI. While the basic program within IPTO continued funding for all areas of AI, SC would seek "generic applications" in four areas critical to the program's applications: (1) speech recognition would support Pilot's Associate and Battle Management; (2) natural language would be developed primarily for Battle Management; (3) vision would serve primarily the Autonomous Land Vehicle; and (4) expert systems would be developed for all of the applications. If AI was the penultimate tier of the SC pyramid, then expert systems were the pinnacle of that tier. Upon them all applications depended. Development of a generic expert system that might service all three applications could be the crowning achievement of the program.

Optimism on this point was fueled by the whole philosophy behind SC. AI in general, and expert systems in particular, had been hampered previously by lack of computing power. Feigenbaum, for example, had begun DENDRAL on an IBM 7090 computer, with about 130K bytes of core memory and an operating speed between 50 and 100,000 floating-point operations per second.[39] Computer power was already well beyond that stage, but SC promised to take it to unprecedented levels—a gigaflop by 1992. Speed and power would no longer constrain expert systems. If AI could deliver the generic expert system, SC would deliver the hardware to run it. Compared to existing expert systems running 2,000 rules at 50–100 rules per second, SC promised "multiple cooperating expert systems with planning capability" running 30,000 rules firing at 12,000 rules per second and six times real time.[40]

Ronald Ohlander managed AI programs for IPTO as the SC program took shape. The navy had sent him to graduate school at CMU in 1971–1975, where he earned a Ph.D. in computer science, working on machine vision with Raj Reddy. In 1981 he finally won a long-sought-after assignment to DARPA, where he inherited a diverse AI program. Vision and natural language understanding were particularly strong; "distributed AI" was a "less focused" attempt to promote AI on multiple processors. Individual support also went to Edward Feigenbaum for expert systems and a handful of other individuals with promising research agendas. Annual support for this area would triple when SC funding came on line, rising from $14 million annually to $44 million.[41] The resources were at hand to force-feed an area of great potential, though Ohlander also faced the risk, as one observer put it, of having more money than scientists.

To launch this ambitious segment of SC, Ohlander published a "qualified sources sought" announcement for expert systems in *Commerce Business Daily* on 21 February 1984.[42] The fifty proposals that arrived in response underwent evaluation by an ad hoc review panel comprised of DARPA colleagues and selected specialists from outside the agency.[43] In the end, Ohlander recommended that DARPA fund six of the proposals. Five of these contracts were for work on specific aspects of expert systems. Stanford University would explore new architectural concepts; Bolt, Beranek & Newman would work on knowledge acquisition; Ohio State University would pursue techniques by which an expert system explained how it arrived at conclusions; and the University of Massachusetts and General Electric Corporation would both pursue reasoning under uncertainty. All these contracts had the potential to provide features that would enhance any expert system, even a generic expert system. To pursue that larger goal, which Ohlander labeled "new generation expert system tools," he chose a start-up firm in Northern California, Teknowledge, Inc. When Lynn Conway saw his list of contractors, she insisted that he add another project in generic expert systems.[44] Ohlander chose IntelliCorp, another Northern California start-up.

The curious circumstances by which two start-up companies located within a few miles of each other came to be the SC contractors for generic expert systems warrant explanation. The common thread in the story is expert-systems guru Edward Feigenbaum of nearby Stanford University. But an even more illuminating vehicle to carry this tale is Conrad Bock, a young man who walked into expert systems at the right time and the right place.

Unhappy as an undergraduate physics major at MIT, Bock transferred to Stanford.[45] Still not happy with physics when he graduated in 1979, Bock cast about for something more engaging. On the recommendation of a friend he read Douglas Hofstadter's *Gödel, Escher, Bach*, a book accorded almost biblical stature within the AI community. Somewhere in its pages, Bock was converted.[46] Soon he was at the doorstep of Edward Feigenbaum.

At the time, Feigenbaum was doing three different things of importance to Bock's future and the future of SC. He was visiting Japan and beginning to develop the ideas that would appear in *The Fifth Generation*.[47] He was commercializing his own work in expert systems. And he was developing a new graduate program in AI. Unlike most programs at prestigious schools such as Stanford which were designed to train Ph.D.s, Feigenbaum's new venture offered a Master of Science in Artificial Intelligence (MSAI). Its explicit goal was to train students in enough AI to get them out pursuing practical applications, but not enough to make them career academics and researchers. It was one more attempt to get this new field out of the ivory tower and into the marketplace.

The program suited Bock perfectly. He completed the course and came on the job market in 1983 at the height of the enthusiasm for AI and expert systems.[48] The job market sought him. Companies were springing up all over Silicon Valley; he had only to choose among competing offers.

He first rejected Xerox PARC (Palo Alto Research Center), Lynn Conway's former home, because its research agenda struck him as too abstract. He wanted something more practical and down to earth. He interviewed at Teknowledge, just down the road from the Stanford campus. At the time the company was looking for staff to help develop system tools. Bock might well have settled there, but he found himself still more attracted to another firm just a few miles away.

IntelliCorp and Teknowledge

IntelliCorp, a Feigenbaum creation, had roots in that other great start-up enthusiasm of the 1970s and 1980s, genetic engineering. The result was what Harvey Newquist calls "biotech-meets-AI."[49] Among the products emerging from the fecund halls of Feigenbaum's Computer Science Department in the 1970s was MOLGEN, an expert system in molecular genetics. It assisted biologists in gene-cloning experiments by suggesting the steps necessary to clone a specific DNA sequence. In 1980 three medical school researchers involved in the project approached Feigenbaum about starting a company. They proposed to make the program available for a fee through time-sharing on a large computer. They would put the program on line, maintain it, and sell access to researchers all over the world. They formed IntelliGenetics in September 1980, gave Feigenbaum a significant block of stock, and placed him on the board of directors. With an infusion of venture capital, they set up mainframes in Palo Alto and Paris and went on line. What they proposed to sell was somewhere between a service and a product.[50]

The market was disappointing; by 1983 IntelliGenetics was experiencing sluggish sales and casting about for a better source of revenue. Edward Feigenbaum suggested that they convert a descendant of his DENDRAL program, EMYCIN, into a general-purpose expert system shell.[51] EMYCIN was the shell derived from his expert system for medical diagnosis, MYCIN. EMYCIN had already been sold commercially as a shell into which other knowledge bases could be plugged. Why not develop it further into a shell for all seasons?

To carry out the plan, IntelliGenetics hired Richard Fikes, an AI researcher with a solid reputation in the computer community. A student of Newell's at CMU, Fikes had worked at Stanford Research Institute (SRI), a think tank once formally tied to the university, and at Xerox PARC. He joined IntelliGenetics because he, like Feigenbaum, wanted to apply AI, to get it out of the laboratory and into practice.[52] His presence and that ambition convinced Conrad Bock to accept an offer from IntelliGenetics and turn down Teknowledge.

The new departure at Intelligenetics produced the Knowledge Engineering Environment (KEE), a LISP expert system shell. KEE came on the market in late 1983, just as SC was getting under way, and, notes Harvey Newquist, "just as IntelliGenetics was running out of money."[53] Feigenbaum recommended that IntelliGenetics apply for SC funds to support the next stage in the development of KEE.[54] This was a step that IntelliGenetics was going to take in any event, but DARPA's interests in a generic expert system were similar to those of IntelliGenetics. This was a potential win-win relationship. Changing its name to IntelliCorp to sound more like a computer firm, the struggling company submitted a proposal to DARPA.

With Lynn Conway's intervention, IntelliCorp won a contract in 1984 for $1,286,781 to develop an "evolutionary new generation system tool." It was evolutionary because it was to evolve from KEE, which in turn had evolved from EMYCIN, MYCIN, and DENDRAL in a straight line from the beginnings of expert systems twenty years before. The specifications of this new system were staggering, a reflection of the optimism of the times (both inside and outside DARPA) and the naivete of a four-year-old start-up company.[55]

The new tool was to support knowledge acquisition and representation (two of the most difficult problems in expert systems), reasoning, and user interface construction. It was to be able to support the building and use of qualitative domain models, that is, to adapt itself to working readily in different realms with different ways of organizing knowledge. It was to expand upon current capabilities to model domain knowledge to include the logical properties of relations between objects, declarative constraints on how objects interacted, multiple hypothetical situations, hierarchical descriptions of objects (nonlinear plans that could move on multiple paths without necessarily following in a single sequence), and justification of derived results (being able to explain how it reached its conclusions).

Furthermore, this tool's problem-solving capabilities were to include truth maintenance, inheritance (i.e., internal learning from the knowledge already embedded), default reasoning into which problems would fall if not directly anticipated in the programmed heuristic, opportunistic control that could ad lib short-cuts where appropriate, automatic planning and scheduling of its processes, a blackboard, and a distributed framework that could handle parts of a single problem in different logical compartments. All these characteristics of what one might call an ideal expert system had been addressed by researchers before. But never before had one program attempted to integrate them all in a single piece of software, to connect them in a single technological system.

On top of developing all this, IntelliCorp agreed to make its program available to other SC contractors and to help them use it. A basic tenet of SC philosophy was to force-feed developments up and down the pyramid. Software developers would be required to run their programs on Connection Machines. Applications projects in image understanding had to use Butterfly architecture. And the developers of KEE had to help architecture designers get the new software to run on their machines and get applications developers to convert it into the specific expert systems they needed. It was not enough that KEE could solve toy problems or work in theory; it had to run on real machines and solve real problems.[56]

As if this project would not tax the young company sufficiently, IntelliCorp was also awarded other ARPA contracts as well. ARPA order 6130 commissioned the company to develop a rule-based language for nonlinear planning based on an underlying truth maintenance system. It was to complement other DARPA research in networking and distributed computing.[57] Furthermore, some SC applications funding went to IntelliCorp to support AirLand Battle Management, an army program analogous to Navy Battle Management, and for work on Pilot's Associate. In all, this second ARPA order provided another $2.8 million over five years, from 1985 to 1989.

Feigenbaum's other commercial venture, Teknowledge, had sprung from outside solicitation—requests for access to his expert systems or proposals to market them. Instead of acceding to these external initiatives, Feigenbaum invited nineteen other computer scientists, mostly colleagues and collaborators associated with Stanford's Heuristic Programming Project, to join him in setting up their own company.[58] Each put up some few thousands of dollars in cash; each expected to do some work, or at least some consulting, for the company; and each would share in the profits.

Teknowledge began with an enormous asset and a great liability. The asset was the expertise of its founders, researchers who were operating in the forefront of a technology with great economic potential. The liability was their lack of business and managerial experience. They surely had the resources to devise and implement useful products, but could they market them? Could they connect their technology to customers? Exemplars abounded in Silicon Valley of researchers turned millionaire. The greatest of them all, Bill Gates, was just then negotiating with IBM to produce the operating system that would make his company, Microsoft, the most dominant player in the computer industry by the end of the decade.[59] What distinguished people such as Gates and Robert Noyce of Intel from the organizers of Teknowledge was that the former spent their entire careers in industry; in fact, Gates had dropped out of Harvard without ever taking a degree. They learned business by doing, and they did it full time. Feigenbaum and his colleagues were amateurs by comparison. Even Danny Hillis was a business professional compared to them.

The one exception within Teknowledge, or at least something of an exception, was Frederick Hayes-Roth, the only initial partner not from the Stanford faculty. Hayes-Roth was then the Research Program Director for Information Processing Systems at the RAND Corporation, the think tank that Keith Uncapher had left to form the Information Sciences Institute at the University of Southern California. He worked on distributed, cooperative, and multiagent computing.[60] A graduate of Harvard University, Hayes-Roth did his graduate work at the University of Michigan, where he took an M.S. degree in computer and communication sciences and a Ph.D. in mathematical psychology.[61] During a two-year postdoctoral fellowship at Carnegie Mellon, he participated in the final stages of the HEARSAY speech understanding research program before it was cancelled by George Heilmeier. Both there and at RAND he had done business with DARPA. He was a logical addition to the group organizing Teknowledge.

At first Hayes-Roth intended to simply consult for Teknowledge on his days off; he came to Teknowledge in part because he had done similar consulting work for Feigenbaum earlier. His boss at RAND, however, felt this posed a conflict of interest. When Hayes-Roth accepted a commission from Teknowledge in 1980 to give the army a two-day seminar on expert systems, his boss told him he could not go. He went anyhow, to be greeted on his arrival with a telegram from his boss saying "If you read this, you are fired."[62] Suddenly, he was a full-time employee of Teknowledge, just the second one the company had. His title was Chief Scientist and Executive Vice President for Technology.

Teknowledge at the time was doing mostly consulting work, providing services for industry and government such as the seminar that led to Hayes-Roth's firing. Such work held no prospect of big money. The big money was in products, such as Intel's chips and Microsoft's operating systems, products of general applicability and wide potential market. But Teknowledge was undercapitalized; it had slight resources to invest in product development. It had produced an IBM personal computer software system called M.1 to exploit some features of the MYCIN program. Hayes-Roth called this in 1987 "the most popular tool for building small-to-medium expert systems"; but priced at a hefty $12,000, its commercial potential was in doubt.[63]

Teknowledge had also developed a more sophisticated system, S.1, to run on VAX and Symbolics machines. When SC began and Ronald Ohlander called for proposals in expert systems development, it appeared an excellent chance to break into the big time, to turn this modest expert system shell into a generic tool of wide applicability. Teknowledge created a subsidiary, Teknowledge Federal Systems, to address problems of interest to the government. Hayes-Roth formed a research division, hired new people, and put together a proposal for DARPA.

Teknowledge proposed "ABE," a multilevel architecture and environment for developing intelligent systems, that is, a tool or shell for creating expert systems.[64] Hayes-Roth likened the concept behind ABE to the neocortex of the human brain.[65] Most of the knowledge humans possess is embedded in the brain. The brain, in turn, is surrounded by a thin layer of neocortex, which organizes the brain's activities and mediates interaction with the external environment. ABE was intended to perform the same function for an expert system. It would, for example, provide blackboards capable of both forward and backward chaining. It would organize data-flow. It would support the importing of existing software and mediate communication between different programs and languages. If it was not quite a generic expert system, it was nonetheless something of a universal tool that could contain and enhance all expert systems.[66]

DARPA executed ARPA order 5255, which originally budgeted $1,813,260 for Teknowledge over two years, with an option to renew.[67] In the first phase Teknowledge was to "design and develop a robust software architecture suitable for building expert systems." The architecture was to be "modular, broad, extensible, suitable for large-scale applications, distributable, and transportable," and it was to feature "reasoning with uncertainty, knowledge acquisition, and cooperative systems."[68] It was, in short, to pursue goals similar to those of IntelliCorp. DARPA was hedging its bet. Though this was not Steve Squires's program, the strategy at work here corresponded to what he came to call the "ensemble model," funding more than one research trajectory to increase the chance that at least one path would pay off. If successful in the first phase, Teknowledge would receive a follow-on contract to transfer the tool to a wide range of applications.

Managing Innovation

If ever a set of ARPA orders exemplified the agency's philosophy of high-risk/high-gain, it was those with Teknowledge and IntelliCorp. Teknowledge, DARPA's first choice, did not even have a product out the door yet, and it was promising to produce in short order a tool of staggering complexity. DARPA was betting on Frederick Hayes-Roth, Edward Feigenbaum, and most of the faculty in Stanford's Heuristic Programming Project. IntelliCorp at least had a product, but it was just transforming itself from a struggling genetic engineering consulting firm into a computer firm taking on possibly the most difficult assignment in the field. The gamble here was based on Feigenbaum (again), Richard Fikes, and KEE. The remarkable thing is not that DARPA chose these two companies for the critical, penultimate technology at the top of the SC pyramid, but that these were the best of the proposals they had to choose from. Teknowledge had a good proposal but no track record. IntelliCorp had a commercial product but nothing close to the generic shell it was promising.

To shepherd these two start-ups on their journey, DARPA assigned J. Allen Sears. Ronald Ohlander, who departed DARPA shortly after getting the Teknowledge and IntelliCorp contracts in place, had recruited Sears and Robert Simpson to help him manage his rapidly swelling AI portfolio. Like Ohlander, Sears was a naval officer, in his case a pilot turned computer scientist. He received a Ph.D. from Arizona State University in 1982 and was detailed to DARPA in the summer of 1984. He assumed responsibility for the speech and knowledge-based systems programs. Simpson, an air force officer, joined DARPA about the same time, after taking a Ph.D. at the University of Georgia in 1985. He took charge of vision-image understanding and natural language.[69]

All three men were typical of the active-duty military officers who did one or more tours at DARPA on assignment from their services. They served beside civilian program managers who came to DARPA from industry or academia, often under the aegis of the Intergovernmental Personnel Exchange Act. All came nominally on two- or three-year tours; sometimes the military officers were extended in these positions, as Ohlander had been.[70] Sometimes the civilians stayed on to become civil servants, as Robert Kahn had done. Sometimes, as in Steve Squires's case, they were civil servants already, simply changing agencies.

The civilians usually came with more powerful credentials in their technical fields; this often meant that they had technical agendas of their own and attempted to shape the field while at DARPA. The military officers came with a better understanding of the clientele for DARPA's products; they were more likely to serve as good stewards and what Hugh Aitken has called "translators" of technology.[71] They could translate service needs for the computer science community and translate technical capabilities to serving officers.

Sears played the latter role. He took over programs for which he had little formal training, but he educated himself on the job and relied on his common sense and management experience to keep the programs on track. Most importantly, he had to recognize good excursions from the plan, to know when a project was making better-than-expected progress and needed more resources to exploit the opportunity. Just as importantly, he had to know when a project was failing and had to be invigorated, redirected, or abandoned.

In the cases of Teknowledge and IntelliCorp, Sears' management style was decidedly hands-off. He dealt mostly with the PIs, Hayes-Roth and Fikes respectively. In contrast to the habits of some program managers, he seldom visited the company laboratories, remaining invisible to the teams working on the projects there.[72] Rather he kept in touch by meeting semiannually with representatives of all the projects he supervised. Usually every six months, sometimes more or less often, DARPA PMs would convene PI meetings, often at remote and physically stimulating resorts such as Hilton Head, SC, or the Shawneee State Park Conference Center in Friendship, OH. But there was little time for recreation, especially at one of Sears' retreats.

These meetings served several purposes. First, they promoted cross-fertilization, one of the goals of SC. They connected researchers working on related projects. At Sears' meetings, for example, researchers in expert systems would meet with their counterparts in speech, natural language understanding, and vision. Each PI made a presentation to the entire group on progress since the last meeting, and everyone got to ask questions. This not only increased the probability that work in one area would stimulate work in another, it also served a second purpose—motivating the PIs to come to the meetings prepared. Research programs seldom produce new and interesting results every six months, but because researchers wanted to avoid embarrassment in front of their colleagues, they usually took pains to find some progress to report.[73]

Third, the meetings provided the PM with an opportunity to see how his or her projects were proceeding. One measure of this was simply what the PIs reported. Seldom, however, would PMs have the time or expertise to know all the fields under their control well enough to judge this for themselves. A Kahn or a Squires might know the whole field well enough to be fully conversant in each of its subspecialties, but most PMs did not. Still, a sharp and attentive PM with adequate technical preparation and a good amount of hard work could use the PI meetings to form a sound opinion of who was producing and who was not. The clues came not so much from what the PIs said; there was a code of silence among them not to embarrass each other in front of the PM.[74] Rather they came from atmosphere, body language, perspiration, and nervousness. One could watch the speaker or watch the other PIs. It wasn't hard to distinguish between the confident, enthusiastic presentation that held the audience and the halting and wordy meandering that lost or confused the audience. In a way, the other PI's were the PM's best guide to what was good and what was not.[75]

Through the final months of Kahn's tenure at DARPA and through the entire term of his successor, Saul Amarel, Sears supervised the development of expert system shells at Teknowledge and IntelliCorp. If neither company could claim any dramatic breakthrough to the generic expert system that SC had originally envisioned, they nonetheless made enough progress to warrant renewal of their contracts for second terms. In October 1986, Sears asked for $5.4 million to continue Teknowledge work on ABE for another twenty-four months.[76] IntelliCorp's contract was similarly renewed on Sears' recommendation for a second phase in spite of the budget squeeze being endured by DARPA in 1986.[77]

But no sooner were those new contracts in place than in blew Jack Schwartz and the winds of AI winter. In September 1987 Schwartz succeeded Saul Amarel as director of the Information Systems Technology Office (ISTO), the new unit formed in 1986 to reunite IPTO and EAO.[78] A computer scientist at New York University and an adviser of long standing to DARPA, Schwartz had no enthusiasm for the job. He took it at the entreaty of his old friend Craig Fields, who was now in the DARPA front office.[79]

The storm that Schwartz visited on ISTO blew up, not within ARPA, but within the larger computer community, especially the commercial AI community.[80] It was driven in part by the collapse of the market for LISP machines as these came to be displaced by more powerful, new work stations.[81] But it also came from the failure of commercial start-ups such as Teknowledge and IntelliCorp to get products out the door at a pace commensurate with the expectations of the venture capitalists who were bankrolling them. Once more AI was finding itself a victim of the unrealistic expectations stirred up by the rhetoric of its more ardent and reckless enthusiasts.

Schwartz brought to bear still another dimension of the problem. His skepticism was based on a deeper and more profound concern about the fundamental nature of AI. In a cold and devastating review of "The Limits of Artificial Intelligence" prepared for the 1987 edition of *The Encyclopedia of Artificial Intelligence,* Schwartz had argued that AI had yet to demonstrate "any unifying principles of self organization," meaning that its "applications must still be seen as adaptations of diverse ideas rather than as systematic accomplishments of a still mythical AI technology."[82] He was especially harsh on expert systems, asserting that "expert systems enhance their pragmatic applicability by narrowing the traditional goals of AI research substantially and by blurring the distinction between clever specialized programming and use of unifying principles of self-organization applicable across a wide variety of domains."[83]

In other words, he believed that expert systems were achieving what success they had by clever programming, not by the application of any general principles. His analysis bode ill for the prospects of achieving a generic expert system of the kind envisioned by the SC program and the contracts with Teknowledge and IntelliCorp. Indeed, Schwartz believed that the same critique applied to AI in general; he concluded that "it may be necessary to develop a relatively large number of artificial systems that mimic particular types of reasoning and mental functions in cases specialized enough to have particularly efficient treatment."[84] In short, there probably was going to be no such thing as a generic expert system, not even the "generic software applications" envisioned in the SC plan. Schwartz's insight was to prove penetrating and robust; it will warrant further exploration later on.

For the time being, it is sufficient to measure the impact of Schwartz on Allen Sears, Teknowledge, and IntelliCorp. He fell on them like a rider out of the Apocalypse.[85] Frederick Hayes-Roth remembers that he came into office with "his knife out."[86] Like George Heilmeier a decade before him, Schwartz put all DARPA's AI programs under a microscope. Perhaps it would be more accurate to say he subjected them to an inquisition. Richard Fikes remembers that on Schwartz's first trip to the west coast after taking over ISTO,[87] he met with each of the DARPA contractors in succession. These were much different encounters than what the PIs were used to with Sears. Schwartz took over a large dining hall in the hotel where the PIs were meeting. He sat at one end of the hall beside Robert Simpson, now Sears' successor as PM for expert systems. Each contractor's PI was summoned to the hall and made to stand alone at the far end. In that awkward setting, the PI had to justify his funding in fifteen minutes or less. "It was Kafkaesque," says Fikes, to whom it was very "clear that he [Schwartz] was there to kill these programs."[88]

Schwartz in fact did not kill the programs outright, as he might have done. But he did make it clear that neither Teknowledge nor IntelliCorp could expect more DARPA support for their expert system tools after their current authorizations expired. Nor could they expect full funding to complete their work. Whatever they were going to produce would have to appear by the end of FY 1990.

The Failure to Connect

The results of SC's investment in generic expert systems can only be described as disappointing, both from DARPA's point of view and from the perspectives of these two companies. At Teknowledge, ABE came in late and deficient. The project's attempt to develop a "federated computer" that would integrate the capabilities of knowledge-processing components independent of platform proved too ambitious. "Intelligent systems," concluded Hayes-Roth, "for the near term at least, will require intelligence on the part of developers to integrate diverse components in effective ways."[89] ABE, in short, would not connect into a single system the ambitious capabilties it had promised.

Inexperienced in developing software of this complexity, the research division at Teknowledge stalled under the pressure and underwent a painful reorganization. Hayes-Roth, the division director, accepted responsibility for the failure and returned to consulting.[90] When the DARPA contract ran out, the division folded its tents and the entire company turned back to the service work whence it had come—consulting and writing customized software. In 1989 it merged with American Cimflex to form Cimflex-Teknowledge, with Hayes-Roth as chairman. The new company was dropped from the stock market in 1993 for lack of trading activity.[91]

ABE turned out to be good, but not marketable.[92] Hayes-Roth notes, for example, that ABE was used as a prototype of CASES, a system that enabled the commander-in-chief of the Pacific fleet to simulate and evaluate alternative naval campaign strategies and tactics. And it was used to model and assemble the Pilot's Associate.[93] But Teknowledge could not sell it commercially, nor could it create a version that would work on SC applications.[94] To make matters worse for Teknowledge, the second contract disallowed funding to translate ABE from LISP into "C."[95] General Motors, one of Teknowledge's major corporate investors, was insisting that the company program its expert systems in "C."[96] And indeed, Kahn had said in the May 1983 version of the SC plan that the program intended "to support the 'C' language, since it is so widely used as a [sic] implementation language."[97] Teknowledge eventually converted ABE to C compatibility, but not with SC funds.[98]

The great advantage for Teknowledge in its SC contract had been experience. The company got an opportunity it could never have afforded otherwise to work on the forefront of expert systems development. Hayes-Roth was a prolific contributor to the literature and he coauthored many of his papers with colleagues at Teknowledge. At the conferences where they presented these papers, they shared views with others working on comparable problems and established contacts that helped them both technically and professionally. Hayes-Roth also got to know many other colleagues through PI meetings. When he and Teknowledge returned of necessity to what is essentially a consulting business, they were better informed, better connected, and more mature in their understanding of the field. ABE never proved to be the product that they had hoped would make them wealthy, but it nonetheless helped the company survive when other AI firms were folding.

The results at IntelliCorp were similar. Like Teknowledge, it found that progress in component technologies did not necessarily translate into commercially viable generic systems. According to Richard Fikes, the company produced a demonstration prototype, "OPUS," incorporating "substantial portions" of the SC agenda.[99] These became part of subsequent versions of KEE, including representation of and reasoning about alternative hypothetical situations, a capability for recording and explaining the justification for derived results, and a significantly enhanced production rule subsystem. But even with KEE already in hand, the production of a true generic tool incorporating all the features promised in its proposal simply proved too daunting.

Schwartz's entry into the equation, just as the second half of the contract was getting under way, made the prospects even bleaker. Fikes left IntelliCorp in October 1987, seeking more stable funding as a research professor at Stanford. Without his guidance, the IntelliCorp research program fragmented. It produced World KEE, essentially an attempt to instantiate a whole series of interrelated capabilities in a single shell. But without Fikes there to guide the program, individual researchers concentrated on solving their part of the puzzle; no one worked on making them connect, so they did not.[100]

Instead, IntelliCorp also lapsed into being a kind of service company. It sold its World KEE to a limited number of customers, especially in Europe, as a kind of expert system shell, that is, as a base on which to hang a custom expert system.[101] But most of the company's work was in designing those customized systems one at a time. Parts of that problem, such as providing blackboards and inheritance and language translation, can become routine when done repetitively. But none of them can yet be automated in a shell.

Like Teknowledge, IntelliCorp is still alive. Most of the other start-ups that began with them in the optimistic days of the early 1980s have fallen by the wayside.[102] Both companies attribute their survival to commercial savvy.[103] They pay attention to the marketplace and they listen to their customers, traits that may have been neglected in the years when they sought generic systems that would be user-independent.

Conrad Bock is still at IntelliCorp, his career yet another result of Strategic Computing.[104] Having witnessed the turn inward from the exciting, optimistic days when DARPA came to IntelliCorp, Bock tends to lump the experience together with his perception of what ails America.[105] While the Japanese, in his view, are prepared to make sustained, long-range commitments to product development, the United States appears to him to be increasingly driven by short-term results and the bottom line. He believes that the Japanese still do the kind of research that he joined IntelliCorp to conduct, but that this is increasingly difficult to find in the United States.

Still, Bock has only happy memories of his work on KEE. He says it gave him the equivalent of a Ph.D. education, exposing him to fundamentals that he never got in his masters program with Feigenbaum, fundamentals that he has been able to apply elsewhere in his research. He has published some of his work and he has personally sold the president of his company on the potential of his contribution.[106] He is part of the human infrastructure that Cooper and Kahn hoped to nurture with their investment in SC.

The Rest of AI

Applications contractors in the SC program did use KEE and ABE. They did not have at their disposal, however, the generic expert system that seemed to be within reach in the early 1980s. They could not exploit "multiple cooperating expert systems with planning capability" running 30,000 rules firing at 12,000 rules per second and five times real time. The retreat from these goals may be traced through the writing of Frederick Hayes-Roth. In an article for the 1987 *Encyclopedia of Artificial Intelligence,* Hayes-Roth said that the then current goal of SC investment in expert systems was to

increase the size of practical rule bases to 10,000 rules or more; increase the speed by two orders of magnitude or more; broaden the set of inference techniques used by the interpreters; improve the methods for reasoning with uncertainty; simplify the requirements for creating and extending knowledge bases; and exploit parallel computing in [ruled-based systems] execution.[107]

Only four years into the SC program, when Schwartz was about to terminate the IntelliCorp and Teknowledge contracts, expectations for expert systems were already being scaled back. By the time that Hayes-Roth revised his article for the 1992 edition of the *Encyclopedia,* the picture was still more bleak. There he made no predictions at all about program speeds. Instead he noted that rule-based systems still lacked "a precise analytical foundation for the problems solvable by RBSs . . . and a theory of knowledge organization that would enable RBSs to be scaled up without loss of intelligibility of performance."[108]

SC contractors in other fields, especially applications, had to rely on custom-developed software of considerably less power and versatility than those envisioned when contracts were made with IntelliCorp and Teknowledge. Instead of a generic expert system, SC applications relied increasingly on "domain-specific software," a change in terminology that reflected the direction in which the entire field was moving.[109] This is strikingly similar to the pessimistic evaluation Schwartz had made in 1987. It was not just that IntelliCorp and Teknowledge had failed; it was that the enterprise was impossible at current levels of experience and understanding.

The other areas of AI research targeted by the Strategic Computing Program had their own stories. Some paralleled expert systems, and some followed very different trajectories. All experienced cross-fertilization; they contributed to and profited from exchanges of component capabilities among areas of research. For example, the blackboard architecture that had first been developed as a part of the CMU Hearsay II project in the 1970s became an important feature of expert systems in the 1980s. Conversely, techniques for reasoning under uncertainty pioneered in expert systems development contributed to the improvement of speech recognition programs. In Squires's model of development, these component improvements contributed to a growing technology base in knowledge-based systems without necessarily achieving all of the specific goals, the "point solutions," identified in the SC plan.

Speech recognition made especially good progress. As recently as 1976, DARPA had withdrawn all support from this field because it did not appear to be making adequate progress. In 1978 and 1979, however, researchers at Carnegie Mellon and MIT concluded that phonetic signals were far more susceptible to analysis than previously believed.[110] The confidence bred of these findings led to two related research trajectories that were pursued simultaneously and integrated by new generation system contractors. One trajectory was in phone recognition. Spoken English has a finite number of discrete sounds, which DARPA reduced to a phonetic alphabet of forty-eight sounds and a single additional category of silence.[111] These sounds can be converted to a stream of data bits distinguishable by a computer program. The simplest speech recognition programs built a database consisting of the signals generated by a single speaker enunciating each of the forty-nine phones discretely, that is, with the words separated by a pause. In the 1980s hidden Markov models were used successfully to automatically derive the probability that spectral representation of human speech corresponded to the vocal articulations of an individual or group.[112]

Even that simple device, however, was unable to tell if the phone combination "[r][uy][t]" should be interpreted as "right" or "write". This is where the second trajectory enters. Speech occurs most often in patterns. "N-gram" statistics came to be used in the 1980s to model syntagmatic word order constraints. As a speech recognition program builds its database, it also accumulates knowledge about the context. It may contain, for example, several records of "turn" and "right" being next to each other, but few if any of "turn" and "write." Thus it might discern from context which word was probably intended by the phone "[r][uy][t]."

At the beginning of SC, it appeared that advances in these two technological trajectories, accelerated by faster and more powerful architectures, might yield speech recognition useful in applications such as Pilot's Associate and Battle Management. But application in those real-world environments meant that many other problems would have to be solved at the same time. For example, real speech occurs not in isolated words but in a continuous stream. In such streams, phones merge together, obscuring the beginnings and ends of words. Furthermore, different speakers produce different sound patterns when saying the same words, such as "tom[ay]to" and "tom[ah]to." Regional dialects compound the problem. Speech with background noise adds additional complications, as does speech under stress. To be truly useful in the military applications envisioned in SC, speech recognition would have to address all these issues and more.

The SC speech recognition program was organized in two new generation systems. Five contractors in eight different programs fed results in

robust speech to Texas Instruments, which performed the new generation integration in support of Pilot's Associate. Eight contractors in nine
projects in continuous speech fed results to integrator Carnegie Mellon
in support of Battle Management. Throughout the program, speech recognition held to its original goal of 10,000-word continuous speech recognition, using speaker-independent natural grammar, moderate noise,
and low stress. This meant that the system could recognize 10,000 words
of natural language spoken by anyone in an environment of moderate
background noise and low stress on the speaker. Such systems were up
and running by the end of SC, being integrated into military applications
and undergoing development for commercial applications.[113] Indeed
this is an area where results exceeded the hard numerical metrics laid
out in the SC plan.

Natural language understanding, a companion program, achieved
comparable results. Its problem at the outset was to translate English
communications into messages understood by the computer, and vice
versa. Like speech recognition, it pursued two simultaneous and
related trajectories: human-machine interfaces and natural language
text understanding. The first was a problem that had occupied AI at
least since the introduction of LISP. In some ways, English is simply another high-level language, whose symbolic code must be translated or
"compiled" into a machine language composed in essence of zeroes and
ones.

While it is reasonably easy to imagine how a Pilot's Associate might
translate a message "Fire!" into a signal that launches a weapon, it is less
easy to see how a computer might respond to a fleet admiral's query
"How far can my battle group sail at twenty knots on its current fuel
supply"? Whether the admiral spoke that message—and thus exploited
advances in speech recognition—or typed it in at a console, the computer still had to translate it into a workable machine language, and
then translate the answer back into English. This involved both the compilers that formed the physical interface between the communicant and
the machine and the program within the machine that translated the
symbols of one into the symbols of the other.[114]

Five contractors worked on various aspects of natural language understanding, feeding their results to two integrators of the new generation
system technology. Each of the system integrators in natural language
was a team: New York University and Systems Development Corporation
(SDC) teamed to integrate text understanding; and BBN teamed with
the Information Sciences Institute to integrate natural language query

processing. The capabilities flowing from these streams of innovation were to feed primarily the Battle Management applications.

In this program, the results were most promising, again exceeding the milestones laid out in the original SC plan. Not only were the metrics of the original plan exceeded in error rate and the size of vocabularies, but new capabilities were introduced based on early progress in expert system development. In the end, the systems to explain underlying reasoning and to automatically acquire new knowledge had to be custom designed for the programs using speech recognition, but these systems nonetheless had powers comparable to what had been predicted in 1983.

Developments in vision were far more disappointing. Part of the reason is that expectations at the beginning of SC were so high. Prior research on computer vision may be grouped in three eras.[115] In the 1950s and 1960s techniques from signal processing and statistical decision theory produced important developments in areas such as Synthetic Aperture Radar, image-enhancement, and terrain-matching, cruise-missile guidance. These were essentially ad hoc inventions, innocent of a conceptual paradigm. In the late 1970s, when DARPA was funding an image-understanding program aimed at photo interpretation, a "signals-to-symbols" paradigm gained currency. It was, however, a theory of low-level, general-purpose vision, limited essentially to two-dimensional analysis.

Optimism that this limitation might be breached grew in the late 1970s, culminating in the 1982 publication of David Marr's pathbreaking study, *Vision,* just as the SC plan was taking shape.[116] As Noam Chomsky had done for speech in the 1950s, Marr refined a new paradigm based on modeling visual cues such as shading, stereopsis, texture, edges, and color to arrive at what he called a "2-D sketch."[117] His theoretical models, combined with the magnified computing power of SC's new architectures, suggested to many researchers that high-level vision was within reach, that a computer would soon employ standard algorithms to distill visual and other signals into machine understanding of what it was "seeing."

By mid-1986, SC was supporting fourteen contractors in thirteen vision projects. Two of these projects fed directly into the Martin-Marietta Autonomous Land Vehicle program: the University of Maryland project on parallel algorithms and systems integration, and the Hughes Aircraft and Advanced Data Systems (ADS) programs on route planning. The other eleven fed into the "new generation computer vision system," which was being integrated by Carnegie Mellon. The new generation system, when completed, was to be transferred to Martin Marietta for incorporation in the ALV.

When a plan was drafted in 1987 projecting developments in the second half of SC, two major trends in vision were apparent. First, a significant effort was being made to integrate new SC technology, especially in computer architecture, into the program. Second, the original claims for what would be accomplished in vision were tempered. No longer did the program plan to achieve by 1993 a capability for "reconnaissance in a dynamically changing environment." Neither did it claim to run programs in 1992 for 3-D vision at one trillion instructions per second or achieve "knowledge-based vision" on a parallel machine of one million processors running at one megahertz symbolic processing rate. Vision itself had proved to be a thorny problem, and the other technologies in the pyramid, such as architectures and expert systems, were not in place to break the impasse. A more detailed examination of the SC vision program will follow in chapter 7.

Thus the record of AI in SC was mixed. Expert systems, the most important segment of AI, the real heart of intelligent machines, failed to produce the generic tool that would have opened up the field. Still, the pursuit of a generic tool stimulated new ideas on the features that a mature expert system should deploy and the ways in which such features might be instantiated. All workers in the field profited from such developments, and custom expert systems grew consistently stronger and more capable.[118] It was these systems that would finally drive the SC applications.

Of the remaining realms of AI, vision proved the most disappointing. It fell farthest short of the goals originally planned, and it proved least responsive to the imposition of vastly increased computing power. Machines could process the data coming in; they just couldn't deploy algorithms that could interpret the images with human speed and accuracy. As with generic expert system shells, significant progress was made in implementing vision for the ALV and other SC applications (see chapter 7). But the hopes for a generic, high-level vision capability were not fulfilled. Indeed, optimism was lower in the early 1990s than it had been in the early 1980s.

Natural language understanding and speech recognition both enjoyed greater success. In these realms computing power did make a significant difference, and linguistics provided conceptual tools that helped as well. The best measure of success here is the one envisioned for SC at the very outset: these developments transferred out of the laboratory and into practical applications, both military and commercial.

Does this mean that AI has finally migrated out of the laboratory and into the marketplace? That depends on one's perspective. In 1994 the U.S. Department of Commerce estimated the global market for AI systems to be about $900 million, with North America accounting for two-thirds of that total.[119] Michael Schrage, of the Sloan School's Center for Coordination Science at MIT, concluded in the same year that "AI is— dollar for dollar—probably the best software development investment that smart companies have made."[120] Frederick Hayes-Roth, in a wide-ranging and candid assessment, insisted that "KBS have attained a permanent and secure role in industry," even while admitting the many shortcomings of this technology.[121] Those shortcomings weighed heavily on AI authority Daniel Crevier, who concluded that "the expert systems flaunted in the early and mid-1980s could not operate as well as the experts who supplied them with knowledge. To true human experts, they amounted to little more than sophisticated reminding lists."[122] Even Edward Feigenbaum, the father of expert systems, has conceded that the products of the first generation have proven narrow, brittle, and isolated.[123] As far as the SC agenda is concerned, Hayes-Roth's 1993 opinion is devastating: "The current generation of expert and KBS technologies had no hope of producing a robust and general human-like intelligence."[124]

Keith Uncapher tells a revealing anecdote on this score. He claims to have repeatedly nagged Allen Newell about never getting involved in the application of AI. Finally, Newell agreed to sign onto the ZOG program, which was developing systems to run the below-decks workings of an aircraft carrier. Though unrealized in practice, the program was technically successful.[125] When Uncapher later asked Newell how much AI was instantiated in the system, Newell said "none." To him, if it was applied, it was not AI. This attitude, which is common among AI purists, invites paraphrase of John Harington's famous observation: AI doth never prosper, for if it prospers, none dare call it AI.[126]

Putting SC to Work: The Autonomous Land Vehicle

Crazy Walking War Machines

Clint Kelly ran SC for three years, yet never learned to use a computer—because, he says, he never learned to type. A thin, gangling man with a gentle temperament and a twangy Kentucky drawl, Clinton W. Kelly III belies the stereotype of the computer whiz who jabbers in technobabble and performs miracles with silicon. He is a manager, not a mechanic. He has a sharp, incisive mind, a solid understanding of the technology, and a true believer's fascination. "I probably never met a technology I didn't like and didn't want to learn more about," he says.[1] He also has a smooth, easy-going manner that makes him particularly accessible and open, characteristics that rubbed off on the contractors who worked for him. He piloted the SC ship while others worked the rigging.

It was not computers per se that got Kelly to DARPA in the first place, but control theory, especially robotics. As a graduate student in engineering at the University of Michigan in the late 1960s, Kelly became interested in synthesizing control mechanisms. He spent two years studying psychology to better understand human behavior, hoping to apply that knowledge to machine behavior. Upon receiving his doctorate in engineering in 1972, Kelly formed his own company, Decisions and Design, to automate intelligence analysis and decision making. In 1980 he joined DARPA's Cybernetics Technology Office under Craig Fields and began a program in legged locomotion, which he considers "the intellectual antecedent of the Autonomous Land Vehicle." He funded a variety of odd mechanical creatures, including a robot that performed backflips. When the Cybernetics Office was dissolved in 1981, Kelly continued his robotics programs in the Defense Sciences Office.[2]

In 1983 Kelly was asked by Robert Cooper himself to help define the applications efforts for Robert Kahn's new computing program. Kelly

was delighted, for it would ensure a prominent place in SC for his robotics ambitions. It also played to his strength. He was a manager, an
integrator, an organizer. He would connect the disparate innovations
arising from Kahn's technology base, synthesizing them in applications
that functioned in the real world. The three applications that were finally
chosen reflected not only the need to court the military services, but
also the diversity of capabilities that might be expected to arise from the
SC research program.

The Pilot's Associate (PA) was intended to be an assistant to the aircraft commander, helping him or her to prepare and revise mission
plans, assess threats, and monitor the status of the plane's systems to
warn of trouble or danger. It would require specialized expert systems
that could cooperate with each other and function in real-time, a speech
system that could understand the pilot's requests and commands under
noisy and stressful circumstances, and compact hardware to deliver the
required computational power (see figure 7.1).

No similar programs then existed within DARPA. After SC was announced, DARPA commissioned several studies by Perceptronics and
other companies to explore the concept of the PA, which was described
only in the vaguest terms in the SC plan. These studies established the
basic form of the PA system. Five cooperating expert systems would be
responsible for mission planning, systems status, situation assessment,
tactics, and interface with the pilot. Most of the planning work for the
program was conducted by air force scientists and engineers at the
Wright Aeronautical Laboratory (AFWAL) at Wright-Patterson Air Force
Base in Dayton, Ohio. The air force had taken a strong interest in AI
by the 1980s, and had its own labs to explore its application. During
1985 and 1986 the air force prepared a description of the planned system
and conducted its own proof-of-concept studies (under DARPA sponsorship). The statement of work for the program "evolved as an almost
philosophical document rather than one of specific tasking defined by
functional and performance specifications," one air force scientist
recalled.

Lacking detailed requirements and specifications, the SOW [statement of work]
became a creative, rather than a restrictive, document. This latitude allowed the
government to buy research towards a goal with the freedom of discovery to
provide direction and objectives defined during the course of the project. The
risks were obvious, but with enlightened management on both sides, the rewards
proved substantial.[3]

Figure 7.1
Pilot's Associate. Adapted from a view graph in Robert Kiggans's file.

Throughout this period, and indeed during the rest of the program, the air force convened a technology advisory board consisting of leading AI researchers in academia and industry, and an operational task force consisting of experienced air force flight crews and engineers, who guided the development of the PA system. A rapid prototyping strategy was also adopted, with early prototypes being used as a baseline to gradually and steadily improve performance. Support from U.S. Air Force Lieutenant General Robert E. Kelly, the vice commander of the Tactical Air Command, ensured that the program was pursued energetically and with relatively minimal direction from DARPA, at least in the early years.

The second SC application, the Battle Management System, shared some features with the PA. It required cooperating expert systems that could absorb and make use of incoming data; a new graphical interface to present information to the user; a natural language system to accept complex queries in normal English; and a speech system that could recognize and respond to a larger vocabulary than that of the PA, although in less stressful and noisy circumstances. There were two programs in battle management. Both were incorporated under the general heading of Naval Battle Management, but they were planned and managed separately. The Combat Action Team (CAT) was also known as Battle Management Afloat because it was installed aboard ship. This was a continuation of the ZOG program on which Allen Newell and his CMU colleagues had worked in the 1970s.

When CMU proposed to continue the work under SC, DARPA sold the project to the navy, especially the Naval Ocean Systems Center (NOSC) in San Diego, California. CMU's plan called for a robust system that would monitor and assess potential threats against a carrier group and recommend possible counteractions. The system would handle new information, make inferences from it, and interact with the human experts who would "teach" it what it needed to know. A Smart Knowledge Acquisition Tool (SKAT) would interview the expert users to define new rules and become better at formulating such rules and improve its rulemaking as it went along. Other systems would include a Spatial Data Management System (SDMS), an interactive graphic display, and the Force Level Alerting System (FLASH), an expert system designed to recognize and announce potential threats.[4]

"Battle Management Ashore" was officially the Fleet Command Center Battle Management Program (FCCBMP, pronounced "fik bump"). Like PA, FCCBMP was to consist of five expert systems. Each would perform a particular task for the headquarters of the Pacific Fleet at Pearl Harbor

(Commander-In-Chief, Pacific Fleet, or CINCPACFLT). First, the Force Requirements Expert System (FRESH) would monitor the readiness of the fleet and assist in allocating its forces according to the capabilities and status of the individual ships. The Capabilities Assessment Expert System (CASES), would compare the relative strength of United States and hostile forces. The Campaign Simulation Expert System (CAMPSIM) would simulate the outcome of different courses of action. The Operations Plan Generation Expert System (OPGEN) would develop operational plans according to specified strategies. Finally, the Strategy Generation and Evaluation Expert System (STRATUS) would assist in developing plans for theater-level strategy.[5]

The most notable feature of the FCCBMP program was its development strategy for "rapid prototyping." A test bed was established at CINCPACFLT in a room near the operational headquarters. The same data that came into headquarters were fed into the test bed, which was carefully isolated so that it could not accidently affect real-world activities. Experienced navy officers and technicians tested it under operational conditions and provided feedback to the developers. The resulting upgrades would improve the knowledge base and interface of the system. New, more powerful computers, ideally machines produced by the SC architectures program, would be added as they became available. By maintaining a close relationship with the end users, the developers could tailor the system to their needs.[6] Requirements for the several expert systems were established by the navy, as opposed to DARPA guessing at what was needed.

As with all of the applications eventually developed by SC, PA and Battle Management consisted of multiple AI technologies. They posed problems of systems integration that had been anticipated by Kahn's prediction of multiple computer architectures. But this was even more complex, for multiple software programs would have to be connected to appropriate platforms and furthermore connected with one another. Connection, in short, would determine the outcome, just as much as soundness of the individual technologies.

Nowhere was this more true than in the third of SC's original applications, the Autonomous Land Vehicle (ALV). Its definition, however, followed a somewhat more circuitous path. Kahn had originally proposed an Autonomous Underwater Vehicle, a development emphasizing sensor analysis and planning. The control mechanisms of the craft would be simple and there would be no obstacles to avoid or terrain to consider. The navy, however, already had a highly classified set of research programs in underwater navigation.[7] So Kahn reluctantly agreed to an autonomous land vehicle for the army. Kelly proposed a walking vehicle such as the human-operated Hexapod then being developed at Ohio State University, upgraded with intelligent capabilities. It would be his "crazy walking war machine," a fellow program manager later recalled with a laugh.[8] Colleagues convinced him that a walker would multiply the engineering problems with little gain in performance. He finally agreed to a wheeled vehicle, though he tried to support both types of locomotion for at least a year.[9]

Working closely with Ronald Ohlander, the AI program manager, and Robert McGhee, a robotics researcher from Ohio State, Kelly outlined the requirements for the vehicle. It would have to be able to plan a route using *a priori* information, execute that plan, and make adjustments when its on-board sensors revealed obstacles. The ALV would have to interpret sensed data, especially visual information, and identify landmarks, obstacles, navigable paths, and so on. The vehicle would require both expert-system and image-understanding technology.

Particularly crucial would be the vision system, the primary capability that DARPA wanted to demonstrate on the ALV. DARPA support of vision research had progressed from signal processing in the 1960s and early 1970s to image understanding (IU) in the late 1970s. The question now was whether it could take the next step to high-level, real-time, three-dimensional IU. For Clint Kelly's purposes, could a computer locate and identify a road? Could it identify trees, boulders, ditches? Could it observe a scene and recognize when and where to move?

This problem is far more difficult than it appears. An image produced, say, by a video camera, first has to be converted into a form that a computer can recognize, a binary collection of ones and zeros. Then—and this is the hardest part—the computer has to form an internal representation of what it sees and extract from that image the information it needs. The DARPA IU program had targeted that problem. Initially conceived as a five-year program, it was still under way in 1994, twenty years later. By 1984 DARPA had spent over $4 million on this effort.[10]

Until 1983, IU had been a laboratory program, fundamental research aimed at conceptual understanding. The belief that it was ready for transition to real-world problems such as ALV flowed from expectations for parallel processing and knowledge-based systems. Knowledge-based systems offered the means of providing the computer with its internal representation.

Information about the world, or at least the part of the world the computer would see, could be programmed into a knowledge base, which the computer would use to interpret the image. If, for example, a computer had the characteristics of a chair stored in its knowledge base, it should be able to recognize a chair when it saw one. The trick was to convert a visual image into a data set that could be matched with the data set listing the chair's characteristics. If a knowledge base had stored the characteristics of a tree and knew the rule that all trees were obstacles, it might be programmed to automatically navigate around any tree that appeared in its path. In the best of all worlds, a computer aboard an ALV might plan its own route over known terrain in pursuit of a predefined mission, adjusting the route when it sensed unanticipated obstacles or received a new mission assignment. It might, in short, combine planning with image understanding.

Parallel processing, it was hoped, would bring such goals within reach. The new architecture supported by SC would provide the brain power for the ALV. Computer vision was computationally intensive, more so than any other subfield of AI. An image, to a computer, is a huge mass of data that must be sifted and sorted quickly. To process the image (that is, to render it in a form the computer can manipulate) takes time; to interpret the information and make decisions based upon it takes more time. Yet a vehicle moving at 60 kph would have to process and interpret an endless sequence of images, and do it in virtually real time.[11]

DARPA estimated that the vision system for the ALV would require 10–100 billion instructions per second, compared to the rate in 1983 of only 30–40 million. Yet to fit on the vehicle the computers had to occupy no more than 6–15 cubic feet, weigh less than 200–500 pounds, and consume less than 1 kilowatt of power. This meant a reduction of one to four orders of magnitude in weight, space, and power.[12] Advances in vision and expert systems would be wasted on ALV without comparable progress on the parallel machines promised by Stephen Squires.

This relationship between the layers of the SC pyramid was just what Robert Cooper had had in mind when insisting on specific applications. Clint Kelly shared his views. The ALV would place enormous demands on the researchers in the technology base. The pressure to deliver the capabilities needed by ALV would "pull" the technology. To make sure that the pull was relentless, Kelly deliberately made the milestones for the vehicle ambitious. As part of the planning process, he and Ronald Ohlander had polled veteran IU researchers, participants in the SC vision program. What, they asked, was a reasonable schedule for the research and what milestones might DARPA lay out. Kelly thought their answers were far too conservative. "I took those numbers," he recalled, "and sort of added twenty percent to them." Taken aback, the researchers objected strenuously. Ultimately Kelly compromised, accepting a less rigorous schedule which he considered too easy but which the researchers still thought was too difficult. Ohlander considered the program "a real stretch," much more so than the other applications.[13]

Working with Industry

After SC was formally announced in October 1983, Kelly and William Isler (who was officially the ALV program manager) set about organizing the ALV program. A single prime contractor would engineer the vehicle itself and incorporate the computing hardware and software produced by the technology base. DARPA expected to award the prime contract to one of the major aerospace companies. These firms had experience in large-scale systems integration and product engineering for defense procurement, and Cooper was anxious to connect them with the DARPA research program. Ideally, the selected company would already have some experience with AI and robotics, especially the remote operation of land vehicles. It would not, however, be on the cutting edge of AI research. DARPA encouraged the formation of teams, especially between the aerospace companies and universities, to promote interaction and technology transfer.[14]

A "qualified sources sought" notice, published in the *Commerce Business Daily* on 27 January 1984, garnered about fifteen responses. Three companies—Martin Marietta, FMC Corporation, and General Dynamics—were invited to submit full research proposals. A team of officials from DARPA and the Engineering Topographical Laboratory (ETL) at Fort Belvoir, DARPA's agent for both the ALV and the SCVision programs, evaluated the proposals and selected Martin Marietta as the prime contractor. Martin had experience in robotics and the construction of remotely operated vehicles, and it had a research site outside of Denver, ideal for testing the vehicle. In August 1984 ARPA awarded Martin a contract for $10.6 million over forty-two months, with another $6 million for an optional phase II for twenty-four months.[15]

To assist Martin, especially in the early stages of the program, DARPA selected the University of Maryland. Long associated with DARPA in the IU program, Maryland employed Professor Azriel Rosenfeld, who had first suggested an ALV.[16] The university already had a practical vision system under development, so its task would be to apply that system to the vehicle to get it running in time for the first demonstration. Maryland would soon form close ties with Martin, which hired many Maryland graduates to work on the ALV.[17] After these initial selections, DARPA contracted with Hughes Artificial Intelligence Laboratory and Advanced Information & Decision System (which soon changed its name to Advanced Decision Systems, or ADS),[18] to develop planning systems for the vehicle. This research was funded separately from Ron Ohlander's technology base program, the emphasis of which was on vision.[19]

As Kelly was organizing the ALV program, Ohlander was doing the same with SCVision. On 20 January 1984, ETL published a "sources sought" notice in the *Commerce Business Daily*, which drew seventy-four responses, over 70 percent of them from industry.[20] Meanwhile, Ohlander began laying out the specific work to be performed for the SCVision program. The major areas of effort would be in Knowledge-Based Vision Techniques and Parallel Algorithms for Computer Vision. In the former, the contractors would develop generic, high-level vision functions, with emphasis on the specific capabilities required by the ALV.

Ohlander broke this objective down into four specific tasks. Task A was to develop a general spatial representation and reasoning system, which would enable the vehicle to construct a three-dimensional representation of its environment. This system would form the basis for the computer's mental image, and be a standard used by all of the SCVision and (ideally) ALV researchers. Task B was to produce a general method of modeling objects, developing a representation for them that could be understood by a computer. Again, this system of modeling was expected to become the standard used by all the SCVision researchers. The last two tasks, C and D, encompassed the recognition and avoidance of stationary obstacles (physical objects, impassible terrain, etc.) and the detection and tracking of moving targets such as vehicles and people.[21]

The other major area, Parallel Algorithms for Computer Vision, integrated the vision research with the new parallel machines being developed by the Architectures Program. This problem was divided into two tasks. The first was the development of a programming environment for the decomposition of IU algorithms. The researchers would figure out a way to break down vision algorithms developed for serial machines into a form suitable for specific parallel machines. The second would develop de novo for specific parallel machines vision algorithms that could then be transferred to the ALV when the new computers were installed.

Road following constituted a third area of effort. Ohlander contracted for a separate autonomous vehicle, a scaled-down version of the ALV, to act as a test bed for the SCVision program. This small vehicle would carry on-board sensors and a control unit transmitting data and receiving instructions via a communications link to remote computers in a lab. It would be used to test not only the road-following system but also other ALV algorithms in cases such as obstacle avoidance and landmark recognition. Results of this development would transition to the ALV no later than January 1986.

The ALV was supposed to be the test bed for the program, but Ohlander was concerned that the ALV contractor would be under such pressure to meet its demonstration milestones that there would not be time to run the experimental software developed by the SCVision researchers. With the ALV being run out of a different office with different program management, Ohlander would have little control over it. He also wanted to see if his researchers could do it better. "It wasn't to outshine them, or anything like that," he later recalled, "but we were . . . used to doing things in universities and with very little money and getting sometimes some startling results. So I . . . had a hankering to just see what a university could do with a lot less money." The test-bed vehicle could be a "showcase" for the research he was sponsoring, the tangible evidence that the money spent on the AI program was paying off.[22]

The competitions held at the end of August produced ten contracts. Carnegie Mellon University was awarded two, for road following and parallel algorithms. It was the logical, even obvious choice, for both these efforts. CMU already had not one but two vehicles in operation, Terregator and Neptune, which had been built for previous projects. Terregator, a six-wheel, mobile platform for testing on-board sensors, connected via two-way radio with a remote computer. Neptune was a smaller, tethered vehicle, also for testing sensors.[23] Chuck Thorpe, who had just completed his thesis in robotics, took charge of the road-following effort at CMU, along with Takeo Kanade, a long-time vision researcher and participant in DARPA's IU program.

CMU was also a logical choice for a contract in parallel algorithms. Its researchers, led by Professor H. T. Kung, were then developing the Warp systolic array for the Architectures Program and were already compiling a library of vision algorithms to run on it. MIT was also given a contract in parallel algorithms for work relating to the Connection Machine. The University of Rochester, which was experimenting with the Butterfly, was given the task of developing the parallel programming

environment. Contracts in Knowledge-Based Systems were awarded to
SRI International for spatial representation, Hughes Research Labs for
obstacle avoidance, ADS for object modeling, and the University of
Southern California and the University of Massachusetts for target motion detection and tracking.

In the spring of 1985 several other organizations joined the program.
General Electric and Honeywell had both submitted proposals that had
not been selected in the initial competition in 1984. Ohlander added
them partly because he had extra money to spend and partly to increase
industry involvement in the program. GE would construct formal, three-
dimensional, geometric models of solids based on two dimensional visual images. Honeywell would attempt to track moving targets. Columbia
University, which had not participated in the initial procurement (its
entry did not arrive on time), offered to develop vision algorithms for
the Non-Von, a parallel machine which its own researchers were then
in the process of simulating.[24]

As in the ALV program itself, the successful bidders were offered contracts for a first phase, in this case for two years, with an option for a
second two-year phase. Unlike the IU program, all of the contractors
were required to produce some tangible results, in the form of working
software, by the end of the first two-year period. It would be highly unlikely that there would be any great breakthroughs in such a short time,
but DARPA would be able to assess the effort of each contractor and
the value of its work before permitting it to continue on to phase II.

Thus, the effort to produce an autonomous vehicle consisted of several major projects administered by separate offices. The ALV program
itself, consisting of the prime integration effort by Martin Marietta and
several lesser technology efforts by Maryland, Hughes, and ADS, was
managed by Kelly and Isler out of the Defense Sciences Office and, after
November 1984, the Engineering Applications Office. The SCVision program, consisting of eleven separate technology base efforts, was managed out of IPTO by Ohlander and, upon his retirement in 1985, by
Robert Simpson. The third major effort, the architectures program to
produce parallel machines for the vehicle, was managed out of IPTO
by Stephen Squires. Connecting these various levels of the SC pyramid
challenged the personal and institutional dexterity of all.

The SCVision program slowly fell into place. The proposals were evaluated at the end of August 1984. A month passed while certain bidders
clarified their proposals. By the time the paperwork was completed and
the ARPA orders that provided the funding were signed, it was already

December and in some cases January. Weeks and even months were spent on negotiating contracts with ETL. CMU and Rochester, whose efforts were more urgent because they were to establish standards for the others, received their contracts in January 1985. The others were brought on board well after: the University of Massachusetts in February, Hughes in March, the University of Southern California in May. MIT and ADS did not receive their contracts until August 1985, almost two years after SC had first been announced. The three latecomers—GE, Honeywell, and Columbia—were not signed on until well into 1986.

Thus, contracting reversed the priorities of SC. The technology base was funded last, though it was supposed to feed the applications, which started first. To compound the problem, the ALV was soon working under an accelerated schedule. In late 1984 DARPA decided to add another demonstration the following May, six months before the first planned demonstration, perhaps to coincide with congressional budget hearings. The requirements for that first demonstration were relatively modest. The vehicle was to move slowly on a generally straight section of paved road for one kilometer. There would be no complicated movements and therefore no planning. All the vehicle had to do was identify the road and follow it. The vehicle was permitted to cheat by driving blind up to 25 percent of the time, using dead reckoning if the vision system made errors or failed in any way. Nonetheless, the added demonstration pressured Martin to get its vehicle running sooner than planned.

Martin quickly assembled its vehicle. Because of the near deadline, it could not afford to take a chance on experimental technology, so most the components were purchased off-the-shelf. On top of an eight-wheeled chassis made by Unique Mobility, Inc., Martin installed a one-piece fiberglass shell for the body. This large, ungainly, box-shaped monster, nine feet wide, ten feet high, and fourteen feet long, housed its sensors in a large Cyclopean eye in the front (see figure 7.2). Driven by a diesel engine, the vehicle turned like a tank, stopping one set of wheels while the other continued to move. The shell contained room for three racks of on-board computers and an air-conditioning system.[25]

The software and hardware architectures were unremarkable, designed for expansion and upgrade as new equipment and software modules became available. The computer hardware consisted of a VICOM image processor and several Intel single-board computers (SBCs), the primary one being based on the Intel 80286 processor. The VICOM was a simple parallel device designed to process and manipulate video im-

Figure 7.2
The Autonomous Land Vehicle (ALV). *Source:* Martin Marietta Corporation, c. 1987.

ages. It was not a particularly good machine, but it was available immediately, and one of the Maryland researchers, Todd Kushner, had extensive experience with it.[26] The SBCs (which were essentially personal computers) performed the calculations needed to navigate and control the vehicle. A single color video camera mounted on top of the vehicle in front provided the sensory input. Unlike the conventional analog television cameras then in common use, this one used several charge-coupled devices (CCDs) to create a digital image for the VICOM. The camera sent three images—red, blue, and green—which the VICOM combined to enhance the features.[27]

For the first demonstration, the vehicle was to follow the centerline of the road to a predetermined point and stop. Three main software systems managed Vision, Reasoning, and Control. Vision used the input from the sensors (in this case, just a video camera) to create a "scene model," a description of the road that the vehicle could use to determine its route. Vision processed and enhanced the images and used an edge-tracking algorithm devised by the University of Maryland to determine its boundaries. From these boundaries, Vision calculated the center points of the road, which a special algorithm plotted on a three-dimensional grid. Vision then passed on the scene model (which consisted of a list of these coordinates) to Reasoning.

At this early stage of the program, Reasoning did little reasoning. It mostly assessed the reliability of the scene model by comparing the current model to those previously given it, looking for discrepancies. If acceptable, its navigator unit estimated the path of the vehicle, called the "reference trajectory." This was something of a guess. Although the Maryland researchers had come up with clever techniques to eliminate superfluous steps and reduce processing time, the vehicle would literally outrun its available data; it moved faster than it could see. If the scene model was deemed unreliable, or if for some reason Vision failed to produce one, Reasoning was permitted to make up its own, calling up stored data to construct an emulated scene model with which it then produced the required reference trajectory. Reasoning then passed on the reference trajectory to Control, which did the driving. Control compared the vehicle's actual location and course (determined using gyroscopic compass and odometers) to the trajectory and calculated the required course corrections. It then fed the necessary commands to the servo mechanisms controlling the engine and the hydraulic pumps that controlled steering and braking.[28]

The demonstration, held in May at Martin's test site near Denver, was successful. The vehicle covered the 1,016-meter course in 1,060 seconds. This was 100 times faster than any autonomous vehicle had driven before, DARPA claimed. It traveled 870 meters under visual control, well above the 75 percent minimum, and produced 238 scene models and 227 trajectories. Reason overrode Vision eighteen times, mostly for intervals of less than a meter.[29]

The ALV engineers had little time to celebrate. In the accelerated schedule, the next demonstration, only six months away, required the vehicle to travel ten kilometers per hour on a straight stretch of road; negotiate a sharp curve at three kilometers per hour; move again on a straight stretch of road at five-to-ten kilometers per hour, stopping at a T-intersection; and then turn itself around 180 degrees to repeat the course in the opposite direction. No vision override was permitted. This exhibition would require somewhat more than the rudimentary road-following system of the first demonstration. Additional sensors would be added to provide the information required. The Vision subsystem had to be capable of fusing the data from these various sensors to produce the new scene models and do so more quickly and accurately than before.

From the Environmental Research Institute of Michigan (ERIM), Martin obtained an experimental, prototype laser range finder that had been developed with DARPA funding. This device sent out a modulated laser beam in front of the vehicle, which scanned both sideways and up and down by means of a rotating mirror. Like radar, the scanner measured the reflection of the beam and calculated the range of whatever was doing the reflecting. Not only could this device locate obstacles and other features of note but, Martin's engineers discovered, it could be used to locate the road itself by identifying its surface and boundaries. The ERIM scanner was installed inside the eye on the front of the vehicle. The Vision system now had to fuse the video and range data to form a three-dimensional image of the road and create the scene model, which now consisted of a series of paired edge-points located in three dimensions. The Reasoning system, too, was more sophisticated than for the first demonstration. A Goal Seeker was added to control the overall operations of Vision and Reasoning based on high-level mission goals and a priori knowledge contained in the new Reasoning Knowledge Base, which included a map describing the road network.[30] The vehicle successfully passed the demonstration in November 1985.

The Tyranny of the Demonstration

While the ALV program moved quickly to field demonstrations, the SCVision contractors began their own work. Carnegie Mellon, eager to start, did not wait for its contract, but began its road-following project in October 1984, upon notification that it had been chosen. Before the month was out, Terregator had made its first runs under the control of its vision system, albeit slowly and with frequent halts. The Carnegie researchers eventually tested several techniques for determining the location of the road, by detecting its edges, for example, or by contrasting its color with that of the non-road areas. By November the vehicle was able to make continuous runs indoors, without the frequent halts. By December it made such runs outdoors in Schenley Park, near the CMU campus. Meanwhile, the researchers experimented with other sensors. In November, the vehicle made its first runs under the guidance of sonar. The following May a laser range finder, just like the one used on ALV, was acquired from ERIM and bolted on the vehicle. The Carnegie researchers encountered problems similar to those of the Martin researchers. The black-and-white cameras had trouble distinguishing the edges of the grey asphalt in Schenley Park, and sometimes mistook tree trunks (with their sharp, firm edges) for roads. The researchers had to use masking tape to mark the roads.[31]

Meanwhile, starting at the end of 1984, the Carnegie researchers took part in discussions with Martin and DARPA officials, planning the future of ALV. While its researchers and engineers were busy preparing the vehicle for its early demonstrations, Martin was planning what the next version of it would look like. The current configuration of hardware and software would be effective only through the May 1986 demonstration. After that, the increasingly complex software would require greater speed and computing power to run in real time. Much of the new hardware was still being developed by the SC Architectures Program, and Martin would have to decide early what systems to adopt. The new algorithms for road-following, obstacle avoidance, route planning, and off-road navigation would, it was expected, come from SCVision.

Early in the program, well before the first demonstration, Martin organized the ALV Working Group, including technology base researchers, to help it define the future architecture of the vehicle. For the remainder of the ALV program, Martin, with DARPA's support and sponsorship, conducted meetings, presentations, reviews, and workshops. The workshops and technology reviews were held annually (quarterly starting in 1986) at a variety of sites around the country, including Key West, Florida; Vail, Colorado; and San Diego, California. These locations were comfortable but, more importantly, they were isolated and removed from the distractions of the everyday world. As at DARPA PI meetings, Martin personnel explained what they were doing and the problems they were encountering. The researchers then gave presentations on their own activities and held extensive, detailed discussions of the issues at hand. The long isolation—the workshops lasted up to five days—produced a sense of intimacy and cooperation, and it helped forge close professional and business relationships, many of which continued beyond SC. This was a goal of SC, going back to Kahn and Cooper's first discussions: connecting academe and industry to tackle common problems and form enduring working relationships.[32]

DARPA and Martin also held more frequent meetings of smaller groups directed toward specific problems and issues—vision workshops, for example, or computer architecture workshops. These were held at the technology reviews, at the ALV test site near Denver, and at the Arlington offices of the BDM Corporation. The ALV Working Group, which acted as a planning and coordination committee, met in January 1985, again in July, and monthly thereafter. The meeting of 26–27 September 1985, for example, included twenty-one people from Martin, Hughes, ADS, ETL, CMU, Maryland, and ERIM. Kelly, Isler, and Roger Cliff of EAO also attended.[33] Ad hoc groups met as needed. On 17 July 1985, BDM hosted a system architecture meeting that was attended by a select group of seven representatives from Martin, DARPA, CMU, Maryland, and the Science Applications International Corporation (who were there because of work SAIC was doing to define ADRIES, a new SC application that would also exploit vision technology).[34]

At these gatherings, DARPA, Martin, and the researchers hammered out the changes to be made in the computing hardware and software. A meeting in October 1985, for example, focused on the computing requirements needed for low-level vision processes. Attendees listened to presentations on several possible parallel systems by their manufacturers, including the Butterfly, Connection Machine, Warp, and PIPE, a machine funded by the National Bureau of Standards. Martin was encouraged but not required to adopt machines developed by SC. Vision researchers, including Larry Davis of Maryland and H. T. Kung of CMU, then debated their merits for the ALV. The meeting was inconclusive. To gather further data on the hardware question, Martin launched a study of all of the possible machines that might be used, analyzing their strengths and weaknesses and the various configurations in which the systems could be combined. This was a lengthy undertaking. There were over 200 machines to be considered, not all of which had yet been released commercially. The evaluation was not completed until 1986.[35]

The hardware question was (by appearances, at least) relatively straightforward; a machine either could or could not perform a particular function satisfactorily. Through standard performance benchmarks and other concrete criteria, it was possible to determine which machine would be best at which task. Furthermore, the task of the hardware was straightforward as well: compute quickly and exchange data with the other hardware components.

Not so with software. The program began with no firm agreement on what components were needed, nor on what each one should do. This was because the ALV program was entering the complex realm of cognition. The early architecture of the first demonstrations, especially the Reasoning system, had been reflexive. Presented with a list of centerline points in the scene model, the navigator and pilot mechanically sought to follow it. The Martin researchers compared these processes to those of the spinal cord and brainstem in humans.³⁶ Future requirements demanded that the Reasoning system act more "intelligently," making decisions based on sensor inputs, map data, mission parameters and goals, and so forth. The Reasoner would have to make such decisions on several levels. It would have to plan an immediate path around obstacles. It would have to plan or replan its overall route to its destination. At the highest level, it would have to decide on the actions required to meet its overall objectives. When, for example, its original plan was unfeasible or superseded by new requirements or events, it would have to use judgment.

To simplify the software problems, Martin defined a "virtual vehicle" consisting of the physical vehicle, sensors, control system, and communications. These well-tested and well-understood systems and their interfaces would remain fixed. The researchers would accept them as a given and would focus their attention on the two key software systems wherein all cognition would lie: Vision and Reasoning.³⁷ Between these two tasks, cognition functions would have to be divided. Where does "vision" end and "planning" begin? Which system should maintain the "world view" of the vehicle, that is, its understanding and model of its environment? More fundamentally, is perception a function of the eye or the brain?

This was not a debate over how vision occurs in humans, but neither was it entirely a debate over the best way to accomplish it in a vehicle. It was also, partly, a debate over who among the contractors should do what. The Vision and Reasoning systems were being developed by different institutions. The contractors most responsible for Reasoning were Hughes and ADS, commercial firms that naturally wanted to expand their own responsibilities in the program and hence their business and future market. ADS was particularly aggressive in pushing the vision-reasoning boundary as close to low-level vision as possible; thus the responsibility for virtually all high-level functions, and overall control of the entire system, would reside in its Reasoning system. ADS also wanted to include the Knowledge Base, containing a priori knowledge and world views, within Reasoning, rather than leave it as a separate component to be accessed by Reasoning and Vision, as had been planned by Martin.

On the other hand, the main vision contractors were universities. Maryland had initially proposed a system architecture that gave Vision a strong and active role in the vehicle's decision-making process. But the Maryland researchers had little interest in vision systems per se and particularly in issues of vehicle control and planning. They were much more concerned with particular capabilities relating to low-level vision—getting the vehicle to reliably recognize the road, for example, or to identify an intersection.[38] Like most academics, they had little desire to expand their own responsibility beyond their particular interests, and they were happy enough to leave the high-level work to ADS. Not so CMU.

Carnegie Mellon had a long tradition of systems work and took a broader systems approach to vehicle development. Its researchers had more experience than those at Maryland in robotic vehicle control, and they did not want to be cut out of the interesting work and relegated to the more mundane tasks of low-level vision. Chuck Thorpe, who represented CMU at these meetings, argued in favor of a more balanced system in which the cognitive functions were distributed among various modules performing autonomously, with none completely controlling the other. The Martin personnel made no judgment at first, observing carefully to see the direction of the debate. Eventually Martin did opt for a more balanced system, as reflected by its adoption of the term "perception" (implying some cognitive function) in place of the more passive "vision."[39] Such debates were neither destructive nor particularly remarkable in themselves, but represented a healthy process by which key technological issues were hammered out.[40]

Nonetheless, the SCVision participants, and their IPTO program managers, were increasingly concerned about the course of the ALV program and their own role in it. Of the three key objectives of the program—to integrate and demonstrate autonomous navigation and vision technology, to serve as a testbed for experiments in the field, and to meet its milestones in public demonstrations—the third was clearly coming to dominate the others. The ALV was the showcase program of SC, much more than the other applications or the technology base. This was in large part thanks to Clint Kelly, who maintained high visibility for the program by issuing news releases and inviting the press to witness the vehicle making its runs. A successful program could give tangible evidence of the value of the emerging technology, and it could maintain the interest of Congress and the services. An unsuccessful program, on the other hand, would have a correspondingly ill effect on SC. ALV simply could not be allowed to fail.

The pressure on Martin's engineers and researchers, intense to begin with, was increased by the addition of extra demonstrations. The pace set by the first one, in May 1985, did not slacken. In December DARPA added a new, yearly demonstration, called the Technology Status Review (TSR), which was intended to be a preview of the technology that Martin was preparing for the upcoming full demonstration. TSRs were less formal and were not intended to be public spectacles. Nonetheless, they kept Martin's shoulder to the wheel.[41]

Martin was having trouble with its road-following system. Quite simply the problem of visual control of navigation was proving far harder than Martin, Kelly, or anyone else had anticipated. The vision system proved highly sensitive to environmental conditions—the quality of light, the location of the sun, shadows, and so on. The system worked differently from month to month, day to day, and even test to test. Sometimes it could accurately locate the edge of the road, sometimes not. The system reliably distinguished the pavement of the road from the dirt on the shoulders, but it was fooled by dirt that was tracked onto the roadway by heavy vehicles maneuvering around the ALV. In the fall, the sun, now lower in the sky, reflected brilliantly off the myriads of polished pebbles in the tarmac itself, producing glittering reflections that confused the vehicle. Shadows from trees presented problems, as did asphalt patches from the frequent road repairs made necessary by the harsh Colorado weather and the constant pounding of the eight-ton vehicle.[42]

Perhaps more alarming to the Martin engineers was the early realization that there would not be one all-purpose, road-following algorithm. Different situations required different programs. The first road-following algorithm that Maryland installed on the vehicle, the "vanishing point" algorithm, had functioned satisfactorily in the lab but not on the road. Under certain conditions the vehicle thought the road had folded back under itself. This algorithm had to be replaced by the "flat-earth" algorithm, so-called because it worked by using a two-dimensional representation of the road and assuming that the road was perfectly flat. The algorithm was quick to run, but it was relatively inaccurate, and, not surprisingly, it worked only on flat ground.

The third program, the "hill-and-dale" algorithm, used a three-dimensional representation of the road. It functioned better on uneven ground, but it did not work on curves. Maryland came up with a fourth algorithm, the "zero-bank" algorithm, which solved this problem; but it ran too slowly on the vehicle's computers and had to be put off until phase II of the program.[43] This appeared to be a case in which enhanced computational power being developed in the Architectures portion of SC might come on-line in time to implement the sophisticated software being developed in the AI portion.

Other problems were caused just by the sheer complexity of the system. By the November 1985 demonstration, 25,000–30,000 lines of code were running in real time on ten different processors. This complexity was blamed for the vehicle's often erratic behavior at that demonstration.[44] The Martin engineers and scientists discovered that systems integration required much trial-and-error. Each new feature and capability brought with it a host of unanticipated problems. A new panning system, installed in early 1986 to permit the camera to turn as the road curved, unexpectedly caused the vehicle to veer back and forth until it ran off the road altogether.[45] The software glitch was soon fixed, but the panning system had to be scrapped anyway; the heavy, 40-pound camera stripped the device's gears whenever the vehicle made a sudden stop.[46]

Given such unanticipated difficulties and delays, Martin increasingly directed its efforts toward achieving just the specific capabilities required by the milestones, at the expense of developing more general capabilities. One of the lessons of the first demonstration, according to the ALV engineers, was the importance of defining "expected experimental results," because "too much time was wasted doing things not appropriate to proof of concept."[47] Martin's selection of technology was conservative. It had to be, as the ALV program could afford neither the lost time nor the bad publicity that a major failure would bring. One BDM observer expressed concern that the pressure of the demonstrations was encouraging Martin to cut corners, for instance by using the "flat earth" algorithm with its two-dimensional representation. ADS's obstacle-avoidance algorithm was so narrowly focused that the company was unable to test it in a parking lot; it worked only on roads.[48]

The pressure of the demonstration schedule also degraded the ALV's function as a test bed for the SCVision program. Originally, DARPA expected that the vision researchers would test their own experimental algorithms on the ALV. Martin did make an effort to fulfill that goal in anticipation that its vehicle would ultimately integrate software that tested successfully. During a workshop in April 1985 Martin announced its plan for the "national test bed," with an elaborate experimentation process to be governed by a science steering committee.[49] Yet the Martin staff and the vehicle itself were kept so busy preparing for demonstrations that they had little time for the difficult and time-consuming process of setting up experiments with risky technology that might contribute nothing to the demonstration requirements.

The vehicle was run frequently, often several times a day, during the weeks prior to the demonstrations. Preparations for the demonstration held in June 1986 began the previous December (not long after the previous showing) with a "best efforts" run, intended to show where things stood and what needed to be done for the demonstration. Serious testing began in January 1987. During the two months prior to the demonstration, over 400 tests were conducted, including 60 runs over the entire test track. By April 1986 the year-and-a-half old vehicle had logged 100,000 miles; before the year was over the engine wore out completely and had to be replaced.[50]

Even had the pressures of the schedule not been so intense, there would still have been a problem with access to the ALV test bed. Denver was not very convenient to the West Coast and even less so to the East Coast, where the bulk of the researchers were. In theory, the researchers would not have to visit the site directly, but could port their code by the ARPANET. But Martin's computers did not run Unix, the standard operating system for the academics. To prepare the software to run on Martin's computers required more time and on-site effort than Martin's personnel could afford to give. In any event, Martin was not connected to the ARPANET, and would not be until at least late 1987.[51]

Researchers grew alarmed at the situation. As a BDM representative noted in July 1985, "Martin Marietta is concerned with issues of integration and meeting demonstration objectives. The university community is concerned that their algorithms . . . might not get integrated at all."[52] The researchers feared that, as Chuck Thorpe observed, they could become merely subcontractors in Martin's effort to achieve its demonstration objectives, and that "the scientific goals of Strategic Computing [would be] subverted in favor of engineering and demos."[53] The emphasis in the milestones on speed was a particular problem. In its quest to get the algorithms and the vehicle to run faster and faster, Martin was forced to ignore research best conducted at low speeds, such as the optimal methods of segmenting an image.[54]

The Carnegie Mellon researchers were particularly restive. They needed ready access to a test-bed vehicle. They had quickly outgrown their current platforms, Terregator and Neptune. Terregator was not suited for its current mission. It was designed for "powerful, go-anywhere locomotion at slow speeds," and it was capable of climbing stairs and bouncing across railroad tracks, though it did not run well on grass. The Carnegie Mellon researchers also had become interested in the problems and possibilities of fusing data from several sensors. Terregator was too small to mount all the necessary sensor equipment, and the communications bandwidth between the vehicle and its off-board computers was far too low to transmit all of the data with sufficient speed. Poor communications was the major reason why the vehicle ran so slowly during its early runs.[55] Aware of the limitations of the current vehicles, the CMU researchers had indicated in their initial proposal that they would need a new vehicle by the third year of the contract (that is, by 1986).[56] They believed they could build such a vehicle themselves at a fraction of the cost of ALV.

The IPTO program managers agreed. Ohlander had long been concerned about whether ALV could adequately perform its test-bed function. Events had borne out his fears. There were definite advantages to having vehicles built at the universities and operated by the researchers. While the ALV acted as a lightning rod, attracting the attention of the press and Congress, a shadow program of IPTO-sponsored vehicles could quietly run experiments using riskier, more radical architectures and components, unconcerned that the inevitable occasional failures might bring bad publicity and scrutiny. At the appropriate time, IPTO could spring the vehicles on the world as the next generation ALV, as the research community believed they surely would be.[57]

In May 1985, therefore, with Ohlander's support, Carnegie Mellon presented a proposal to construct vehicles they called "ALVan" (later called "Navlab," for NAVigation LABoratory). The vehicles would be based on a commercially purchased Chevrolet van that would be modified by William Whittaker's Civil Engineering Lab. Carnegie proposed to build two vehicles with an option for three more, the idea being that the extra vehicles would be sent to other sites for the use of the other SCVision researchers. The first two vehicles could be produced for less than $1.2 million; the additional vehicles would cost $265,000 each. Carnegie would supply the vehicles with video and acoustic sensors and basic control hardware and software similar to Martin's "virtual vehicle." The other research sites could then install and test their own hardware and software configurations.[58]

The New Generation System

In its proposal CMU also offered to perform the role of system integrator. This reflected another of Ohlander's key concerns, that of facilitating the process of technology integration, of connection. As program manager and one of the architects of SC, he had long pondered the problem of how to transition the technology-base program into working applications. This was, after all, what SC was all about. The challenge was partly technical. SCVision included a dozen contractors, most of them working on somewhat narrow, well-defined pieces of the puzzle—

developing methods for geometric reasoning, for example, or object recognition, or obstacle avoidance. Each of the contractors had its own hardware, software languages, protocols, and ways of doing things. There were virtually no universal standards. How could their components be connected with one another in a single working system? It was as if someone were trying to build a car using mechanical and body parts intended for a dozen different makes and models. To make matters worse, the components were all experimental; no one knew for certain if each part would work at all, let alone connect with the others.

The problem of integrating ALV technology was complicated by the culture of academe. Research was traditionally performed independently by the various institutions. Even within a given institution, groups working on different projects might not communicate with each other and coordinate their efforts. Partnerships among two or more universities were uncommon, and between universities and industry they were rarer still. The kind of partnerships that DARPA had in mind for SC—between universities and aerospace defense firms—was virtually unheard of. Academics, who performed the great bulk of basic research in computer science, wanted the freedom to experiment and explore new ideas. Such research is inherently risky, and most of the ideas will fail. Indeed, failure is an essential part of the process. Development contractors and program managers, on the other hand, generally abhorred experimentation and risk. They could not afford to fail. Their goal was a product that met specifications and performed as expected. They had timetables and budgets to meet. They dared not experiment with uncertain technology that could delay or hinder their efforts. Furthermore, the end user wanted a reliable product that performed to specifications. The technology had to be safe, tried, and predictable. The imperatives of this community matched poorly with those of academe.

The challenge came down to program management. The original SC plan, predicated on the notion of moving technology out of the labs and into the real world, did not say how to do it; it did not specify the process. It could not. No one yet knew what that process would be. DARPA had little experience with such efforts and none on such a scale. ILLIAC IV, as has been seen, was hardly a model to follow.[59] The new methods of procurement specified by Cooper, and later by the Competition in Contracting Act, made Ohlander's task doubly difficult. In a previous era, he would have held informal discussions with the researchers at an early stage in the planning process, so that contractors understood what was expected of them and what they would contribute. Such a casual approach was no longer possible. The potential contractors were all competitors now.

Ohlander could gather the researchers together to help write the SC plan, but he could not discuss specifics with them about who would do what. The researchers themselves would have been very reluctant to talk, especially with their potential competitors. Ohlander could not be certain that the successful competitors would work well together. He had never worked at all with some of them, especially those in industry, such as GE and ADS. All he could do was to specify in their contracts that they were required to participate in open discussions and exchange information freely with the other contractors. They would have to trust in his ability, as the source of funds, to twist arms if necessary. Running SCVision in two phases made this easier, as it allowed DARPA to cut off recalcitrant or underperforming contractors after only two years.

As SCVision was getting underway and the participants were coming under contract, Ohlander turned his attention to the problem of how to coordinate their efforts. He first recognized that there had to be one locus for the program, one contractor responsible for the integration. Martin Marietta was supposed to do this, but by the time of the first ALV demonstration in May 1985 it was clear that it could not. Martin's engineers had neither the time nor the inclination to play around with flaky, untried algorithms. A better solution would be to have a contractor within the SCVision program do this, one whose special, assigned task was to make the various components fit into one system. This would require money, Ohlander realized, because it was a time-consuming task. The SCVision and ALV programs, as then structured, provided no funds for integration, and they did not specifically designate any contractor to do the work.

Carnegie Mellon was the logical place to perform the integration work. Its technological philosophy emphasized the development of functioning systems instead of abstract capabilities. Integration, of both technology and research efforts, was a part of the school's culture, and it had a strong influence on its graduates, such as Ohlander himself and Frederick Hayes-Roth. The various CMU labs and projects—the Vision Lab, the Civil Engineering Lab, the Mobile Robotics Lab, the Warp Group—had the reputation of cooperating closely, without the jealousy that sometimes infected departments and labs at other schools; the feud between the AI Lab and the Laboratory for Computer Science at MIT was legendary.

Indeed, for a long time CMU did not divide up its research program into separate, independently-funded projects. In the earliest days, the various research efforts were funded from essentially a single, annual grant from DARPA. Later, as DARPA funding came in multiple, targeted contracts, the computer scientists still spread the wealth, the funded researchers allocating money to worthy projects. The pressure for accountability and tight funding ultimately changed these informal methods, but the spirit of cooperation lived on, thanks to leaders such as Alan Perlis, Nico Haberman, Herbert Simon, Raj Reddy, and above all Allen Newell.[60] Thus Red Whittaker's Civil Engineering Lab had gladly volunteered the use of the Terregator for Takeo Kanade's vision researchers and was looking forward to the opportunity to build a new vehicle for the SCVision program.

In its integration proposal, Carnegie offered to develop an overall architecture for an autonomous navigation system that would be used as a standard by all the SCVision researchers. This architecture would be developed in consultation with the other researchers and with Martin. Furthermore, CMU would "develop an understanding of the research being done at other sites . . . [and would make] appropriate tools . . . available to those sites." When software modules developed by the other contractors were ready for transfer, CMU would integrate them into the overall system. CMU would periodically release the entire system, especially to Martin Marietta for use on the ALV.[61] Ohlander requested $1.2 million for the eighteen-month effort.[62]

CMU's proposal immediately met resistance from Clint Kelly. From the start, Kelly did not think there was money enough for more than one test-bed vehicle. Funding for additional vehicles might be taken from ALV; so might attention. Kelly looked to Martin to manage a common test bed. At a meeting with the technology-base contractors in July 1985, Martin proposed to host two vehicles. One, for demonstrations, "would sacrifice generality in algorithms as necessary to meet demonstration objectives"; the other would be a true experimental vehicle, "a national resource for the university community."[63] The proposal made sense conceptually, but in practice Martin was stretched to the limit by its demonstration commitments. In the end, Kelly and EAO reluctantly accepted the idea of a test bed managed by the technology base.

The DARPA Director's Office was harder to convince. By the summer of 1985, Charles Buffalano had assumed de facto charge of DARPA in the place of the soon-to-be-departing Robert Cooper. Buffalano shared Ohlander's concerns about the problem of technology transition, and he wanted to see a plan specifying how IPTO would organize the process.[64] Soon after forwarding the Carnegie proposal, Ohlander retired from the Navy and left DARPA, moving to ISI, where he could continue to work for DARPA, albeit indirectly. Buffalano and Kahn prevailed upon him to prepare such a plan. Kahn himself was preparing to leave DARPA without having formalized a plan to manage the program he created, and Lynn Conway had proved ineffective at mobilizing the SC process she envisioned for connecting the disparate parts of the program. Throughout the summer of 1985 Ohlander discussed his ideas with the DARPA AI program managers, Alan Sears and Bob Simpson, and with the DARPA community, which was then still in the process of getting SC up to speed. By the early fall he had completed a draft of the Strategic Computing Technology Integration and Transition Plan.[65]

The plan applied not just to SCVision but to each of the four SC programs in AI. Each program was centered on a concept called the New Generation System (NGS). This would be a functional system connected to the related application but having generic capabilities—a generic vision system that could run on ALV, a generic speech system that could be applied to PA, and so on. The NGS would be an integrated system that incorporated the technology produced by all the SC contractors. One contractor or team of contractors would be charged with developing the NGS; the others within each AI subfield would be "Component Technology" (CT) contractors, whose role was to provide pieces for the final system. The NGS contractors had the task of testing those pieces and arranging their transfer into the NGS. They would do this by establishing and disseminating standards in software and by creating a system architecture that would promote the integration task.

The NGS was not designed to replace the applications, but to complement them. Indeed, the NGS integrator and the applications contractors were to have compatible systems themselves, to promote the transfer of technology from the NGS to the application. Free from the tyranny of the demonstration schedules, the NGS contractor could experiment and tinker with the system, exploring ideas and possibilities and leaving the applications contractors to focus on their narrow performance requirements. Thus the NGS system would represent the true state of the art. When ready (after about two years, Ohlander thought), it would be transferred whole to the application. The process of development would be ongoing. The system would continue to evolve and improve, and every two years or so another NGS would be transferred.[66]

An attractive aspect of this plan from DARPA's point of view was that the NGS contractor would help organize the program, working with the

CT contractors and greasing the wheels of transition and integration. To ensure that they obtained the component technology they needed, the NGS contractors were to make bilateral agreements with each of the CT contractors and the applications contractor(s), setting schedules and standards. This would relieve the inexperienced and overburdened AI program managers, who between them oversaw dozens of SC programs, to say nothing of the ongoing basic efforts.

The Integration and Transition Plan was never formally implemented or even finished. Buffalano and Kahn both left in the fall of 1985 without formally adopting the plan. Sears and Simpson adopted it in principle and organized their programs according to its principles. Interested institutions were invited to compete for the position of NGS integrator, and as early as August 1985 the contractors were tentatively selected. In natural language understanding, two teams were selected for the NGS: NYU and SDC, for a NGS for text understanding; and BBN and ISI, for a New Generation query system for database and expert system access. Carnegie Mellon and Texas Instruments received contracts for New Generation Speech Systems. IntelliCorp and Teknowledge were NGS contractors for New Generation Expert Systems. And Carnegie Mellon, of course, was assigned the New Generation Vision System.

In the end, Ohlander's NGS plan proved more effective at defining a problem than solving it. The difficulty was inherent in the SC plan. How would advances in Kahn's technology base "transition" to Cooper's applications? This difficulty transcended the disjunctures between technology push and demand pull. It also included sequencing, the simultaneous development of applications and the technology breakthroughs that were supposed to make them possible. Perhaps the greatest difficulty, the one that Ohlander's plan most clearly addressed, was research focus. Integration of research efforts was easiest when the projects proceeded on similar or parallel trajectories, when they shared common methodologies, standards, and infrastructure. DARPA's attempt to create a common LISP programming language is a good example. But DARPA also insisted on casting a wide net, on hedging its bets, on supporting all promising lines of research.

By definition, research was an excursion into the unknown. No program manager could be sure in advance which trajectory would reach the target and which would stray off course. The greater the dispersion of trajectories, the higher the probability that one would succeed. But that dispersion bred incompatibility and complicated problems of integration. The generic software systems examined in chapter 6, for example, attempted to incorporate all the capabilities being developed in the other AI contracts—blackboards, forward and reverse chaining, inheritance, and so forth. They failed. What was worse, they never attained a form that could be usefully transmitted to applications.

The experience with ALV mirrored what was going on elsewhere in the SC program. The applications failed to connect with the technology base. Instead, applications extemporized ad-hoc, off-the-shelf solutions to meet demonstration deadlines. Meanwhile, the many research projects in the technology base rose and fell on their own merits. Mutually incompatible, they seldom achieved integration, let alone transition. Some few followed a trajectory that mapped closely with the SC vision. Speech recognition, for example, achieved its metric goals in part because it ran on more powerful architecture made possible by advances in VLSI and other infrastructure. And it was integrated successfully in a New Generation Planning System and went on to application in the PA. The more common experience, however, looked like that of the ALV.

The ALV and Its Shadow

By early 1986 Carnegie Mellon was running tests on its new Navlab vehicle. (There was only one; the second vehicle, and the three that were optional, were all canceled, victims of the severe budget crisis that year.) The van was equipped, Thorpe likes to say, with "onboard power, onboard computers, onboard sensors, and onboard graduate students."[67] Designed for running tests at slow speeds only, the vehicle had a top speed of only twenty miles per hour. During its early tests it ran at ten centimeters per second—"a slow shuffle." Unlike ALV, it could be driven to a remote test site by a human driver and then switched to automatic. Like the ALV, the vehicle mounted a video camera and an ERIM range finder.[68]

The system installed on the Navlab, which the Carnegie researchers called "Codger," was a variation of the blackboard architecture developed in the CMU Hearsay-II program of the 1970s. In this system, the "whiteboard" did not control the functioning of the whole; rather it connected the various components. Each module ran continuously. If it needed data (a scene model, or map information), it would submit its request via the whiteboard and wait quietly for another module to respond. This system, the researchers thought, would be particularly effective when the various modules ran on multiple machines, as on the

ALV.[69] By November 1986 they successfully demonstrated to DARPA officials the Park road-following system, so named because it was run in Schenley Park.

Meanwhile, Martin Marietta struggled with ALV. Much of 1986 was spent trying to improve the performance of the road-following algorithms. After the November 1985 demonstration, the perception system had been fully overhauled. The effort was largely successful. The new system performed much better, and the demonstration of June 1986 went so well that a few weeks later Kelly held another one for the press.[70] Then Martin personnel began to integrate and test software to enable the vehicle to avoid obstacles, a requirement for the October 1986 TSR and the 1987 demonstration. Also during the summer, Martin completed plans for the installation of the computing hardware for the second version of the vehicle, planned for 1987. The Mark II was to receive two VICOM computers for stereo vision, a Butterfly for navigation-reasoning, and a Warp parallel machine with a SUN front end for image processing.[71] This was a clear bid to make an SC application run on technology developed lower down the pyramid.

By this time relations between the aerospace giant and the SCVision researchers had warmed considerably. Simpson noted that after a period of stand-offishness on the part of Martin's personnel early in the program, they became much more willing, even anxious, to work with the researchers, having discovered by then just how hard their task was.[72] Martin worked especially well with CMU, which took seriously its responsibility to deliver technology to the ALV. Personnel from both organizations met to discuss needs and opportunities; and Martin went so far as to enroll one of its employees in Chuck Thorpe's program, so he could serve as a liaison and keep the company abreast of Carnegie's research.[73] Carnegie returned the favor. Martial Hebert, a Carnegie professor, spent time in Denver. When the new Warp machines arrived in the summer of 1987, Martin reported, "We have received outstanding support from the CMU Warp team, headed by Professor H. T. Kung, and greatly appreciate their assistance during our recent visit."[74]

More importantly, technology finally began to filter up to the ALV from the technology base. During the summer of 1986 Martin received a path planner, an obstacle detection routine, a sensor calibration package, and utility software from Carnegie.[75] Later that year Carnegie began installing on the ALV its Generalized Image Library, an extensive collection of basic image processing functions. This effort was completed in early 1987. Specialized drivers were also installed to permit CMU's code to run without alteration, and Martin held discussions with Carnegie about the possibility of transferring the Navlab's "whiteboard" to the ALV.[76]

By 1987 the ALV was much improved. During the winter it was entirely overhauled, receiving a new engine, chassis, and shell, and later its Warp and SUN computers. The hardware now supported Unix, C, and the SUN Operating System, compatible with whatever software the technology base could deliver. The software architecture was much more sophisticated than that of the early days of 1985, so that it was becoming much easier to swap modules in and out of the system. For example, the Reasoning system, which formerly had consisted of a "monolithic block of code," was redesigned into a group of separate Unix processes not unlike the knowledge sources of a blackboard system—on which, indeed, the system was modeled. The knowledge base was also broken up into two parts to permit easier modifications of the knowledge.[77]

These improvements produced better performance. In the summer of 1987, a team from Hughes came to Denver to test its planning and obstacle-avoidance software. For the first time an autonomous vehicle drove itself cross country according to a route it selected itself, based only on digital map data and sensory input. The vehicle successfully drove around gullies, bushes, rock outcrops, and steep slopes.[78] In November the ALV performed quite well at its formal 1987 demonstration. It achieved a top speed of 20 km/hr as required, and averaged 14 km/hr—four more than was required. The vehicle maneuvered around plastic trash cans placed in the road, at a speed of up to eight km/hr.[79]

Yet the vehicle's days were already numbered. On 6 November 1987, the day after the triumphant demonstration, a panel of DARPA officials and technology-base researchers visited the Denver site to review the ALV program and to make recommendations about phase II, due to begin the following spring. Doubtless reflecting the opinions of the other researchers, Takeo Kanade of CMU, while lauding Martin's efforts, criticized the program as "too much demo-driven." The demonstration requirements were independent of the actual state-of-the-art in the technology base, he argued. "Instead of integrating the technologies developed in the SC tech base, a large portion of Martin Marietta's effort is spent 'shopping' for existing techniques which can be put together just for the sake of a demonstration."[80] Based on the recommendations of the panel, DARPA quietly abandoned the milestones and ended the ALV's development program. For phase II, Martin was to maintain the vehicle as a "national test bed" for the vision community, a very expensive hand servant for the researchers.

For all practical purposes, therefore, the ambitious ALV program ended in the fall of 1987, only three years after it had begun. When phase II began the following March, Martin dutifully encouraged the researchers to make use of the test bed but attracted few takers. The researchers, many of whom were not particularly concerned about the "real-world" application of their technology, showed more interest in gathering images from ALV's sensors that they could take back to their labs.[81]

The cost of the test bed became very difficult to justify. Phase I of the program had cost over $13 million, not a large sum by defense procurement standards, perhaps, but large by the standards of computer research. Even with the reduced level of effort during phase II, the test bed alone would cost $3–4 million per year, while associated planning, vision, and sensor support projects added another $2 million.[82] Perhaps most disappointingly for Kelly, the army showed little interest in the program, not yet having any requirement for robotic combat vehicles. One officer, who completely misunderstood the concept of the ALV program, complained that the vehicle was militarily useless: huge, slow, and painted white, it would be too easy a target on the battlefield.[83] In April 1988, only days after Kelly left the agency, DARPA canceled the program. Martin stopped work that winter.[84]

As it turned out, the ALV was not gone for long. In 1990 Congress consolidated all military robotics efforts into a Tactical Warfare Programs Office within the Department of Defense. Interest in robotic vehicles grew after that, especially in the wake of the Gulf War in 1991, when "smart weapons" and "unmanned air vehicles" (UAVs) received much publicity. Starting that same year, Lieutenant Colonel Erik Mettala, a DARPA program manager, launched several programs to develop unmanned ground vehicles (UGVs), including Demo-I for teleoperated vehicles and, in 1992, Demo-II for autonomous vehicles. Like the ALV, Demo-II was oriented towards demonstrations. Unlike ALV, it was explicitly directed toward military application, and it had the full backing of the army.[85]

In a sense, too, ALV lived on in Navlab. After the cancelation of Martin's effort, Carnegie Mellon continued with its work, continually improving its vehicle. Thorpe and his researchers deliberately made their task as difficult as possible. They sought vehicle autonomy under bad lighting conditions and over rough, complicated ground as well as on the highway. By 1991 Carnegie had outgrown Navlab. With funding from Metalla's program, they constructed Navlab-II based on an army high mobility medium utility vehicle (HMMV). This vehicle, which has driven up to 62 miles per hour on the highway, employed neural nets, a technology that SC discounted. Navlab-III, a CMU faculty member's Honda Accord, is controlled by a commercially available Sparcbook notebook computer. By 1995 there were five Navlab vehicles.[86]

The ALV experience offered some important lessons for DARPA. The concept of a generic new generation system proved elusive. As both Martin Marietta and CMU discovered, there would be no ideal, all-purpose vision algorithm, even for road-following alone. As late as 1991, seven years after the start of the ALV program, the CMU researchers reaffirmed that no single perception system could address all possible road configurations, let alone all outdoor environments. The perception system had to be tailored to specific circumstances, conditions, and tasks. In the area of program management, DARPA discovered the difficulty of integrating technology in an orderly fashion, especially software produced by a variety of contractors.

After early and apparently sincere efforts to organize the SCVision effort to promote the transfer of component technologies into the new generation system, CMU decided that it could develop whatever technology it needed without help. Software components, it found, worked better when custom-developed in-house according to the actual conditions and requirements. Components produced elsewhere were hard to integrate. Furthermore, if they were designed under lab conditions (as almost all are), they did not work well anyway.[87] As for the component technology contractors, they could market the fruits of their labors independently—or not at all. Ultimately very little of the technology base was either integrated or transitioned. Bob Simpson, the AI program manager, considered making it a stipulation for continued funding that the component technology contractors provide technology for integration, and that Carnegie integrate it. In the end, however, he thought better of it, allowing Thorpe full control over CMU's new generation system.[88] The individualistic research culture was too entrenched for DARPA to overcome; connection was too difficult.

Other lessons were less technical. The ALV program demonstrated that the military services had to support applications programs. The other SC applications—the battle management programs and PA— were developed in close coordination with the prospective users and received strong financial and programmatic support from the navy and air force. The FCCBMP program in particular was a model for transition to the services, and was the most successful of the original SC applications. CINCPACFLT used the test bed as it would a functioning program, and it worked closely with DARPA and the Naval Oceans System Center to make the improvements it needed to develop a practical, working system. ALV was conceived, funded, and managed with little interest or support from the army, the expected end user of the system. The program became an expensive white elephant for DARPA. DARPA learned from this experience: The UGV program was conducted—and funded—jointly with the army. The technology was connected with the user.

Finally, at the most fundamental level, ALV cast doubt on Robert Cooper's belief that applications would drive SC through technology pull. It was never entirely clear exactly how this would happen, since the applications programs started simultaneously with the technology base. Inevitably, they had to begin with off-the-shelf technology, looking to upgrade with new components as the technology base advanced. But the tyranny of the demonstration preoccupied Martin Marietta's engineers, and the risk of an embarrassing public failure frightened them away from many of the innovations arising in the technology base. Instead of pulling technology, the original ALV stifled it and used up funding that might have been better used pushing technology through the base.

Part III

8

ISTO: The Middle Years of Strategic Computing 1985–1989

Reconciling Form and Function

By the end of 1985, SC was moving into second gear. Over $100 million had already been spent on the program. Most of the initially planned efforts had been funded and organized, and the contractors were onboard or soon would be. Ninety-two projects were underway at sixty different institutions, about half in universities and government labs and half in industry.[1]

The founders had much to be proud of, but few of them were still around to enjoy the success. Indeed, only a handful of those who had participated in the planning process in 1983 were still at DARPA two years later: Al Brandenstein, Steve Squires, Craig Fields, and Clint Kelly. The original leaders of the program—Robert Cooper, Charles Buffalano, Robert Kahn, and Lynn Conway—as well as key program managers, such as Ron Ohlander and Paul Losleben, had all moved on.

In January 1986, Dr. Robert "Cliff" Duncan became the eleventh director of DARPA. A U.S. Naval Academy graduate and navy fighter pilot in the 1950s, Duncan held advanced degrees in aeronautical engineering. He had served stints at NASA on the Apollo project and at Polaroid, where he oversaw the development of the SX-70 camera. Ready to retire from Polaroid in 1985, Duncan aspired to be deputy director of NASA. Instead, he was offered the top job at DARPA, an agency little known to him. Asking around, he received glowing reports about it, especially from ex-director George Heilmeier, who said it was the best job in Washington, not excepting the presidency. Duncan accepted the position.

IPTO, too, had a new director. Early in 1985 Robert Cooper had begun his search for a replacement for Kahn, who was due to leave DARPA in September. Cooper consulted a group of senior computer scientists, mostly members of the DARPA research community. The group recommended several candidates, including Dr. Saul Amarel, an AI researcher and longtime professor at Rutgers University who was then on sabbatical at Carnegie Mellon. After a string of rejections by other candidates, DARPA approached Amarel. Having spent fifteen years as chairman of the Computer Science Department at Rutgers, the professor was looking for a change. He was keenly interested in the office and in SC, though he had some doubts about the job, particularly about the level of funding he would have to work with. Before agreeing to serve, Amarel held a long, closed-door meeting with Buffalano at a neutral site, Keith Uncapher's Information Sciences Institute in Los Angeles. The search for a new IPTO director had been long and unavailing, and Buffalano needed to land Amarel. He agreed to fund IPTO at $192 million. Amarel accepted. Arriving at DARPA late that summer, a few weeks early, he took charge of IPTO immediately upon Kahn's departure in September 1985.[2]

A soft-spoken man of quiet dignity and unfailing manners, this former officer of the Israeli army had entered the computing field in the 1950s, when he received a Ph.D. in electrical engineering from Columbia University. Amarel was interested in AI from his earliest days in the field, especially in control theory and machine learning. He spent eleven years at RCA Labs in Princeton, New Jersey, where he founded the "Computer Theory Research Group." In 1969 he moved over to Rutgers University in New Brunswick, where he founded the school's computer science department, and where, except for brief sojourns elsewhere (including DARPA), he has remained ever since. Amarel continued his AI work and kept in close touch with the AI communities at Carnegie Mellon and Stanford.

For four years during the 1970s Amarel was a member of the Jasons, an organization that met during the summers at La Jolla, California, to study various issues of interest to national security. In 1975 the Jasons were commissioned by Heilmeier to investigate and report on DARPA's AI program. After interviewing the leaders of the field—Marvin Minsky and Patrick Winston of MIT, John McCarthy and Edward Feigenbaum of Stanford, Allan Newell of CMU, and others—Amarel prepared a spare, two-page report indicating rather vaguely that AI was a promising field that had achieved some interesting things and would continue to do so in the future.[3]

Amarel was unusual in being an outsider at DARPA. Hitherto, IPTO directors had been selected from among those already serving in the office, often the out-going director's handpicked protégé who had apprenticed as deputy director. This ensured some continuity, not just of management procedures, but of values and ideals. In this way, the vision that had guided the office from its founding, that of human-computer collaboration, was maintained and perpetuated.[4] Amarel's apprenticeship was limited to the two months preceding Kahn's departure. Nonetheless, he brought ideological continuity to the office. He was not only an AI researcher, but a firm believer in the IPTO vision. The computer, he believed, should be an effective assistant to humans, not merely a low-level technician that followed explicit instructions. Interactive computing via time-sharing and graphics, networking, and AI—all of these were ultimately directed toward this goal. The goal of AI itself, he believed, was to mechanize the human decision-making and problem-solving processes, not necessarily in the same way humans do them, but in ways that computers can.[5]

Amarel was also a firm believer in SC even before he arrived. It was not just the overall approach of the program that he liked—simultaneous efforts in hardware, software, and applications—but the systems approach taken within the individual fields of the technology base. He believed in connection. Vision researchers, for example, had traditionally focused on narrow, self-contained problems in areas such as low-level vision, representation, and modeling. They did not work together to produce an integrated, functioning system. In short, they did not connect. SC was attempting to break that mold. Unlike some in IPTO, Amarel considered it important to continue work on the applications as well as on the technology base, not so much as goals in themselves but as means to an end. Like Cooper, he believed that applications could focus disparate research projects and pull the technology to fruition. The overall goal, in his mind, was to develop working AI systems. "I felt the Strategic Computing Program was a unique opportunity . . . to do something important in AI," he later recalled. "It had to be done carefully and well. It had to be taken seriously."[6]

Amarel brought to DARPA a number of concerns about the program and a number of ideas for changes. One concern was the model for technological transition and integration established by Ohlander, especially as applied to the SCVision program. Ohlander's model called for the component technology contractors to work separately on individual pieces of the vision system, which would then be integrated into the working New Generation System (NGS) by the integrating contractor, CMU. CMU would then deliver the system for integration into the ALV.

Amarel had little faith in the process of transition and integration. He believed that priority should be given to one contractor to develop an entire system. Rather than work on their own projects and leave systems integration to someone else, component contractors should work directly on a unified system.

In the case of vision, for example, the eleven subcontractors for the Carnegie Mellon "New Generation Computer Vision System" should be working on CMU's projected system, not on interesting subproblems that CMU would be left to assimilate. Similarly, CMU's program should be tailored to fit the Martin-Marietta ALV, not adapted after the fact to fill a niche on the demonstration vehicle. Amarel therefore encouraged CMU to focus on NavLab while ignoring the integration of technology developed elsewhere.[7] This shift in managerial emphasis was perhaps good for NavLab, but not so good for ALV.

Amarel also had doubts about another basic tenet of SC: the quest for generic AI systems. While targeted for specific applications, these systems were nonetheless expected to fit a variety of tasks. By 1986 it was becoming clear that this goal was overly ambitious. Knowledge-based systems in particular were difficult to apply outside the environment for which they had been developed. A vision system developed for autonomous navigation, for example, probably would not prove effective for an automated manufacturing assembly line. "There's no single universal mechanism for problem solving," Amarel would later say, "but depending on what you know about a problem, and how you represent what you know about the problem, you may use one of a number of appropriate mechanisms."

Amarel wanted to focus less on generic systems and tools (such as expert system shells and development tools) and more on task-specific systems, such as vision systems designed for navigation in particular environments. In expert systems, too, he downplayed the importance of shells, and he promoted instead the design of optimal architectures suited to different classes of problems. This perhaps explains why the work performed by Teknowledge and IntelliCorp on their new generation expert systems tools received so little priority or attention during the next two years. To increase the general utility of such systems, however, Amarel encouraged the development of tools that permitted the transfer of designs and systems from one problem area or environment to another.[8]

On the architectures program, Amarel thought that too much effort was going into hardware and not enough into system software, algorithms, design principles, and compiling techniques. In a subtle criticism of Squires' growing tendency to emphasize the speed and power of his machines, the new IPTO director professed a desire to see conceptual work performed on architecture taxonomies and models of computation before becoming too involved in benchmarking and performance evaluation. As in AI, he thought that the design of the architectures themselves should more closely reflect the tasks to which they would be applied. And finally, he wanted to increase efforts in developing parallel software specifically for AI tasks such as vision, natural language, and speech.[9]

The greatest concern of the new IPTO director was the structure and organization of SC. The program had been scattered across DARPA quite deliberately so that pieces of it resided in five separate offices. No one was formally in charge. The previous arrangement, the uneasy triumvirate of Kahn, Conway, and Buffalano, had not worked well. Now even that was gone. Clint Kelly appeared to exercise authority in the program unofficially, almost by default,[10] but there was no formal structure. It was not even clear who had signature authority for program funds.

It appeared to Amarel that the division of the program was leading to a lack of coordination and wasteful duplication. Nowhere was this more evident than in the showcase vision/ALV programs. In addition to Martin Marietta, EAO was funding work at the University of Maryland and Hughes for basic vision systems, route planners, and obstacle avoidance systems. Martin itself was subcontracting directly with ADS for additional planning and obstacle-avoidance software. Thus EAO was funding its own research in the technology base, while IPTO was funding its own application in autonomous navigation, the NavLab. In a period of tight budgets, this was insupportable.[11]

Some of Amarel's discontent with his inheritance at IPTO reflects the natural shift in emphasis that occurs with changes of institutional leadership. But some of it reflects the uncertain administrative legacy of SC's founders. Cooper and Kahn never reconciled their differing views of the program. The pyramid never mapped on the time line. The simultaneous launching of applications programs and the research base to support them forced applications developers to work with off-the-shelf technology while holding places and expectations for base technologies that did not yet exist. No method existed to compensate for predicted developments that failed to materialize or unexpected developments that suddenly appeared. In short, the projects did not connect.

In addition to the conceptual dilemma that dogged SC, no scheme of program management had taken root. Lynn Conway's visions of rational process had departed with her. Ronald Ohlander's new generation scheme informed work in his portfolio but eroded in the hands of his successors and failed to convince Amarel. Steven Squires's elaborate scheme of technological development, including gray coding and the maturity model, gave coherence to the architectures program but escaped the comprehension of most of his colleagues. Amarel thus found himself in charge of a program of enormous potential but uncertain process. He undertook to impose his own order upon it.

Soon after his arrival at ARPA, Amarel began to campaign for a reorganization of SC. After several false starts, he prevailed on Duncan in March 1986 to fuse IPTO and EAO into one office that would be responsible for both technology base and applications work. By April 15 Amarel and Kelly had agreed to merge their operations in an Information Science and Technology Office (ISTO). Amarel would be Director, while Kelly was to be Executive Director, the "chief operating officer." Robert Kiggans was named Deputy Director with additional responsibility for programs relating to networking (see figure 8.1).[12]

At the same time a new organization was established for SC itself. Kelly became DARPA Special Assistant for Strategic Computing, reporting to the DARPA director. The title was perhaps consolation for the fact that his formal position in ISTO was, as Amarel later admitted, meaningless.[13] Within the agency at large, however, Kelly had considerable power. He acted as DARPA's official representative and liaison on SC matters with Congress, the military services, academe, and industry; he also developed plans, priorities and budgets; he worked closely with Amarel to administer SC; and he chaired an SC Steering Committee, consisting of all DARPA office directors with SC funds. Finally, he met twice a year with an SC advisory committee, appointed by the DARPA director and consisting of representatives from academe, industry, and the military services.[14]

Responsibility for SC funding decisions was divided among the office directors and the special assistant. The special assistant established overall funding plans and priorities. The office directors proposed plans and budgets for the efforts performed in their offices, and they were responsible for spending the SC funds allocated to them. However, they had to get the special assistant's signature before they could spend anything.[15]

This arrangement suited the interests of both Amarel and Kelly. In practice, Amarel would focus on the technology base, while Kelly de-

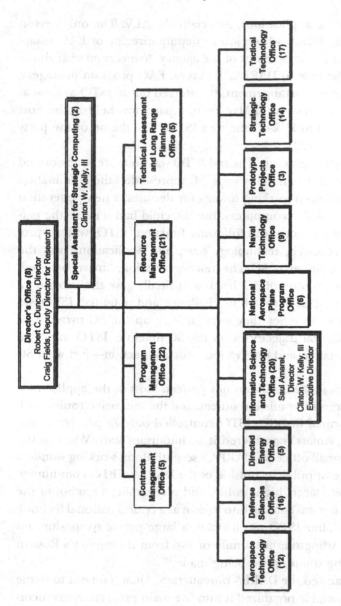

Figure 8.1
DARPA organizational chart, September 1987. *Source:* DARPA telephone directory

voted himself to the applications, especially the ALV. The only person left out was Craig Fields, whose job as deputy director of EAO disappeared. He became chief scientist of the agency, "concerned with all the disciplines of relevance to DARPA."[16] Several EAO program managers, unhappy with the new arrangement, transferred out of ISTO as soon as they could. Nonetheless, on 6 May 1986, a party was held at the Fort Myer Officers' Club to kick-off the new ISTO. The theme of the party was "a new beginning."[17]

And it was a new beginning. The old IPTO was no more, transformed by SC, its own creature. In many ways, SC represented the culmination of what IPTO had been working toward for decades: a new generation of intelligent, interactive computers. But the child had remade the parent in its own image. Form was following function. ISTO, with its programs of basic research, technology base, and applications, was the institutional embodiment of SC. The new organization, Amarel believed, could realize the promise of SC, because it finally gave the program a structure and a process to match its challenge and potential. ISTO was a conduit of ideas, whether they were moving up the SC pyramid or connecting horizontal trajectories on the SC timeline. ISTO was organized for coordination and integration—for connection—just what SC needed.

The organization of SC still was not perfect. Most of the applications were scattered across five mission offices, and the microelectronics work was being performed in DSO. ISTO controlled only 75 percent of the SC budget. Still, Amarel considered it an important start. Whereas the old IPTO was a small office on DARPA's seventh floor, working wonders with little fanfare or publicity outside of the DARPA/IPTO community, ISTO, like SC, was large, high-profile, and very public, a symbol of the fact that computing was coming into its own as a critical national technology. Just a week after ISTO was formed, a large public symposium on SC was held in Arlington, just a mile or two from the agency's Rosslyn office, showcasing the progress being made.[18]

Having reorganized the DARPA bureaucracy, Amarel set out to shape the ISTO program. He organized it into five main program areas incorporating both the basic and SC programs, each of which represented about half of the ISTO budget.[19] "Computer systems" focused on basic computational power. It included Steven Squires's multiprocessor architectures; efforts in microelectronics, especially the ongoing programs in VLSI design and manufacturing; and a new program in software technology. This latter effort, run by the newly recruited Dr. William Scherlis of Carnegie Mellon University, sought improvements in the vast array of software systems used by the DoD, with specific emphasis on the programming of parallel machines being developed by Squires. The second main area, "machine intelligence," focused on AI programs, both SC and basic efforts (such as the image understanding program). It also included a $7-million robotics program inherited from EAO. A third area, "prototype applications" (later called systems integration), addressed the IPTO portfolio of SC applications, that is, ALV and AirLand Battle Management. It also covered several of Kelly's larger robotics projects, such as the six-legged Adaptive Suspension Vehicle (the "Hexapod") under development at Ohio State.

The fourth main program area, "automation technology," subsumed the VLSI program. It continued to shorten the design cycle of basic computing technologies such as VLSI chips, circuit boards, and system modules. It also developed scalable VLSI design rules, promoted rapid prototyping of both chips and complete machines (especially multiprocessors being developed by Squires' program), and supported the use of computer-aided-design (CAD) tools for such devices. The MOSIS effort received funding from this program area, both basic and SC money. Ultimately automation technology was expected to be a model for the rapid procurement of critical parts by the defense and intelligence communities.[20] The fifth and final ISTO program "C³ [command, control, and communcation] technology" (later called "Networking/C³") focused on network development.[21]

To support these thrusts, Amarel recruited a number of new program managers. By 1987 ISTO counted fourteen technical and managerial officers, including eleven program managers. In contrast, IPTO, in the early summer of 1983, had numbered only six such officers, none of whom remained.[22] In addition to Scherlis, Amarel brought in Lieutenant Colonel Mark Pullen for computer systems; Lieutenant Colonel Russell Frew, to run the AirLand Battle Management Program; and Lieutenant Colonel John Toole in automation technology. Pullen was a particularly important if little-heralded addition to the office. An army officer with a Ph.D. in computer science and extensive training on parallel systems, Pullen was efficient, organized, and deft at handling paperwork. He brought an important measure of bureaucratic stability and organizational ability to the architectures program. Together with Squires, he helped make it the powerhouse of ISTO in the late 1980s.

Amarel also set out his vision for the new office, which was generally in keeping with that of the old IPTO. He reaffirmed that ISTO's mission was to support the needs of the DoD. Computing, he said, should "facilitate, assist, and amplify the activities of military personnel." The technology should be transparent to the user, flexible enough to meet the users' needs under a variety of conditions, and continually evolving according to the changing requirements and advancing capabilities of America's military adversaries.

The specific goals of ISTO's program were to advance American military capabilities directly and to develop the technology base that permitted the building of usable computer systems. For the first goal, he identified three ways in which advanced computers could support the military. One was to assist soldiers, sailors, and air personnel to operate military systems under critical time constraints. Sometimes the machines would take primary responsibility for action, as in the ALV; sometimes they would advise and assist the human, as in the Pilot's Associate. Amarel attached the highest priority to computer systems such as Battle Management that could support planning and command functions, again under critical time constraints. Finally, computers could help to design, manufacture, and maintain defense systems, and train the people who were involved in the work. This included the development of complex software systems and simulators. Amarel attempted to drive the entire ISTO program, both basic and SC, toward these goals.

Although ISTO's primary client would be the military, Amarel also emphasized the civilian benefits of its efforts. "It is important," he argued, "that strategic planning at ISTO be guided by a broader view of national goals for advanced computer technology—in addition to the more specific, defense-oriented, goals." What was good for the country was also good for defense—and vice versa.[23] Indeed, in 1987 Amarel proposed a new national initiative called "computer-aided productivity" (CAP), which was explicitly directed toward the civilian economy. Building on ISTO's program in Automation Technology and a planned new effort in Manufacturing Technology, the ten-year initiative would seek to improve design and manufacturing processes in industry by applying AI techniques and high performance computers. Like SC, it would consist of technology base and application components, but unlike SC it would be a multi-agency effort. Amarel suggested a total budget of $400 million, of which $125 million would come from DARPA, an equal amount from other agencies, and $150 million from industry via cost sharing.[24]

Reconciling Vision and Budget

Amarel's great ambitions for ISTO in general, and SC in particular, foundered for want of funds. Amarel never received the level of funding he had been promised by Buffalano, partly because of political forces beyond DARPA's control, but also because of the changing priorities and outlook of the DARPA leadership. When Amarel arrived at DARPA in August 1985, Buffalano was gone and James Tegnalia was Acting Director, pending the arrival of Robert Duncan. Tegnalia was a "bombs-and-bullets" man, far more interested in weapons systems than computers, more concerned with applications than basic research. He focused on military hardware projects, such as the new Armor/Anti-armor program mandated by Congress to improve the relative technological advantage of the American armored forces over those of the Soviets. He immediately began withholding money from the IPTO budget, including SC funds. For example, he reneged on a commitment Buffalano had made in the summer of 1985 to restore funds for an MIT contract. In fact, he redirected both the restoration and an additional $2 million that Buffalano had allocated from unspent SC money.[25]

Although some of Tegnalia's actions were based on his own priorities for DARPA, some appear to reflect a more widespread skepticism about IPTO in general and SC in particular. At a review of the SC program held in late 1985, shortly after Duncan's arrival, Amarel was challenged to "clarify the goals, approaches, and coherence of the [SC] program." The SCVision program appears to have been a particular concern of Tegnalia's. In postponing a decision on Columbia University's proposed effort in vision, Tegnalia noted, "I think we have too much in that area now!"[26] This concern doubtless accounts for the long delay in approving Carnegie Mellon's NavLab project. Thus, Amarel found that his first task as director of IPTO was to justify SC and petition for the release of the funding Buffalano had promised him. Though most of the funds were eventually returned, as late as January 1986, $5.5 million of SC money budgeted for the current fiscal year was still being withheld.[27]

Tegnalia would later continue to delay signing off on requests for SC funds and to favor the applications efforts over the technology base research. The blockage became so pronounced that Senator Jeff Bingaman, DARPA's chief supporter in Congress, publicly expressed alarm in the spring of 1986. To make matters worse, the Director's Office delayed allocation of funding for SC beyond the upcoming FY 1987, making it very difficult to conduct long-range planning for the program.[28]

As it turned out, these were only the first of Amarel's problems and by no means the worst of them. Shortly after his arrival in Washington, Congress, in a budget-cutting mood after several years of military expansion and growing budget deficits, ordered a reduction in R&D funding for the DoD. DARPA lost $47.5 million.[29] Directly on the heels of that decision, in December 1985, President Reagan signed the Gramm-Rudman-Hollings Act, which mandated automatic, across-the-board spending cuts to balance the federal budget by 1990. The law specified a budgetary reduction of $11.7 billion for FY 1986, already underway; half was to be absorbed by the DoD. The impact on DARPA was compounded by President Reagan's decision to exempt expenditures for military personnel and the Strategic Defense Initiative, whose shares of the reductions were absorbed by the rest of the department.[30]

The combined cuts devastated DARPA. The technical offices immediately lost $121 million. The research offices—Information Processing Techniques, Engineering Applications, Defense Sciences, and Directed Energy—took the brunt of the reductions, averaging 27 percent each, while the mission offices—Strategic Technology and Tactical Technology—fared far better. IPTO and EAO (the future ISTO) together absorbed 62 percent of the total cuts.[31] In January the combined IPTO/EAO budgets for 1986 (SC and basic programs) stood at $244.5 million; by July the budget of the newly formed ISTO was $169.4 million, a total reduction of $75.1 million, or 31 percent.[32]

The SC portion of the ISTO budget also fell by almost a third, from $123.5 million to $83.4 million.[33] The same was true for the SC program as a whole. The original plan had called for $95 million to be spent on the program in 1985, rising to $150 million the following year and holding steady after that. As late as April 1985, Kahn had optimistically proposed raising the funding levels above that $150 million baseline to allow for inflation and new starts, reaching a high of $284 million by 1991. Largely because of the transfer of gallium arsenide and other strategic programs to the Strategic Defense Initiative Office (SDIO), however, the actual expenditure for fiscal 1985 was only $68.2 million, a reduction of 28 percent.[34]

For the following year, the administration had asked Congress for $142 million for SC. The Senate Armed Services Committee recommended cutting $42 million out of the appropriation, but a coterie of sympathetic senators, led by Bingaman, Edward Kennedy, and John Glenn, succeeded in restoring the funds. As late as January 1986 it appeared that DARPA would get the full request.[35] Two months later, as a result of all the cuts, the allocation was down to $116 million. Further-more, $12 million was earmarked by Congress for a new computing cen-ter at Syracuse University. Only $104 million was effectively available for SC.[36]

Amarel hoped to get the SC budget back up to $124 million in 1987 and to $145 million by 1989, but DARPA's priorities were changing. The Reagan military buildup was effectively over. DARPA's budget was level-ing off. Its leaders had other interests besides SC. The Director's Office, under Tegnalia's influence, continually raided the computing budget for pet projects such as Armor/Anti-armor (a $40 million program in 1987),[37] taking away all of Amarel's discretionary funding. Duncan him-self was particularly fond of a new program initiated by his predecessor to develop the National Aerospace Plane, a vehicle that would speed air travel (and perhaps even serve as an orbital vehicle) by flying into the stratosphere. This program consumed $100 million in 1987.[38] Amarel spent much of his time during his two years at DARPA explaining and defending computer research to Tegnalia and Duncan, often in vain. In 1987 the allocation for SC dipped further to $102.3 million. Not until 1988 did the SC budget rebound to $132 million. This would be the highest yearly allocation SC ever received; afterward, the High Perfor-mance Computing program split off from SC, taking a good chunk of money with it. But this was well after Amarel was gone.[39]

The impact of the cuts on SC was profound. Some insiders believe that they account for the failure of the program to achieve more than it did. The cuts eliminated any reserve funding for new starts or opportu-nistic spending and forced a slowdown in many ongoing efforts. The architectures program suffered the most, its 1986 budget dropping from nearly $40 million in January to less than $25 million by August. By the following January, its FY 1987 budget was down to an anemic $20.7 mil-lion. The AI program was cut also, though far less drastically, from $21.5 million to $19.8 million for 1986. It then rose slightly the following year to $22.9 million.[40]

Different program managers reacted differently to the situation. Some, such as the AI program managers, preferred to spread the pain by cutting all programs equally. Carnegie Mellon, for example, acquired one autonomous vehicle instead of five. Stephen Squires was much more selective. Not wishing to delay the launching of his program, especially the stage-3 prototyping efforts, some of which had already achieved great momentum, he cut back instead on the efforts to develop generic soft-ware, development tools, and performance modeling. Thus, while the program to produce prototype architectures and scalable modules was cut by less than 4 percent, the rest of the program was cut 86%.[41] Squires also performed what he later described as a "ruthless triage." He categorized and prioritized projects by type of contractor: universities, commercial firms, or defense industries. He first terminated all defense industry projects that did not involve some other part of the DoD. He then shut down marginal industrial projects and finally marginal academic projects. In this manner, he sought to preserve projects having the greatest generic potential in a broad range of applications.

Victims of the pruning included Columbia University's DADO and NonVon architectures and Texas Instruments's optical crossbar. Squires, however, did manage to save some programs by obtaining funding from the Strategic Defense Initiative Office in exchange for his assistance with the office's troubled architectures program, "a $20 million seat on the board," he called it. SDIO provided $15.2 million in FY 1987, raising the contribution annually for a total of $209.2 million by FY 1992. This money funded some software and benchmarking projects of interest to both DARPA and SDIO, as well as some of the stage-2 and stage-3 prototyping efforts, including those by IBM, Encore, Carnegie Mellon, and Harris Corporation.[42] If SC was not a cat's-paw for the Strategic Defense Initiative, as its critics had once claimed, it at least became a helpmate.

In spite of this external funding, the budget cuts severely slowed the architectures program. Stage-1 and stage-2 efforts were cut or delayed, with a resulting slowdown in the development of the second-wave architectures. The purchase of full- and reduced-scale prototypes was severely curtailed, arresting the development of software for these experimental systems; the importance of this loss can hardly be exaggerated. Some of the applications programs, such as the Battle Management and ALV programs, failed to get the parallel machines they had been expecting. The microelectronics program also suffered. All the programs run by the Defense Sciences Office, including those in wafer-scale integration, optoelectronics, and advanced packaging—totaling about $6 million— were cut for 1987.[43]

Strategic Computing at Age Four

Amarel returned to Rutgers University in the fall of 1987 after a two-year leave of absence. For all his achievements, the tour at DARPA had been a disappointment. Instead of inaugurating a new era of increased focus and efficiency for DARPA computing research, Amarel had poured his time and energy into a rearguard action to protect the program's funding. Decisions about what to save and how to save it devolved from his office to the program managers. All the while the ability to run a coordinated, integrated program eroded. And, unbeknownst to Amarel, the destabilization of the program was producing a new center of gravity within SC.

In spite of these centrifugal forces, Amarel had held the program together during unprecedented adversity. Most importantly, he had formulated a new organizational arrangement, ISTO, that survived his departure. Just repairing the damage done by the old IPTO-EAO schism would have been a significant contribution. But Amarel also left 153 separate projects under way in 91 institutions. Funding for these activities spread smoothly across DARPA offices in just the way that Cooper had envisioned at the outset. The original programs were nearing the end of their first phases. As might be expected, some had much to show for their efforts, others less so.

The applications programs were among those demonstrating tangible progress. After the schedule slipped in 1986, Martin Marietta had overhauled the ALV and was preparing to install the Mark II architecture and algorithms that would carry it successfully through its obstacle avoidance tests in 1987. Elements of the ALV's vision and reasoning systems were transferred to two teleoperated vehicles being developed under the Advanced Ground Vehicle Technology (AGVT) program, one based on the tracked M113 being developed by FMC, the other a Commando Scout by General Dynamics. Both vehicles were run successfully on the Denver test site during 1986.[44]

Work also progressed satisfactorily on the Naval Battle Management programs. Program manager Al Brandenstein's emphasis on making the Fleet Command Center Battle Management Program (FCCBMP) a functional test bed run by the navy was paying off. During 1985 the computing hardware was installed at Commander-in-Chief, Pacific Fleet (CINCPACFLT) at Pearl Harbor, Hawaii, and the test bed itself was completed and run up in March 1986. Since 1984 work had been proceeding on one of the five cooperating expert systems of FCCBMP, the Force Requirements Expert Systems (FRESH), which would monitor the Navy's ships and help allocate them for various assignments according to their readiness.

In June 1986, FRESH Prototype One was installed in the test bed, and two months later it was demonstrated successfully to DARPA and the navy using the IRUS natural language generator developed by BBN, SAIC, and the Naval Ocean Systems Center (NOSC). At that time the IRUS vocabulary recognized 5,000 words, including proper names and naval terms; it successfully comprehended and responded to queries in terms usually used by the operations staff. By 1987 the enhanced system was performing in ninety minutes tasks that would usually take the CINCPACFLT staff fifteen hours. By then the navy employed the system routinely to monitor the readiness of its ships in the Pacific. Of the other four expert systems, work on CASES was begun in August 1987 (nearly a year behind schedule, probably because of budgeting problems); work on CAMPSIM, OPGEN, and STRATUS would not be started until the following year.

Meanwhile, the Combat Action Team (CAT) was installed for testing on the *Carl Vinson* in early 1986. Fleet exercises in May convinced the navy that it was indeed a useful system. An improved version was installed in August and the *Carl Vinson* was sent out on a six-month tour of duty in the western Pacific. The system showed real promise in spite of its inability to operate in real time.

Its successes notwithstanding, CAT also revealed some of the difficulties researchers faced when working directly with the services. System developer Carnegie Mellon established a good relationship with NOSC, which sent its personnel to Pittsburgh for extended stays to exchange information and ideas. Twice a year the latest version of CAT went from CMU to NOSC for adjustments and additions to its knowledge base. But much of the knowledge domain was classified, and the Carnegie researchers did not have—or want—a security clearance. Thus, the researchers were not allowed to see the contents of the knowledge base they were attempting to develop for the navy. They had to guess at what kind of information would be required and use false data in shaping the latest version of the system. When NOSC returned the upgraded system to CMU, it stripped out all classified data, thus preventing the developers from seeing what the information was and how it was being used. This was hardly an ideal method for developing a practical, functioning expert system.[45]

The PA program had gotten off to a slower start than either the ALV or the naval battle management program. Following the program definition phase, when it was decided exactly what the PA system should do, two contractor teams were selected in February 1986. Lockheed-Georgia led one team composed of eight companies and universities. McDonnell Aircraft headed the other, three-company team. Lockheed was awarded

$13.2 million, and McDonnell $8.8 million; both companies contributed 50 percent of the cost in anticipation that the work would give them an advantage in the competition for the lucrative, next-generation, Advanced Tactical Fighter contract. The program was managed at DARPA first by John Retelle, then, after 1987, by Llewellyn Dougherty; the work itself was supervised closely by the Air Force Wright Aeronautical Laboratory (AFWAL). By the fall of 1987, both teams were busily preparing for their first make-or-break demonstration, to be held the following March.[46]

By 1987 four new applications programs had been added to the original three. They bestowed upon SC a mixed blessing. Appearing later in the development process, they naturally exploited new and more complete understanding of what the technology base could really deliver. They provided, in short, a better match between promise and payoff. On the other hand, they necessarily diluted the funds available to develop the original applications. In a very real sense, they diminished the likelihood that those stalwarts of SC would come to fruition.

The first new application, AirLand Battle Management (ALBM), took its name from the army's war-fighting doctrine in Europe. Like the naval version, it applied expert-system and natural-language technology to the problem of command and control on the battlefield. If successful, it would reduce by 97 percent the time it took an army corps staff to prepare a battle plan; instead of three days, the plan would require just two hours. In an apparent bid to please the customer, DARPA salted the proposal with the clever acronyms that the military services so often employ. FORCES and STARS were the two major software systems. As in Navy Battle Management, FORCES, the planning system, would consist of three related expert systems: MOVES(C), to help a corps' staff plan its maneuver; MOVES(D), to do the same for a division staff; and FIRES(C), to plan a corps' artillery fire support. TEMPLAR, a Tactical Expert Mission Planner, to be developed by TRW and the Rome Air Development Center, would assist NATO with air missions in Central Europe. STARS operated on a different level, providing an expert system development tool specially tailored to the army's needs.

Unlike the ALV, ALBM attracted immediate interest from the army, perhaps because it met an obvious service need. It could not only help in conducting military operations, it also could be used for training as well, a capability of particular interest to the army. In 1985 the army agreed to support and cofund the program with ARPA. An RFP was published in January 1986, and by the fall the MITRE Corporation, Cognitive

Systems Inc., Advanced Decision Systems, and Lockheed Electronics Company were selected to work on various aspects of the system. As with the navy battle management program—and unlike ALV—the army played an active part in the program and worked closely with the contractors using the "rapid prototyping" development strategy.

AI "cells" were established at Fort Leavenworth, Kansas; Fort Sill, Oklahoma; and elsewhere to compile the expert knowledge needed for the system. The various prototypes were made available for testing by the army as soon as they were ready. One such experimental prototype expert system, ANALYST, developed by MITRE for use by the 9th Infantry Division at Fort Lewis, Washington, received enthusiastic reviews. Fruitful cooperation with the army was further abetted by the assignment of Major Michael Montie, who moved to ISTO in the summer of 1986 from his previous assignment to the Tactical Technology Office. In addition to being a career service officer familiar with the needs and culture of the army, Montie also came from a DARPA office accustomed to dealing directly with the military services and meeting their needs.

Two other new applications continued previous DARPA and service programs, applying image understanding to intelligence analysis. The Advanced Digital Radar Imagery Exploitation System (ADRIES), jointly funded with the army and air force, analyzed the data produced by Synthetic Aperture Radar (SAR). These high-resolution radar images could reveal tactical targets on the battlefield, but they generated more data than intelligence officers could process in real time. ADRIES addressed the problem with INTACTS, to screen the SAR imagery for military targets, and MISTER, to recognize exactly what the targets were, using the model-based vision techniques being developed by DARPA's Image Understanding program. Like the ALV, this program was to draw upon the results of developments in vision, knowledge-based systems, and planning, and it was expected to apply the new parallel architectures to the problem, including Warp, Butterfly, and Connection Machines.

As in the battle management programs, ADRIES emphasized getting system prototypes into the hands of the eventual end users, who would test them and recommend changes and improvements. Begun in 1983, before SC was announced, the program completed its first phase in March 1986 with the development of an overall system architecture and early prototypes of the five main processing subsystems. Both INTACTS and MISTER were due to be completed in 1991.[47]

A second follow-on application began in July 1985. SCORPIUS, the Strategic Computing Object-directed Reconnaissance Parallel-processing Image Understanding System, screened aerial and satellite reconnaissance photographs for objects of interest such as ships, tanks, buildings, and depots. In this phase, it concentrated on identifying submarines at naval bases and bombers at airfields. A continuation of programs begun as early as 1981, SCORPIUS (like ADRIES) was expected to benefit heavily from the object-recognition techniques developed by the IU program and SCVision, and by the development of parallel architectures, such as the Butterfly and the Connection Machine. In keeping with the spirit of SC, the SCORPIUS program was not expected to deliver a finished, working product, but merely demonstrate the feasibility of automating the process of photo-interpretation. With joint funding from the CIA's Office of Research and Development, Major David Nicholson ran the program out of the Tactical Technology Office.

The last applications project added to SC, "smart weapons," had about it the look and feel of James Tegnalia. It applied machine intelligence and advanced computer architectures to autonomous navigation. Unlike ALV, however, the goal of the program was explicitly military. These machines would locate and attack targets far behind enemy lines, just as Computer Professionals for Social Responsibility had claimed that ALV would end up doing. One "smart weapon," the Autonomous Air Vehicle (AAV), would fly into enemy territory and launch both "smart" and "dumb" bombs at a target. The second, called the intelligent munition (IM), was the smart bomb itself. Both platforms required massive computing power in a small package, much smaller than the 6–15 cubic feet required by the ALV. SC promised a "soupcan computer" that would deliver 50–100 MFlops in a volume of less than 20 cubic inches.[48] The program began in the summer of 1986 with the award of contracts to seven teams headed by major defense contractors such as General Dynamics, Hughes, Lockheed, Martin, and Northrop.

The technology base contractors also reported significant progress. In January 1986 Carnegie Mellon had demonstrated a speech system that could recognize 250 words spoken continuously regardless of the speaker. By 1987 this system could cope with a 1,000-word vocabulary with 95 percent accuracy, operating at 10 times real time on a parallel system. Texas Instruments produced a robust 200-word connected-speech-recognition system that was installed on F-16 fighters for operational testing by pilots.

In Natural Language, BBN's IRUS system, with a vocabulary of 4,500 words and 1,200 domain concepts, received a favorable response in 1986 when it was installed in a Battle Management Program test bed. By 1987

BBN IRUS was becoming JANUS, a system designed to understand vague and confusing queries. JANUS was successfully demonstrated in the Air-Land Battle Management Program in the fall of 1987. Meanwhile, New York University and Systems Development Corporation cooperated on the PROTEUS system, intended to comprehend written text. In 1986 the first version successfully processed naval casualty reports (CASREPS) about equipment failure, in this case air compressors, and work began on PROTEUS-II. Advances in Vision allowed Carnegie Mellon's Nav-Lab vehicle to run up and down Schenley Park at low speeds. In expert systems, Teknowledge completed the first prototype of ABE during 1987, while IntelliCorp, having implemented its state-of-the-art Truth Maintenance System in 1986, was working to incorporate it into KEE.

Parallel architectures made even more dramatic progress. The Warp and Connection Machines stood out among the first generation of parallel machines, but other projects were also doing well. Generic software and programming tools advanced more slowly, but showed some promise. Carnegie Mellon's Mach operating system for multiprocessors sought to restore the simplicity and flexibility for which Unix had once been noted. Like Unix, it would adapt to a variety of platforms and break down a program into separate parts (known as "threads") that could be executed independently and concurrently. By 1987 Mach was running on the VAX, the Encore Ultramax, the RP3, and other machines, raising hopes within DARPA that it would be available in 1988 as a substitute for Unix.[49]

Meanwhile, Squires and Pullen were working on the next wave of architectures. Rather than an incremental advance of the first generation, Squires sought new technology demonstrating different computational models. He aspired to nothing less than parallel machines performing at the level of traditional supercomputers. He focused on single-chip implementation of processors, memory, and communications and on improved packaging both to dissipate heat and to limit size and weight for defense applications. In accordance with the Maturity Model, this next generation underwent theoretical exploration even while the first wave prototypes were being built. For example, Professor Charles Seitz of the California Institute of Technology, the designer of the hypercube topology for parallel systems, worked on a new routing chip to speed communications within a machine. Squires anticipated the "second wave" prototypes of this design in mid-1989 and commercial versions a year later.[50]

By 1987 Squires realized that many of the parallel architectures initially intended for symbolic processing, including the Warp and Connection Machine, had great potential for general-purpose and numeric applications. He therefore encouraged upgrading of first-generation machines to adapt them to this purpose. Even as General Electric was manufacturing the PCWarp, Carnegie researchers were at work on a new version in which each processor was implemented on a single, custom-made VLSI chip produced by Carnegie's new partner, Intel.

The "integrated Warp," or iWarp, aimed at gigaflops performance while occupying just a tenth of the volume of Warp. DARPA also provided funds for Thinking Machines Corporation to upgrade the original Connection Machine into the second model, the CM2, which was introduced in April 1987. While retaining the same basic architecture, the CM2 had twice the clock speed (8 MHz), it used 256K memory chips in place of the old 4K chips, and it sported improved processors with floating-point capability for improved numerical computation. A communication system provided much more flexible and programmable interconnections among the 64K processors, and the data vault vastly increased the machine's storage capacity. By 1988 CM2 achieved a sustained 2–10 gigaflops on large problems. DARPA placed at least sixteen CM2s with various organizations over the next two and a half years, although some of the machines were paid for by the navy or SDIO.[51]

DARPA also solicited new architecture proposals for second-wave systems and succeeding generations. A broad agency announcement (BAA) published in April 1987 garnered thirty-eight responses. Fifteen contracts worth $55.7 million were awarded by September 1988. Contracts went to ten different institutions, including seven universities and three industrial organizations.[52]

The microelectronics program, gutted by the transfer of GaAs research to the Strategic Defense Initiative, had turned its focus to connecting technologies. It was supporting preliminary programs in optoelectronics that promised accelerated communication, computer memory access, and processor throughput, in both silicon and GaAs circuits. In packaging technology, DARPA had concluded that it could leave "conventional packaging" to the VHSIC program and turn its attention instead to the "high density packaging structures" required by the "Soupcan computer." SC had enhanced infrastructure for the DARPA community by distributing machines at a total savings of $3 million, expanding access to the ARPANET for SC researchers, supporting the introduction of Common LISP programming language to universities and industry, and implementing VLSI designs through the MOSIS program. Chips of 1.2 micron CMOS were already delivered; still smaller devices and wafer scale integration were in the works.[53]

By the time Amarel left DARPA in the fall of 1987, SC was proceeding as its creators had envisioned. Difficulties were becoming apparent, particularly in the effort to achieve generic AI systems and algorithms, but researchers were achieving favorable results in the attempt to apply the technology to specific tasks. Speech recognition and natural language understanding appeared particularly promising. The architectures program was making especially good progress; the quest for scalable parallel systems appeared to be paying off and promised to go far, if the money held out.

More important was what SC had set into motion: building up a technology base by interesting industry and the military services in machine intelligence technology; beefing up the research programs, not just at DARPA's traditional circle of universities, but in an expanded community that now numbered twenty-eight schools; and supporting the training of graduate students and researchers in industry. In fall of 1988 DARPA estimated that it had helped award 120 master's degrees and 100 doctorates in computer science.[54] Finally, SC had transformed IPTO from a small, obscure office operating in the shadows into the large, visible, high-profile Information Science and Technology Office, the institutional embodiment of SC. This was the legacy that Amarel passed along to his successor, Dr. Jacob T. Schwartz.

Waiting for the Wave

Schwartz, from New York University (NYU), harbored a very different perspective on SC and the IPTO/ISTO philosophy. He had trained as a mathematician, but, in his words, "got sidetracked . . . into science and never recovered." His many interests included programming-language design, computer architectures (especially multiprocessing), and, contrary to what many people believe, AI and connectionism. Schwartz first met DARPA when applying for SC funds for a parallel machine he was helping to develop at NYU, the Ultracomputer Project. He became acquainted with Kahn and later with Craig Fields, then acting director of the agency, who invited him to join the SC advisory committee. Schwartz assisted Fields with the search for a successor for Amarel. When Field's chosen candidate backed out suddenly, Schwartz found himself being recruited. He accepted the position "out of personal regard" for Fields, whom he admired deeply. Unlike the DARPA regulars, he always referred to his superior as "Dr. Fields" and never as "Craig."[55]

Like Amarel, Schwartz was an outsider. He entered DARPA as an office director without any apprenticeship or inside experience. Unlike Amarel, however, Schwartz spurned the belief that ISTO, with its relatively small budget (SC notwithstanding), could transform the world as its program managers often claimed. Just before coming to ISTO, Schwartz was told by a high IBM official that the industry giant had to spend a billion dollars a year just to keep up with Japanese developments in mainframe computers. Next to such industry expenditures, Schwartz believed, the government programs were but "a drop in the bucket."[56]

More fundamental was Schwartz's basic philosophical disagreement with the ISTO and SC approach to technological development. Schwartz called this the "swimmer model." DARPA swims in a fluid medium, choosing a destination and struggling toward it. Perhaps the water will be rough and rife with currents, but the swimmers remain focused on their objective and continue until they achieve it. Thus DARPA, having decided that it wanted to achieve machine intelligence, was swimming toward that objective, regardless of the currents flowing against it.

Schwartz contrasted this outlook with the "surfer model." Surfers are far less active agents. While swimmers move toward their goal, surfers ride the waves that come along. DARPA's role, as Schwartz saw it, was to watch closely for the waves, the technological trends and opportunities that promised to carry DARPA's relatively modest funding to significant results. In other words, Schwartz would focus on those areas in which his money could have the greatest leverage. He liked to call this approach "waiting for the wave." While swimmers could easily be swept out to sea, challenging forces they were powerless to overcome, surfers would always make it back to shore.[57]

Schwartz's views transformed ISTO policy. Saul Amarel had always considered ISTO's SC to be an integral part of the ISTO program, a complement to basic research, but he distinguished SC and basic money in his program and budget reviews. Schwartz ignored the distinction in his own program reviews, treating all funding allocated to ISTO, from whatever the source, as a single pot of money.[58] By the summer of 1988 the director had reorganized the ISTO program into four major areas.

The first, with the largest share of the funding, would eventually become High Performance Computing. It included essentially all the architectures, microelectronics, parallel software tools and algorithms, and infrastructural efforts supported by ISTO. The second area was AI &

Robotics and Applications, which incorporated all programs relating to machine intelligence (both basic and SCI), including the application efforts such as AirLand Battle Management and SCORPIUS (ALV had been cut by this time). The third area was Networking/C^3, which supported networks, database technology, and other programs more explicitly directed towards military communications. The final area was Manufacturing. Schwartz himself was much interested in the application of computing technology to manufacturing, especially in computer-aided design (CAD);[59] he promoted a small program ($6.2 million in 1989) to promote the use of CAD in various applications.[60] If the SC pyramid and timeline were not already in a shambles, Schwartz's reorganization rendered them thus.

Schwartz ignored SC. In his view, it embodied the Swimmer Model, and its emphasis on machine intelligence pitted it against the current. SC, according to Schwartz, was an optimistic concept sold to Congress in the hysteria over Japan's Fifth Generation. Schwartz objected to AI for pragmatic reasons. He did not share the philosophical misgivings of such leading critics as Hubert Dreyfus and Joseph Weizenbaum; theoretically, he believed, AI was possible and promising. Practically, however, it was nowhere near ripe. Computing still lacked the power to run the necessary algorithms fast enough. Even more importantly, the fundamental concepts that underlay the field needed refinement.

In particular, the current algorithms did not scale; new ideas and new approaches were needed. DARPA could not guarantee such new, fundamental ideas in AI merely by throwing money at the problem. Schwartz compared the situation to infant mortality under the Chinese emperors: however much money they spent on the "best doctors in the universe," the emperors could not prevent their children from dying from diseases they did not understand. "If you don't have the idea," said Schwartz, "you can't create it by a crash program."[61] Schwartz also likened the quest for AI to rowing to the moon. It is possible to get to the moon, but not by rowing. "No matter how many galley slaves you put on the galley and no matter how hard you beat them, you are not going to row to the moon. You have to have a different approach."[62]

When Schwartz turned his jaundiced eye to DARPA's AI programs, he found what he had expected to find: dead ends. He traveled around the country visiting the contractor sites and forcing the AI researchers to justify their funding. Researchers at Carnegie Mellon recall Schwartz asking them in advance of his visit to Pittsburgh to show the progress they had made in the first three years of their SC funding. They prepared a detailed presentation quantifying their progress in speech recognition, autonomous navigation, and natural language processing. At the end of the half-hour briefing, led by the distinguished professors Allen Newell and Raj Reddy, Schwartz announced that he saw no improvement whatsoever. The astounded researchers, proud of their efforts, asked what he meant, and he replied that he "didn't see any new ideas." They tried to explain the theories and ideas underlying their work, but to no avail. Schwartz wrote an unfavorable evaluation of the CMU program. One researcher concluded, "We were going to get an 'F' regardless of what we did at that point."[63]

Not all researchers recall such a heavy hand. Edward Feigenbaum, for one, thought he was treated fairly by Schwartz and maintained his friendly relations with the director.[64] Nonetheless, the budget axe fell heavily on AI and robotics at DARPA. In Amarel's last year, 1987, ISTO's combined basic and SC budgets for those programs totaled $47 million. By 1989 Schwartz's budget in those areas was less than $31 million.[65] The basic program in AI was reduced from the planned $30.6 million to $21.8 million, a cut of $13 million from the year before.[66] In spite of his disastrous review at Carnegie Mellon, Schwartz increased funding for the Speech program, which was making important progress in its efforts to create standard benchmark tests. He also raised the budget of CMU's NavLab, in part because of the tremendous prestige of Professor Takeo Kanade.[67] The other programs, however, fared less well.

Knowledge-Based Systems (Expert Systems) represented what Schwartz considered the flawed AI paradigm in its purest form. His visit to the sites of these programs, recounted in chapter 6, made his audience at CMU look warm and sympathetic by contrast.[68] Amarel had steadily increased funding for expert systems, from $3.6 million in 1986 to $5.2 million in 1988. Schwartz slashed that figure to $3 million by 1990.[69]

Meanwhile, Schwartz began funding several new areas in machine intelligence, reflecting his desire for new approaches. Neural Modeling drew on new, biologically based paradigms from neuroscience, reflecting the ISTO director's growing interest in the revival of connectionism.[70] Another program, Machine Learning, applied case-based reasoning (the particular expertise of AI program manager Bob Simpson) and neural net research. A third program, Computational Logic, attempted to measure progress in AI subfields by developing benchmarks, and to explore the value of AI techniques for conventional software.[71]

Schwartz also cut funding for the applications programs in ISTO, eliminating the ALV entirely. AirLand Battle Management appears to have received his support, partly because the army was so enthusiastic about the program, and partly because it was already in the process of transitioning over to the army and would be out of DARPA by 1990 anyway. SCORPIUS, too, was on its way out, with ISTO funding for that program, only $1 million in 1989, to be cut completely by 1990. Funding was increasing for a program in "advanced battle management."[72]

Schwartz's cuts received strong support from the DARPA director's office, especially deputy director of research Craig Fields. The AI program managers, Allen Sears and Robert Simpson, now found few in either ISTO or the front office who shared what Simpson considered the "AI vision."[73] Indeed, Fields pressed Schwartz to make further cuts in the AI and applications programs.[74] Raj Reddy, the president of the American Association for Artificial Intelligence, recalled getting "a call for help" from Simpson and Sears in November 1987.

"We are being asked some tough questions by the front office," they said. "What are the major accomplishments of the field? How can we measure progress? How can we tell whether we are succeeding or failing? What breakthroughs might be possible over the next decade? How much money will it take? What impact will it have? How can you effect technology transfer of promising results rapidly to industry?" They needed the answers in a hurry.[75]

These questions resemble those asked by George Heilmeier a decade earlier. They bespeak a hard-minded pragmatist, skeptical of the utopian visions and failed predictions that had infiltrated DARPA from the AI community. Schwartz, like Heilmeier, wanted proof, performance, and products. Programs that failed to produce would be trimmed to free up funding for those that could.

Robert Simpson took special pains to resist Schwartz's attack on the AI programs. He argued forcefully for AI at every opportunity, both within DARPA and outside the agency. He compiled a list of more than eighty AI systems that were either operational or prototyped. He insisted to all who would listen that the impact of AI technology was out of all proportion to the amount of money spent. "Give me one B1 bomber or one B2 bomber," he said, "and I'll fund AI research for the next decade."[76] These arguments had little impact on Schwartz, but they did convince Craig Fields to restore much of the funding for the basic AI program for 1991.[77]

In Schwartz's view, architecture, not AI, rode the wave in the late 1980s. He believed that high-performance computing, especially parallel architectures, held out more promise than any other SC program.

Schwartz himself had a strong research interest in that subject. At NYU, he had formulated the key concept behind the parallel Ultracomputer project called the "omega network," which provided more pathways between processors and memory than other parallel topologies. In 1984 he chaired a panel for the Office of Science and Technology Policy on computer architectures that endorsed DARPA's SC efforts in that area, though it emphasized the importance of general-purpose architectures. "It remains difficult to derive any specific supercomputer designs from the requirements of artificial intelligence," the panel reported, and "the division of supercomputers into two distinct genuses of 'scientific machines' and 'artificial intelligence machines' is artificial."[78] It was during the two years of Schwartz's directorship at ISTO that the drive for general-purpose high performance computing came to outweigh the quest for machine intelligence.

Schwartz transformed computer research at DARPA, not just by slashing artificial intelligence but by elevating architectures. He cared little for the conceptualizations undergirding the SC pyramid or even the time lines, because in his view AI could not yet support machine intelligence. Funneling advances up the pyramid availed nothing if none could rise above the AI ceiling. Better, therefore, to invest in technology that was progressing and ride that wave for all it was worth. If that meant scrapping SC, then so be it.

As best he could, Schwartz promoted Squires's architectures program, providing it with what little extra funding he had and reorienting much of the rest of the ISTO program to support it. Schwartz claimed in retrospect that this transformation had already begun when Schwartz arrived at DARPA. Squires had gained influence in the waning days of Amarel's tenure and used it to advance architectures over other program areas. For example, the BAA published in April 1987 was intended to promote the software component of the architectures program, but Squires used it instead to launch the second generation architectures, funding hardware prototypes and general systems efforts such as Intel's Touchstone project.[79]

Squires's success to date fueled his optimism and ambitions. His first-generation machines were surviving in the marketplace. The second-generation machines promised to carry the parallel paradigm further. And new ideas and technologies were appearing on the horizon that would ultimately form a third generation. Squires now concluded that general systems (as opposed to the specialized, task-oriented systems envisioned by the original SC plan) could achieve teraFLOPS capabilities (one trillion floating-point operations per second) by the early 1990s. Increasingly, this tantalizing prospect became his goal. A month after Schwartz's arrival at ISTO, Squires proposed a teraOPS computing technology initiative. This five-year program would follow the methodology of SC. Simultaneous efforts at all levels of technology from microelectronics to applications would feed advances up the pyramid while applications pulled the technology base. This looked like SC, phase II, without the AI. Phase I was nearing completion. This could potentially serve as a follow-on initiative which might deflect congressional inquiries about the achievement of machine intelligence. Squires proposed a budget of $30 million in 1988, rising steadily to a sustained level of $110 million in 1991.[80]

Plans, however, were no substitute for funds. By 1988 the architectures program was desperately short of money, having fared badly during the budget cuts of the Amarel era. Squires had hoped to spend $40 million on his program in 1986, though $10 million was potentially to be held pending the final cuts mandated by the Gramm-Rudman Act. The cuts proved to be worse than expected. The architectures program received only $24 million for 1986. Amarel and Squires had hoped to get it back up to $33 million in 1987 and $39 million in 1988, but by 1987 these numbers were down to $20 million and $25 million, respectively. Essentially, Squires' budget had gone flat at the very time his program appeared on the verge of a breakthrough.[81]

Schwartz sought to support Squires's efforts with the limited budgetary flexibility he had. Redirecting funds from the applications and the AI program, Schwartz put the money into parallel algorithms and software development tools, funded VLSI accelerators, promoted network access to parallel machines, and provided some infrastructural funding to purchase machines. That left little for redirection. The new parallel machines developed in Squires's program had to get into the hands of users, but only $5 million could go to this purpose in 1989, half the previous year's investment. And funding for parallel machine development itself had stagnated around the $20 million range. In a program review of August 1988, Squires estimated that he needed $32 million, including $17 million for the development of second-generation machines and $10 million for the purchase and distribution of first-generation machines. Money still flowed from SDI, including $3.8 million for computing infrastructure in 1988, but Squires doubted the reliability of that source. SC alone could not support the program he envisioned.[82]

Strategic Computing 2

At the end of 1989 it appeared that SC might disappear altogether. SC never had an obvious ending point, and the work in software, AI in particular, was obviously not going to achieve any headline-grabbing breakthroughs anytime soon. Fields expressed concern about this as early as the fall of 1988, as the program was entering its sixth year and was receiving special attention from Congress.[83] There was growing pressure on DARPA to show some result for SC. Congress wanted products, not promises.

In addition, there was continuing financial pressure on SC. Its allocation for FY 1988 was higher than expected, at $132 million; but then it declined again, so that by 1990 it had dipped down below $100 million.[84] Fields feared that it would appear that DARPA was "being paid twice" for SC and ISTO's basic program, especially intelligent systems (the basic AI program).[85] The distinction had always been somewhat fuzzy; now, with increased congressional scrutiny, it had to be made more explicit. SC remained a programmatic entity within DARPA, but no mention was made of it either in the director's statement or subsequent testimony. It appeared to be sinking into oblivion.

Steve Squires and Craig Fields rescued it. Fields had done quite well for himself since his appointment as Chief Scientist of the agency in 1986. The following year both Duncan and Tegnalia left DARPA, Duncan to become Director of Defense Research and Engineering, Tegnalia to return to private industry in advance of the new, restrictive ethics laws that would have shortly constrained his employment options. For a number of months, Fields had virtual charge of the agency—it was during this period that Schwartz was hired—until the arrival of the new director, Raymond Colladay, in early 1988. Colladay remained for little more than a year, leaving in May 1989. Fields then assumed the directorship, the first insider to rise to that office since Steven Lukasik (1971–1974).

Fields had strong ideas about the direction of the agency, and he was the first DARPA director since Cooper to take a keen interest in SC. Indeed, many credit him with revitalizing the program and reinfusing it with a sense of mission. Certainly he saved it from oblivion, prodding Schwartz to continue it as a distinct program, and finally in 1989 compelling Schwartz to appoint Steve Squires the Director of Strategic Computing.[86]

That spring, Squires set out to prepare the SC annual report, the fifth in the series. Instead, he ended up preparing a plan for a second phase of the program, to run from 1990 to 1995. This remarkable document, which became known as "SC2," followed the spirit of the original 1983 plan in a number of ways. It identified parallel computing as a key to machine intelligence. It retained the SC pyramid and timelines. It reemphasized the importance of integrating the technologies developed at the various levels of the pyramid, stressing the steady, purposeful flow of technology. Parallel architectures enjoyed somewhat enhanced visibility and significance in the achievement of machine intelligence, but did not displace it atop the pyramid—yet. Technology flowed up the pyramid, as in the original plan; it also flowed down. Not only would basic concepts and enabling technology support the applications programs, for example, but ideas developed from systems integration would be returned to the basic researchers. Saul Amarel and others had considered the one-way flow of knowledge a basic flaw of SC.[87]

Yet SC2 was a much more sophisticated document than the 1983 plan had been, reflecting the experience gained in the first five years of the program and the vision and principles that had been applied in the SC Architectures program. It emphasized a broad systems approach to the development of computing technology. SC programs were to be much more interconnected than heretofore. All SC areas were to create products. These would be consumed by other parts of the program or transitioned out to commercial or military use. The products could be either component technologies, such as microelectronics devices or software modules, or supporting technologies and infrastructure, such as software standards or design environments. The component technologies could be either modular, a complete unit ready to be plugged into a larger system, or scalable, independent systems that could grow from simple demonstrations to large, complex applications.

SC2 also made up for the original plan's most glaring omission; it articulated a coherent strategy for technological development:

• Seeking general results, to avoid premature specialization

• Structuring projects to provide incremental results

• Encouraging projects to use results from other research sources via "open program management," that is, to obtain ideas and technology from anyone who had something useful to offer, and in turn to offer all research results to anyone who might find them useful

• Encouraging interdisciplinary collaboration

• Employing "mixed strategies" incorporating a variety of approaches, from among which certain approaches would be chosen to be continued according to a judgment of their maturity

• Developing and specifying a transition strategy for each project and area

• Organizing funding decisions in an investment-oriented pattern; that is, each effort had to be an investment that, if successful, would advance the goals of the program as a whole

• Selecting and organizing the projects to minimize DARPA's funding obligation, involve other government organization, and mobilize the private sector

• Making each program area responsible for putting its results into forms suitable for use by others

• Establishing producer-consumer relationships

• Continually reviewing projects and programs to ensure their productivity and value to SC

Two key ideas informed this strategy: the Maturity Model that Squires had developed for the architectures program, and the concept of "producer-consumer relationships," or the technology "food chain." The latter concept likened technological development to a vast food chain. When a "producer" introduces a useful technology into the environment, it gets devoured by a "consumer," who then produces a new product which itself is consumed, and so on in a continuing process. For example, a researcher might develop new fabrication design tools that could enable a microelectronics researcher to implement a new gate technology, which in turn is consumed by a microchip designer to devise a new processor that is consumed by a computer architect for a new parallel system; and so on to the very end user, a soldier in the field, a nuclear physicist, an aerospace engineer, or even a schoolteacher.

If a producer creates a product that no one wants to consume, the technology dies out. If no one produces a technology that consumers need for their own products, the process of technological advance grinds to a halt for the lack of the enabling technology. This model challenged DARPA to fund those products that would find consumers, and to find products for which there were hungry consumers to keep the process going. Everyone was to connect their work with another level of the food chain. Both of these fundamental ideas, the Maturity Model and the Technological Food Chain, required careful, active management. They became the new catechism for computer development in the early 1990s at DARPA.

SC2 also differed from the original plan in certain other respects. Networking joined SC, filling the vast and improbable lacuna left by Barry Leiner's decision in 1983 to avoid the visibility and politics of SC. In the ensuing years, networking had grown to rival architecture as the most dynamic technology in the computer field. But DARPA no longer led the development. ARPANET disappeared in 1989, absorbed by NSFNET, a growing number of regional networks, and an emerging network of networks called Internet. NSF, not DARPA, was pioneering the Internet. A group at the European Laboratory for Particle Physics near Geneva, Switzerland, introduced in 1990 a new message protocol, HTTP, that would facilitate access to the Internet and spark an exponential growth in the coming decade.[88] DARPA had to reinvest in networking if it wanted to play a role in this explosion and claim some of the credit.

In another major shift in emphasis, SC2 removed "machine intelligence" from its own plateau on the pyramid, subsuming it under the general heading "software." This seemingly minor shift in nomenclature signaled a profound reconceptualization of AI, both within DARPA and throughout much of the computer community. The effervescent optimism of the early 1980s gave way to more sober appraisal. AI did not scale. In spite of impressive achievements in some fields, designers could not make systems work at a level of complexity approaching human intelligence. Machines excelled at data storage and retrieval; they lagged in judgment, learning, and complex pattern recognition.

Furthermore, the availability of unprecedented computing power did not, as many researchers had hoped, solve the problem. Instead of aiming for machine intelligence as an immediate goal, researchers turned increasingly to programming that would exploit the new architectures to solve immediate problems in the real world.[89] The banner of machine intelligence was quietly taken down from the DARPA flagpole, replaced by the more mundane but realistic pennant of software engineering. DARPA did not quite give up on machine intelligence, but it shifted its focus to nearer term goals. If SC had been an AI program, it was no more.

Finally, SC2 divested the program of the military tone and style that had characterized the original plan. The Cold War was ending. Ronald Reagan had ridden off into the sunset in January 1989, his defense build-up of the early 1980s steadily eroded over the course of his second term. The Berlin Wall fell in the summer of 1989; the Soviet Union collapsed shortly thereafter. Japan now appeared more threatening to U.S. interests than world communism, and economic competition replaced the arms race as the desideratum of national security. In this political environment, DARPA shifted its focus to dual-use technologies, those with both military and civilian applications.

These changes, however, took place behind the scenes. SC2 was never released. No formal explanation about the record of SC and its future was ever made to Congress. Instead, Fields quietly reoriented SC, molding it to suit the changed political landscape. Those modifications saved many SC projects and the bulk of the program's funding. Eventually they would also cost Fields his job.

In spite of the subsiding Cold War, Fields connected SC more tightly with the military services. He placed increasing emphasis on obtaining memoranda of understanding (MOUs) and memoranda of agreement (MOAs) with the services, committing them to financial support of the program. This bought security for the programs by giving the services a greater stake—and a greater say—in the work.

Under Fields's direction DARPA also grew more aggressive in transitioning technology into the services and industry. This campaign responded to increasing political pressure to show results, not just in SC, but in defense R&D generally. In 1986 the president's blue ribbon panel, the Packard Commission, had recommended that DARPA play a greater role in developing and prototyping not just experimental technology but products for deployment to the field.[90] DARPA's charter was changed for the first time in its history to conform to the new mandate, and the agency hastened to comply, throwing together a program in "prototype projects" and redirecting current programs.[91]

Pilot's Associate fell victim to this reform, even though many observers felt it was proceeding well. A successful demonstration, Demo 2, in March 1988 appeared to promise a bright future for the project. Although one of the two contractor teams, led by McDonnell Douglas, made a poor showing at that demonstration[92] the other team, headed by Lockheed, performed spectacularly. "Lockheed demonstrated a system that met the computer science and Artificial Intelligence challenges," recalled Captain Carl Lizza, an air force scientist participating in the program. "More significantly," he continued, it "clearly showed the power and promise of the technology in a credible operational environment." The demonstration was "the pinnacle of the program."[93]

Shortly afterward, to the shock of Captain Lizza and other participants, DARPA ordered the air force to redirect Pilot's Associate away from concept demonstration, experimentation, and risk, and toward short-term, low-risk development using available computing systems. No longer did the program have the latitude to "explore, create, postulate, and fail," Lizza recalled. Now it had to deliver a fully functional product ready for deployment in an existing aircraft. At the same time, the program migrated out of SC and into the DARPA Prototypes Office, to be paid for with 6.3 (advanced development) funds.[94] Politics clearly drove this decision, not just the Packard Commission report and budgetary considerations.[95] Still, the redirection undermined any possibility of getting a worthwhile, working system out of the program.

Under pressure to produce and intimidated by the difficulty of achieving real-time performance, the Lockheed engineers and air force scientists opted for conservative technology. In particular, they switched from a loosely coupled blackboard architecture, similar to that of the NavLab and the later version of the ALV, to a tightly coupled, shared-memory architecture not unlike that of the ALV in its early days. The final product, demonstrated in June 1992, was officially an "unqualified success," but insiders such as Captain Lizza labeled the program a failure.[96] "The behavior of the cockpit displays," said Lizza, "was anything but intelligent or intuitive." No "pilot's associate" system found its way into operational use by the end of the decade.

The final strategy for protecting SC was not to discuss it at all. DARPA published no annual reports on the program after 1988. In his congressional testimony at the annual budget hearings in March 1990, Fields did not once mention SC by name.[97] In a piece of bureaucratic legerdemain, SC simply disappeared in the early glow of a new program just then dawning on the horizon.

In 1990 the high performance computing initiative (HPC) captured the attention of the White House and Congress. It promised and delivered vastly greater funding than SC ever could. HPC embodied many of the goals and methods of SC, but it profoundly redirected them. In a way, it replaced SC2. In another way it terminated SC.

9

The Disappearance of Strategic Computing

SC did not quite end; it simply disappeared. As a line item in the DARPA budget, it vanished in 1993, but long before that it had faded from public view. The annual reports ceased in 1988. DARPA directors reduced the prominence of SC in their congressional testimony through the late 1980s and into the 1990s. SC dropped from the DARPA telephone directory when the Information Science and Technology Office (ISTO) was broken up in 1991. SC had virtually vanished at least two years before its decade was done.

In another sense, however, SC did not really disappear. Rather it was transmuted into High Performance Computing (HPC). In this transformation, SC lost its soul; it abandoned the quest for machine intelligence. Steve Squires's architectures program formed the heart of the successor program. It carried with it those technologies that promoted processing speed and power—number crunching. Left behind in the fading dream that had been SC was the failed promise of AI. What Cooper had once hoped would be the demand pull of applications in machine intelligence instead became a "reverse salient" that retarded development in other fields. If intelligent programs would not run on massively parallel machines, then find applications that would.

These applications came to be called "grand challenges." They were problems collected to fit the solution that Squires had already invented. Ends were now imported to match the means at Squires's disposal. Soon ends and means were indistinguishable.

None of this happened by design. Rather an independent research trajectory collided with SC. In the wake of their convergence, Squires jumped ship, taking with him the SC architectures program, its ancillary technologies, and the long-neglected but suddenly fashionable field of networking. Renamed High Performance Computing and Communication, this juggernaut sped down the information superhighway. SC was left to limp into the sunset on the back road of machine intelligence.

The slow erasure of SC from public and congressional view raises the same question that historian Peter Galison addressed in his 1987 study *How Experiments End?*[1] Students of the research enterprise tend to focus on the hopeful beginnings. What paradigms guide investigators to their projects? How is research shaped by the questions it seeks to answer? Who gets funding and why? Knowing how good research begins, one might presumably come to understand the keys to success.

Most research does not succeed, however, at least not in the sense of achieving startling results. Even when a research project produces a revolutionary breakthrough, it does not necessarily coincide with the end of the program. Sometimes great achievement appears well before the project runs its course. More often the project reaches the bottom of a dry hole with nothing left but to move on and warn the community in a final report not to drill again at that place. Seldom is it easy to know when the end has come. Usually, it is when the money runs out.

The eschatology of research is especially important for DARPA. By some estimates, as many as 85 percent of DARPA projects fail.[2] This is the price of high-risk research. It means that DARPA must be especially nimble in recognizing the dry hole and pulling up stakes. But a long-term, highly visible project such as SC, announced with great fanfare and promising dramatic achievements, cannot be closed down overnight. As it became clear in the later 1980s that SC would not achieve its goals, some graceful termination had to be found. The merger with high-performance computing provided a perfect screen. To understand how SC disappeared within HPC, it is necessary to trace the trajectory of supercomputing, whose meteoric ascent in the early 1990s would mask the decline of SC. Like the program it replaced, HPC had roots in 1983.

The FCCSET Initiative

SC was not an isolated phenomenon. It was, in fact, only one manifestation of a larger trend toward greater federal involvement in the high-end computing industry. The government had promoted computer technology from the earliest days of electronic computing. Even before DARPA entered the field with J. C. R. Licklider's initiatives in 1962, the federal government had supported such landmark projects as the World War II ENIAC at the Moore School of the University of Pennsylvania,

MIT's Whirlwind, and the early mulitprocessor ILLIAC IV, developed at the University of Illinois.³ Yet all of these efforts were to develop single, one-of-a-kind machines, or specific components or technologies. Not until the 1980s would there be coordinated efforts to promote the whole field of computing. SC was the first such program. Within a decade, it would be dwarfed—and supplanted—by the High Performance Computing and Communications Initiative (HPCC), a much larger effort spanning the federal government.

The architectures program of SC was driven—at least at first—by the needs for greater and cheaper computing power for machine intelligence applications. Yet AI researchers were not the only ones who needed high performance. While sensor interpretation, data sorting and fusion, and other such command-and-control functions had driven much of military's involvement in computing, there was always a strong demand for superior number-crunchers.

Throughout history the demand pulling computer development has been, not for machines that could "think," but for machines that could calculate. Blaise Pascal's early machine in the seventeenth century was intended to help his father compute taxes. Charles Babbage designed his Difference Engine, and later the Analytical Engine, to calculate astronomical tables for the Royal Navy. Herman Hollerith's machine tabulated census tables, and Lord Kelvin's harmonic analyzer calculated tides. One of the first electronic computers, ENIAC, was created to calculate ballistic tables for the artillery in World War II, and it was also put to work on problems of nuclear physics for the Manhattan Project.

In the decades after World War II, the scientific and engineering communities came to appreciate the value and possibilities of high-end computers. By the 1970s scientists and engineers were becoming increasingly dependent on computers. They demanded more and more power. To accommodate this demand, manufacturers began to create the so-called conventional supercomputers, ranging from pioneers such as Control Data Corporation's CDC 6600 and IBM's 7094 in the late 1950s and early 1960s, and stretching through the succession of Cray machines that came to define the field. The Cray X-MP, delivered in 1982, set the industry standard on the eve of SC.⁴

Supercomputers brought about great advances in the sciences and engineering; but by the 1980s they were becoming obstacles. There was growing concern over the future of scientific computing. The impending physical limitations on, and escalating costs of, conventional, von Neumann-style systems were as much a concern to the scientific community as to DARPA. Researchers, now fully aware of the value of supercomputers in solving hitherto unsolvable problems, feared that the machines—and thus the researchers themselves—would soon reach the limit of what they could do. The researchers also feared that the rising costs would soon put supercomputing out of their reach. Access to super-computers was already considered a critical problem.

There were only about sixty supercomputers in the United States; by 1983, Cray, the industry leader, had installed only forty-six machines in the United States.[5] The cost of time on the machines that were available was prohibitive, especially for academic researchers. Few universities could afford a good, high-end machine; most were lucky to be able to purchase VAX minicomputers. In 1983 only three universities possessed a supercomputer.[6] Even if a researcher could gain access to a machine, at a government lab for example, his or her academic budget did not buy much time. Increasingly, researchers were going to Europe, where access to supercomputers was easier and cheaper.

Researchers were not the only ones concerned by the problem. Government officials, too, worried that their agencies would not obtain the computational power to accomplish their missions, whether nuclear weapons research in the Department of Energy (DoE), intelligence analysis at the National Security Agency (NSA), or engineering design for space vehicles at the National Aeronautics and Space Administration (NASA). Their greatest fear, however, was that the Japanese National Superspeed Computer Project would capture the supercomputing market and destroy the American industry, as had happened in consumer electronics. Such a scenario would place technology critical to American security interests in the hands of foreigners, which was unacceptable to many in the government.

In 1982 and 1983, just as SC was taking shape at DARPA, three other federal organizations initiated comparable steps to address the supercomputer problem. The National Science Foundation (NSF) working group on computers in research suggested establishing ten university supercomputer centers with network links to other schools.[7] A panel of the Federal Coordinating Council for Science, Engineering, and Technology (FCCSET) recommended that federal agencies design and establish supercomputer centers and networks of their own for long-term needs.[8] The Panel on Large Scale Computing in Science and Engineering (known as the Lax Panel), sponsored by the National Science Board, recommended a coordinated national program to increase access to supercomputers.[9]

By 1984 federal money was flowing. NSA, NSF, and DoE all received funds to build "supercomputer research centers" or buy time on existing machines. In August 1984 NSF's Office of Advanced Scientific Computing purchased time from three existing supercomputer centers for the use of researchers, and began the process of selecting five new centers. NSF also increased support for its own network, NSFNet, which would possess the necessary bandwidth to allow researchers to access the machines remotely. NSA announced the establishment of another supercomputer research center in Landover, Maryland, in November 1984. DoE received $7 million from Congress to expand access to supercomputers for energy research scientists and $7 million more to establish a supercomputer center at Florida State University. NASA began a five-year, $120-million program to build a supercomputer to design and test commercial and military aircraft.[10]

These initiatives were part of a widespread pattern of government support for computer research, a pattern that had yet to achieve focus or coordination. In fiscal year 1983, the federal government invested $173.4 million in R&D for advanced computer development. By far the largest portion, $93.6 million, came from the Department of Defense, and two-thirds of that total came out of DARPA. NSF was next, with $37.8 million, and then NASA with $20.3 million. The Departments of Energy and Commerce together contributed another $21.7 million. In FY 1984 total funding would rise to $226.9 million.[11] The primary supporters of basic research were NSF, DoE, and DARPA. NASA and the military services offered some support, but generally for mission-specific programs. In the Department of Commerce (DoC), the National Bureau of Standards (NBS) focused on standards, metrics, and benchmarks, while the NSA and the Central Intelligence Agency (CIA) focused on intelligence-related activities.[12]

The lack of cooperation among these federal agencies virtually guaranteed that some needs would be overlooked while other suffered from duplication of effort. Each agency followed its own agenda and pursued its own projects with its own budgeted money. One estimate listed thirty-eight separate parallel architectures projects sponsored by seven separate organizations by 1985. A third of these projects were jointly sponsored by two or more offices; the rest were pursued independently.[13] Frequently, individual program managers shared ideas and results on a formal or informal basis with others working on related projects, but such contacts took place at a relatively low level.[14] As of 1983 there was no unified government strategy or approach to the development of, or investment in, supercomputing. And while there was a growing sense that the federal government should take action, there was no agreement as to what that action should be.

The first step toward coordination came in response to Japanese computing initiatives. DoE and NSA convened a conference at Los Alamos National Laboratory in New Mexico in August 1983 to discuss the threat posed by Japan. The meeting, which included 165 representatives of academia, government, and the supercomputer industry, debated the problems of supercomputing technology and the issue of public support for the industry. Some, including Nobel Prize–winning physicist Kenneth Wilson, insisted that the United States had to develop a mass market for supercomputing, but there was little consensus on the form of federal involvement.

NSA Director Lincoln Faurer left the meeting "a little taken aback by what we [the computer users] perceived as a lack of a sense of urgency on the part of the government people in attendance."[15] James Decker of DoE presented a draft of recommendations to be made to FCCSET, but it called for government action only in the most general terms. Richard DeLauer, the Undersecretary of Defense for Research and Development, held out little hope for federal money, while Robert Cooper told the attendees that DARPA's new SCI would concentrate on AI and symbolic processing; conventional supercomputing, he said bluntly, had little utility outside of the national laboratories.[16] The Frontiers of Supercomputing conference highlighted the problem but failed to generate a solution. That role fell to FCCSET.[17]

That same year, FCCSET began laying plans for the proposals that would eventually become the High Performance Computing Initiative. FCCSET was an interagency committee established by Congress in 1975 under the auspices of the Office of Science and Technology Policy (OSTP) to coordinate activities of the executive branch of government in technical realms. In the spring of 1983 FCCSET established three panels to address supercomputing. Two of the panels (which were later merged into one) dealt with procurement of and access to high-speed numerical supercomputers; they were chaired by a representative of the DoE. The third, the Computer Research Coordination Panel, addressed symbolic processing and artificial intelligence. Not surprisingly, Robert Kahn of IPTO chaired this panel. The meetings were attended by representatives of the DoE, NSA, NASA, NSF, DoC, and CIA; Stephen Squires and Craig Fields also sat in.

Although chartered to focus on symbolic processing, Kahn's panel soon came to address the general problems of architecture relevant to both numerical supercomputing and symbolic processing.[18] In a June 1985 report the panel emphasized the importance of parallel processing (which it called "very high-performance computing" [VHPC]) to achieve great gains in both numeric and symbolic computing. It noted that a strong federal program to develop such computers already existed. For the then current FY 1985, the panel estimated that over $100 million was being spent by five departments and agencies for VHPC. Apparently considering this level adequate, the panel did not call for an organized program or for specific new funding levels. It did recommend greater coordination among the various agencies, however, and increased emphasis on the problems of infrastructure, technology transition, and the training and retention of qualified researchers in universities and the government. The panel declared its intention to remain in existence to function as a coordinating body, reporting regularly on the state of the field and providing an institutional mechanism for interagency communication.[19]

It is important to note that the content of the report bears a strong resemblance to SC, at least as Kahn had initially visualized that program. It emphasized infrastructure and scalable parallel architectures, and it called for development programs to seek generic technology having the widest possible application. The report even included a rendering of a technology pyramid almost identical to that of SC, except that at the level of "functional capabilities" it included "scientific computing" as well as AI, and it did not specify any applications (see figure 9.1).

Finally, the report proposed a version of the maturity model Squires was then developing for the SC architectures programs, describing the development of computing technology in four stages: basic research, exploratory development, advanced development, and production engineering. Throughout its long history, culminating in the HPCC initiative of the 1990s, the FCCSET approach to government support of computer research retained the imprint of DARPA's SC. HPCC was a government-wide version of SC without Robert Cooper's influence.

This connection notwithstanding, Kahn's FCCSET panel failed to adopt the tone of urgency and promise that had pervaded the SC plan of 1983. Rather than a call to action, Kahn's panel issued a plea and a set of guidelines. Perhaps believing that SC would solve the problem, Kahn produced a less provocative document than others that were then flowing from the FCCSET initiative.[20]

The FCCSET Computer Research Panel continued to meet for another year, with Kahn remaining as chair at Amarel's request after leaving

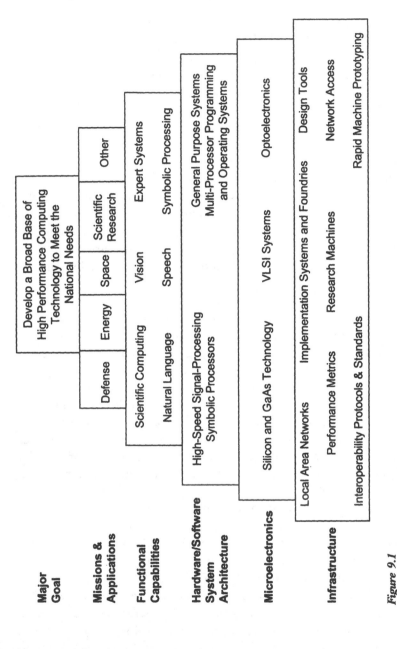

Figure 9.1
FCCSET "Advanced Computer Research Program Structure and Goals." Adapted from "Report of the Federal Coordinating Council on Science, Engineering, and Technology Panel on Advanced Computer Research in the Federal Government," June 1985, figure. 1.

DARPA in September 1985. After March 1986 the panel went dormant for a year.[21] In these early years, DARPA's influence on FCCSET was stronger in shaping the institution than in promoting action. DARPA, in short, seems not to have appreciated the need for a government-wide initiative.

Connecting the Supercomputers

DARPA also appears to have played a less-than-prominent role in another sea change taking shape under the FCCSET umbrella. Networking came to the attention of FCCSET at just the time it was losing its salience at DARPA. Barry Leiner's decision to keep his networking research programs outside of SC had denied them a visibility and source of funding that he felt better off without.[22] DARPA would have to scramble in the 1990s to reconnect computing and communication to blend with the trajectory taking shape in FCCSET in 1985.[23]

The FCCSET Committee on High Performance Computing tracked industrial R&D and commercial developments in the United States and overseas, especially component manufacture (e.g., chips) and computer networks. To study the latter problem, it established in 1985 the networks working group, which published a report in February 1986 calling for the interconnection of existing federally supported telecommunications networks.[24]

By 1986 Senator Albert Gore, Jr., of Tennessee had become fascinated by the possibilities of the "information superhighway," a term he popularized to suggest the analogy with the interstate highway system. Under Gore's leadership, Congress in August called on OSTP to study and report on the state of the networks in the United States. FCCSET's response to Senator Gore captures in microcosm the ways in which the DARPA-ISTO model of technological development was both transferred and transformed in the pursuit of HPCC. DARPA representatives played prominent roles in the flock of panels and committees FCCSET created on the road to Senator Gore's information superhighway. And they infused the process with ISTO concepts such as gray coding and the Squires-Scherlis concept of a food chain.

Coordination on the scale being attempted by FCCSET, however, surpassed the comprehension of a DARPA program manager or even an office director. Entire agencies, with their different operating styles and political agendas, had to be coordinated. Policies could not be formulated and handed down; they had to be negotiated. And the entire process played out in a far more public, and more political, arena than the offices and meeting rooms of DARPA. The institutional form of the enterprise was the committee; its currency was the report.

To conduct the study, FCCSET formed a panel, which held a workshop in San Diego in February 1987. Soon after, President Ronald Reagan's science adviser, William Graham, decided to expand the mandate of the panel to focus not only on networks themselves but the context in which they would be used. In March he chartered the FCCSET Committee on Computer Research and Applications to complete the study. The committee created three subcommittees: the Subcommittee on Scientific and Engineering Computing, chaired by James Decker (later David Nelson) of DoE; the Subcommittee on Computer Networking, Infrastructure, and Digital Communications, chaired by Gordon Bell (later William Wulf) of NSF; and Kahn's old panel, now revived and called the Subcommittee on Computer Research and Development, under Saul Amarel (later Jack Schwartz). This last panel included a strong contingent of members and observers from ISTO, including Squires, William Scherlis, and Robert Simpson. Kahn participated in its meetings as well. In addition, Squires served on the Subcommittee on Science and Engineering Computing, while ISTO program manager Mark Pullen represented DARPA on the Subcommittee on Computer Networking, Infrastructure, and Digital Communications.

The FCCSET committee held a series of workshops involving hundreds of computer scientists and engineers in industry, government, and academe. It also formed several working groups to explore various technical aspects of the problem of advanced computing, including a "TeraOPS technical working group." Amarel's panel focused on the development of advanced computer architectures. The group held two workshops and received from the NSF Advisory Committee for Computer Research an Initiatives Report in May 1987, which reviewed the field and recommended investment in parallel systems and software technology. In September the panel held a workshop on advanced computing technology at Gaithersburg, Maryland, attended by 200 people.[25]

During the spring of 1987, Graham asked for specific proposals concerning networks and computer research. On June 8 FCCSET briefed him at the White House. Paul Huray spoke as chairman of OSTP, and Decker, Amarel, and Bell presented the findings and recommendations of their respective committees. They proposed a "national computing initiative," a plan developed by Amarel, Squires, and Scherlis to achieve

TeraOPS systems in the 1990s through scalable parallel processing, and a "national research network" to achieve high-bandwidth networks with a gigabit communications capacity. These two proposals ultimately would form the basis for HPCC.[26]

With Graham's approval and some further study, OSTP issued its landmark report, *A Research and Development Strategy for High Performance Computing.*[27] This document (for which Squires had been the Executive Secretary) emphasized the growing importance of advanced computing in the United States, and warned that the Japanese and the Europeans threatened the American lead in high-performance computing and networking. It proposed a strategy for bolstering these areas, based on the coordination of existing federal programs, an augmentation of this government effort, and the close cooperation of government, industry, and academia to develop "a shared vision of the future."[28] It emphasized the importance of industrial financial participation—the federal money was intended to leverage private investment, not supplant it—and the transfer of technology and research results to the private sector for commercialization. All of these principles resonated with SC.

The strategy itself involved simultaneous efforts in four areas. High Performance Computing Systems focused on the development of advanced parallel computers through a process similar to that in the SC architectures program, whereby the government would purchase prototype systems and make them available to the research community. The Advanced Software Technology and Algorithms area would promote advances in all areas of software technology, from programming tools and operating systems to compilers and applications. In particular, application software should be directed to solving "grand challenges," problems that were of special significance to—and difficulty for—the scientific and engineering communities. The National Research and Education Network (NREN) would replace existing networks in a three-stage program, culminating in a three-gigabit transmission capability within fifteen years. Finally, Basic Research and Human Resources, an infrastructural effort, would support fundamental research in computer science and promote the training of personnel to ensure a strong technology base.[29] The major difference between this plan and SC was that it left out AI and stressed networking.

The report did not spell out exactly how these goals were to be accomplished. It was only a strategy, a guideline that would require further analysis and planning before implementation could be attempted. It did propose a funding plan, beginning at $140 million in the first year and

rising to $545 million by year five, for a total of $1.74 billion. About half
the money would go to basic research and HPC development; the rest
would be applied to networking and research in computational science
and engineering. This money would be on top of then current funding
for high performance computing within the various agencies, which was
estimated at about $500 million per year.[30]

Graham signed the report and passed it on to Congress. The document garnered considerable interest in Washington. Senator Gore held
hearings on the subject in August 1988 and then asked OSTP to prepare
a second report detailing a plan by which the strategy could be implemented.[31] Suddenly the proponents of a federal initiative to promote
high-end computing were a long step closer to their goal.

The Gap between Conception and Legislation

The 1987 OSTP report—and, even more, the cooperative effort to prepare it—energized the participating agencies to talk more and press forward their own in-house programs. It was during this time, the fall of
1987, that Squires and others in ISTO began laying plans for the Tera-
OPS Computing Technology Program, building on the ideas presented
in the National Computing Initiative proposal. The program would be
launched from the second wave of parallel systems then being developed
under SC, with the goal of achieving TeraOPS performance (a trillion
operations per second) within five years. The SC program had demonstrated the feasibility of a number of architectural models:

• Shared-memory multiple instruction stream/multiple data stream
(MIMD) multiprocessors like the RP3

• Message-passing MIMD multicomputers such as the Cosmic Cube and
the Hypercube;

• Data parallel single instruction stream/multiple data stream (SIMD)
multicomputers such as the Connection Machine

• Systolic multicomputers such as the Warp

• Hybrid systems consisting of a heterogeneous combination of processors coupled together, such as the MOSAIC

Squires concluded that all of these types could be scaled up to a TeraOPS
capability. The key would be to focus on advances in the components
themselves—processors, memory, and coupling components (i.e., communications switches)—to improve the performance of the basic mod-

els. Efforts also would be undertaken in wafer-scale integration, three-dimensional chip-packaging techniques, gallium arsenide or "cooled silicon" chips, optoelectronics, and so forth.[32]

Reflecting the SCI approach, just then mirrored in the FCCSET plan for HPC, the proposed program would include four components. One would focus on research in "microtechnologies" (i.e., microelectronics), including new packaging concepts, feature design, and materials. Another area, called "microsystem design," would focus on design rules, tools, and frameworks for rapid design and prototyping of computing systems. A third area, systems architectures, would scale up selected first- and second-generation SC systems to achieve sustained 100 gigaops (.1 TeraOPS) by the early 1990s. "Systems will be selected for scale-up based upon evidence that they are appropriate vehicles for attacking selected grand computational challenges of defense significance." The final component, "driving applications," would develop prototype software for applications chosen for significance to national defense and usefulness in understanding computational issues. Once the feasibility of such a prototype had been demonstrated, it would then be turned over to mission-oriented programs for advanced development.[33]

In essence, this program was SC without AI. It was a body without a brain. It aimed for thinking machines that could not think. The critical defense applications that Squires had in mind focused on scientific and engineering problems, such as aircraft design, circuit design and simulation, modeling of material properties, and advanced composites and alloys. The only SC-type applications problems were image analysis and the analysis of data from large sensor arrays.[34] The FCCSET process and the success of the early SC parallel systems in numerical computing and modeling had convinced Squires that DARPA's efforts in high performance computing should be directed toward scientific and engineering applications, not machine intelligence. ISTO director Jack Schwartz, himself skeptical of AI applications, supported the TeraOPS plan enthusiastically and contributed some ideas of his own.[35]

During the next year Squires continued to study the technological trends and prepare "roadmaps." By the fall of 1988 he had a clear idea in his own mind of how TeraOPS systems could be achieved during the 1990s. In a detailed briefing that he presented both within and outside of DARPA, he laid out his complex ideas for the process.[36] By this time Squires had narrowed the taxonomy of parallel architectures to four basic types: multicomputers, systolic systems, data parallel systems, and multiprocessor systems. He argued that all could achieve TeraOPS capabilities. The key was to build the machines with modules that could be mixed and matched to achieve any desired capability. Thus the systems could be scaled up to a TeraOPS merely by adding additional modules.

These modules would consist of three-dimensional packages of stacked hybrid wafers, each of which incorporated 1,000 advanced components (microprocessors, memory, and coupling devices). The exact mix and design of the components themselves would vary according to the kind of computer architecture being constructed. The modules could be stacked into cube-shaped "multi giga supercomputers," which could themselves be combined into TeraOPS systems. How many modules were required to achieve this capability depended on the power of each processing node. At one extreme was a massively parallel system of a million nodes, each capable of a million operations per second. At the other extreme was a hypothetical single node capable of a trillion operations per second, which Squires considered impossible. Squires calculated that the optimum configuration, in terms of risk, was a system of a thousand nodes, each capable of a billion operations per second.

Such systems could not be produced at once. Squires expected that they would require several generations, each of which would have ten times the capability of the previous one. Advances would occur in all technologies in a carefully managed development cycle. Squires represented this cycle as a triangular "scalable parallel computing spiral," integrating design, manufacture, and application (see figure 9.2). Squires proposed a budget considerably greater than that of a year before: more than $1.1 billion over seven years, peaking at $210 million in 1993. More than half of this money would go toward prototype development and manufacturing; the rest would be applied to component design tools, system software, and facilities.[37]

By this point ends and means appear to have changed places in Squires's mind. TeraOPS had become the holy grail. The great, beguiling challenge was how to get a machine to switch a trillion times a second. What such a machine might do receded to secondary importance. Squires became obsessed with connecting switches, not with connecting the resulting supercomputer to users in the real world. It was simply assumed that if he built it, they would come.

In expectation of greater funding, DARPA published another BAA at the end of February 1988, calling for proposals in parallel architectures, preferably leading toward the goal of a TeraOPS system.[38] By June forty-

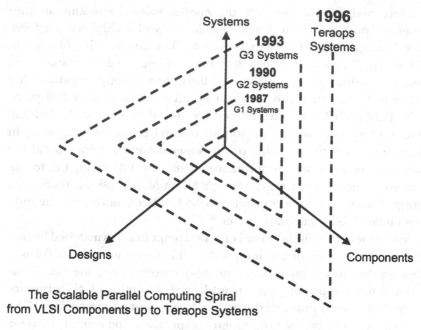

The Scalable Parallel Computing Spiral
from VLSI Components up to Teraops Systems

FOR OFFICIAL USE ONLY

Figure 9.2
"The Scalable Parallel Computing Spiral" G1 = 1 Gigaops; 2 = 10 Gigaops; G3 = 100 Gigaops. Adapted from DARPA "TeraOPS Computing Technology Program" slides, "Introduction, 28 September 1988.

two responses had been received. Similar activity was taking place in the various agencies. While the White House showed little interest in the idea of a national initiative, Congress proved much more receptive. In October 1988 the first bill on the subject, "The National HPC Technology Act," was introduced in the Senate, followed by both House and Senate bills the following year.[39]

Outside government, interest in high performance computing also gained momentum. By March 1989 about 400 conventional supercomputers had been installed in the United States, more than a six-fold increase since 1983.[40] And, as Squires had anticipated, researchers were finding applications. For example, the Cray X-MP at the Pittsburgh Supercomputing Center, one of those established with NSF funding, supported the work of 1,700 researchers from 1986 to 1989.

In 1988–1989 researchers performed projects in physics, biology and biochemistry, astronomy, mathematical sciences, and geosciences. One project modeled the growth of the *Bacillus Subtilis* bacterium; another explored problems in the design of gallium arsenide chips; a third developed images using two-dimensional Markov chains. The Aluminum Company of America studied the design of improved aluminum cans, and two political economists studied the history of congressional voting patterns by analyzing two million roll-call votes taken over 200 years. The Pittsburgh Center encouraged the use of its facilities by holding two-week summer workshops to give researchers intensive training in scientific computing.[41] As interest in supercomputing grew, so did the demand for improved performance. When the Pittsburgh Center installed its new, state-of-the-art Cray Y-MP in March 1989, it reported a surge of proposals from researchers who had been waiting for the hitherto unavailable computing power.[42]

Excitement over the potential of supercomputing was matched by concern over the state of the industry. At this time the concern focused less on the advancement of the field—far from hitting the "wall," the conventional supercomputer manufacturers continued to introduce steadily improved products during the 1980s—but on the survival of the industry itself in the face of foreign competition. The case of the semiconductor industry was a powerful warning.

In the late 1970s Japanese companies dumped DRAM chips onto the American market below cost, capturing the market. By the time Congress had passed antidumping laws, the American production of DRAMS had been virtually wiped out. U.S. computer manufacturers, including those making supercomputers such as Cray and ETA, became entirely dependent on Japanese suppliers. The Japanese subsequently cut back on production, which, combined with American tariffs, caused the price of the DRAM chips to skyrocket, to the detriment of the American supercomputing industry. The Japanese firms also captured the market in bipolar logic chips (the kind favored by U.S. supercomputer manufacturers) and then (some Americans suspected) refused to export the latest state-of-the-art components to the United States, although the Japanese denied it.[43]

Further loss in the semiconductor industry was avoided by legislative action and especially the development of government-encouraged consortia such as Sematech, but the damage had been done and the danger was clear. Americans could not rely on shutting out the Japanese from American markets and technologies; active steps had to be taken to promote and secure the competitiveness of the American industry, both technologically and commercially. This would require funding, but, more importantly, centralized leadership to guide the fractured American industry.[44]

On 8 September 1989, OSTP issued *The Federal High Performance Computing Program*, a plan for implementing the strategy outlined nearly two years before. As in the previous document, the plan called for efforts in four areas: high performance computing, software and algorithms, networking, and basic research and human resources. Yet this plan now called for a coherent program of which these areas would be subsets. The federal part of the program would be conducted by the individual departments and agencies, which would have varying degrees of responsibility within each area, under the general supervision of OSTP through the FCCSET Committee on Computer Research and Applications. A High Performance Computing Advisory Panel, consisting of government, industry, and academic leaders, would monitor the progress of the program, its continued relevance over time, the balance among its components, and its outcome.[45]

For three of the components of the program, the plan designated one or two lead agencies or departments, although all would continue to play greater or lesser roles in that area according to its relevance to the agency's mission. The DoD would lead the High Performance Computing Systems area, which included research in advanced architectures, system design tools, the transfer of technology to facilitate the construction of prototypes, and the evaluation of the systems that emerged. Naturally, DARPA was responsible for high-risk research and prototyping, while the other agencies would continue to purchase first production models of new systems. NASA would lead the advanced software technology and algorithms component, focusing on the development of software components and tools, computational techniques, application efforts, and high performance computing research centers.

NSF and DARPA together would develop and deploy the "national research and education network," which would supplant the existing Internet. DARPA would be in charge of gigabit technology, while NSF would take the lead in deploying and maintaining the network. As proposed in the HPC strategy, the effort would be conducted in three phases, with the third phase achieving one to three gigabit/second capability by the mid- to late-1990s. Finally, no agency would have particular charge of the basic research and human resources component, though DoE and NSF would play leading roles.[46]

Several other agencies would have lesser roles in the program: the National Institute of Science and Technology (NIST, formerly the

National Bureau of Standards) and the National Oceanic and Atmospheric Administration, both in the Department of Commerce; and, to a lesser degree, the Departments of Transportation and Health and Human Services.

The plan called for new federal funding on top of what the agencies already spent on HPC-related activities, estimated at almost $500 million in FY 1992.[47] The first year would require $151 million, of which $55 million would go to High Performance Computing Systems, $51 million to Advanced Software Technology and Algorithms, $30 million to the National Research and Education Network, and $15 million to Basic Research and Human Resources. The proposed funding level rose incrementally to nearly $600 million by year five.[48]

A key element in this program was the "Grand Challenge" problems. These were scientific and engineering problems that were considered of particular significance to the American economy and defense. Their complexity required exceptional computing capabilities far beyond what was possible in 1989. They included the prediction of weather, climate, and global change; superconductivity; the design of semiconductors and new drugs; and research in fluid dynamics and nuclear fusion. These challenges posed particular problems for the Advanced Software Technology and Algorithms effort, but they also drove the rest of the program. For example, many were expected to require TeraOPS computing capability. The grand challenges thus functioned as applications of SC by exerting a technology "pull."[49]

Fields and Boehm in the Last Ditch

That same fall of 1989, SC entered its final active phase. In November ISTO received its third director since 1986. Barry Boehm had worked in interactive computer graphics with the RAND Corporation in the 1960s and had helped define the ARPANET in the early phases of that program. He then moved to TRW Inc. to work in software, cooperating with a university team on a DARPA-funded software environment project called Arcadia. In 1987–1988 he helped Squires and Scherlis identify the software issues that should be explored in the HPC program then under study. In 1989 Boehm was about to retire early from TRW and was planning on moving to a university, probably the University of Southern California. Squires and Scherlis prevailed upon him to take the position of ISTO director, left vacant by Schwartz's departure in September. Boehm became Director of SC in addition to director of ISTO. Squires continued as Chief Scientist of ISTO, a post he had assumed shortly before Schwartz's departure. Boehm leaned heavily on Squires, who worked with the program managers (especially the newer ones) to help them conceptualize their programs and fit them into a coherent whole.[50]

Fields wanted to restore SC as a coherent, integrated program, reversing Jack Schwartz's insistence on blending SC and basic funding. Boehm agreed. He planned a new strategy with Fields and then held a series of off-site meetings with the office personnel before his arrival. Out of these conferences came the "DARPA ISTO Strategic Plan" in January 1990. Squires's vision, ideas, and experience permeated the plan. At the highest level, the plan espoused allegiance to traditional DARPA policy. It would support American national security goals by adhering to three principles: (1) focus on high-leverage areas, primarily high-risk, high-gain, longer-range programs of critical importance to DoD; (2) accelerate high-leverage technology by creating critical-mass programs, stimulating synergy between programs, and keeping programs fresh by continual evaluation, redirection where necessary, and strategic use of reserve funding; and (3) accelerate technology transfer into DoD missions, by stimulating awareness of the technology among the military services, stimulating the use of the technology within the ISTO community through user test beds and joint programs, and stimulating external behavior by eliminating barriers to the use of the technology (i.e., facilitating its commercialization and availability), arranging cooperative funding, and so on.[51]

As the plan became more specific, the hand of Steve Squires came into view. A "technology maturity model" would determine the support directed to any given technology (see figure 9.3). Furthermore, an "information technology infrastructure model" would define the relationship of the various elements of the ISTO program (microsystems, computing systems, networks, software, etc.) and how they contributed to a particular mission or objective. This model reprised the "food chain" first laid out in SC2, the idea that all programs are defined by the relationship between the suppliers and the customers of the technology.[52] Each program would contribute to "paradigm shifts," that is, major changes in the nature or use of the technology. The model emphasized the relationships between the various levels of technology, from manufacturing and development tools and device technology to computer components to full computing systems, networks, software, and so on up to the final end product. Finally, the plan set out the criteria for the ISTO basic and SC programs. This reflected DARPA Director Craig

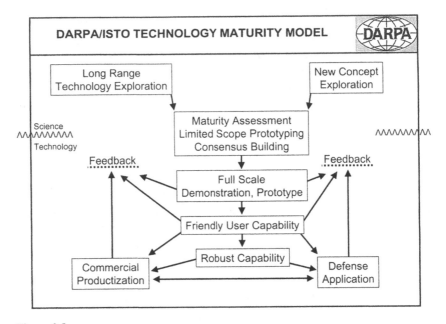

Figure 9.3
DARPA-ISTO Technology Maturity Model. Adapted from "DARPA ISTO Strategic Plan," January 1990; Boehm collection.

Fields's desire that the two parts of ISTO's program be better defined and delineated.[53]

Within the SC program, the three remaining SC applications were winding down and transitioning out to the military services. These programs were relatively successful. In the Fleet Command Center Battle Management Program (FCCBMP), for example, two of the components of the system, the Force Requirements Expert System (FRESH) and the Capabilities Requirements Assessment Expert System (CASES) were fully operational by the late 1980s, and the navy was using them extensively. More importantly, through its experience with FCCBMP, the navy came to appreciate the value of using such systems to ease its command-and-control burden. FCCBMP became an integral part of the navy's new Operations Support System (OSS), a more comprehensive system encompassing several computer-based C^2 programs. The development of OSS was begun in 1987.

The navy also absorbed the new development methodology that DARPA introduced it to with FCCBMP, that of relying on evolutionary development using rapid prototyping. Instead of defining the require-

ments of the end product, establishing the specifications (milspecs) for
it, and then seeking to build a production machine, the navy would instead build a basic prototype and gradually improve it, adding or changing capabilities, functionalities, and hardware as experience and
requirements dictated. This, perhaps, was SC's most important contribution to the navy.[54]

Two applications programs were relatively new but already underway
when Boehm arrived at ISTO. One was WHISPER, a program to apply
speech recognition technology to the analysis of voice intercepts; in effect, the computer would analyze radio and telephone messages. The
other was the Survivable, Adaptive Planning Experiment (SAPE). This
program was designed to prepare plans for nuclear war and update them
even in the face of an attack. The Single Integrated Operations Plan
(SIOP), the master plan for conducting a nuclear war, generally required eighteen months to revise; SAPE sought to reduce that time to
three days. The program also sought to reduce from eight hours to three
minutes the time required for retargeting strategic weapons; it would
permit retargeting during the course of an enemy strike. SAPE, begun
in 1988, had been moved to SC by March of 1990. It was the only SC
program explicitly directed toward strategic conflict.[55]

Boehm later recalled inheriting an embarrassment of riches. Some of
the expensive applications programs were transitioning out, and funding
for SC remained strong. Though SC was barely visible in the DARPA
budget, masked within the broad funding category "strategic technology," and DARPA Director Craig Fields failed to mention it in his congressional briefings in 1990, the program still rode a wave of support
from previous funding cycles and a general sentiment within Congress
that DARPA was underfunded, given its many contributions to national
security. Boehm asked other offices for proposals but elicited little response. Ultimately, he chose to fund four small applications programs
within ISTO. SC, which Cooper and Kahn had intentionally spread
widely around DARPA was now limited to the two technology-base offices, ISTO and the Defense Sciences Office.[56] The most plausible explanation for this constriction of SC influence was the growing realization
in the new decade that 1993 was fast approaching. Soon SC might be
called to account for the $1 billion it would have distributed in its ten-
year history.

Boehm appears to have chosen his new applications with just this calculation in mind. These projects were well-defined, realistic attacks on
concrete problems, alive to the limitations that AI had demonstrated in

earlier, more ambitious SC applications. One program, TIPSTER, resembled WHISPER except that it dealt with written messages. It sought to apply natural language and message understanding technology to the analysis of documents received by an intelligence agency, although it was expected to have considerable commercial application as well. A third program, Research And Development Image Understanding System (RADIUS), took after ADRIES and SCORPIUS; it sought to apply image-understanding technology to the analysis of visual images, also for the intelligence community. An Intelligent Interactive Image Understanding work station would call up the image on a light table and then work with a human specialist to analyze it. More limited than SCORPIUS, this system worked interactively with humans instead of attempting to replace them.[57]

The final application, the Dynamic Analysis and Reprogramming Tool (DART), would help the military services plan and organize the transportation for a deployment. Such work was remarkably time consuming, even when performed by highly skilled and experienced planners. The planners determined the available transportation resources (ships, aircraft, etc.), decided how they should be loaded for most efficient on- and off-loading, how they should be routed, and then where problems were occurring. Most of the work was performed manually and required days, weeks, or even longer for a major move. Steve Cross, a new AI program manager in ISTO, likened this process to "monks transcribing manuscripts by candlelight."[58] DART applied expert systems to reduce this time to hours and to increase flexibility by generating more than one plan. Cross sought to give this system more functionality than the previous SC applications of expert systems, generating dummy plans for deception and reusing elements of previous plans.[59] In 1990 Operation Desert Shield used DART to move VII Corps from Europe to the Persian Gulf.

With Fields's approval, Boehm also restored the funding that Jack Schwartz had cut from AI. Boehm had been skeptical of AI prior to coming to DARPA, but he soon concluded that good technology could come out of the program. Furthermore, he received visits from many leaders of the field—"the Feigenbaums, the McCarthy's, the Roger Schanks"—convincing him that DARPA funding was crucial to the AI community and to the survival of the AI field. The new emphasis was more pragmatic and less ambitious than the original AI program, but it nonetheless represented an attempt to salvage the useful work that had been done and restore some lustre and morale to a program that had once been the essence of SC.[60]

In the early spring of 1990, the reorganized SC technology base consisted of five areas: machine vision, speech processing, applied mathematics, high-definition display technology, and distributed-parallel systems. Though progress had been made in the area of vision, there had been no breakthroughs and no comprehensive vision system. Rand Waltzman, one of the new AI program managers, sought to establish a standard vision environment which would be defined by the image understanding system being developed by RADIUS. A key emphasis would be on setting standard interfaces to allow components of the vision system to be integrated—the goal that Ohlander and Simpson had long sought to achieve.[61]

The speech program, under the management of Charles Wayne since 1988, incorporated some of the efforts of the now defunct SC natural language program. It consisted of fourteen projects conducted by eleven institutions. Wayne's most important contribution in this area was the definition of a standardized library of spoken sounds that could be used to test and evaluate systems developed by the research community. At annual meetings that Boehm characterized as "bake-offs," DARPA would furnish the participants with speech samples with which they would test the accuracy of their systems. By the fall of 1989, the Sphinx system recognized a speaker-independent, thousand-word vocabulary with 94 percent accuracy at near real time. The Dragon Dictate system, by Dragon Systems, demonstrated the capacity to understand 30,000 words at 30 words per minute after adapting to a given speaker.[62] The program could not yet recognize or understand speech in a noisy, stressed environment; nor could it use natural grammar. But Wayne set his sights on understanding spontaneous speech with a 5,000 word vocabulary in real time by 1993.[63]

The Distributed-Parallel Systems area included such key projects as Mach, which was continuing to prove an important development. By 1990 it was becoming a standard operating system for distributed and parallel systems. At least a dozen computer platforms supported it, ranging from conventional machines such as SUN work stations and the DEC Vax, to conventional supercomputers such as the Cray Y-MP, and to the parallel machines such as the BBN Butterfly, IBM Multimax, and the IBM RP3X. At least 150 sites were using it.[64]

These remaining AI and applications programs, however, occupied the pinnacle of an SC pyramid from which the base had disappeared. If the architectures, microelectronics (now called "microsystems"), and computing infrastructure areas were still a part of SC in 1990, it was in name only.[65] Indeed, as a program review held in March 1990 makes clear, these areas and networking were now effectively organized in a separate HPC program. Since 1987 Squires had increasingly conceptualized these areas as an integral program, and, working closely with the microsystems and infrastructure program managers in ISTO and DSO, succeeded in forging the sort of cooperation and coordination that SC as a whole had never quite achieved. Furthermore, Squires' objectives for the HPC program were avowedly dual-use (in keeping with the tenor of the FCCSET initiatives), while SC was being directed toward more explicitly military objectives.[66]

By this time the second generation systems were well under way and were continuing their technical advances. If the AI programs were still struggling to achieve the goals set out in 1983, the Architectures program was forging far beyond what had been expected. Thinking Machines' CM2 was achieving multigigaflop performance on scientific computation at military and government labs. The Army Engineer Topographic Laboratory placed a Connection Machine on a radiation-hardened trailer to support its image exploitation. The full 64K-processor system was measured at a sustained 7 gigaflops; in March 1991, TMC claimed the speed record for high-performance computers, with a peak speed of 5.2 gigaflops. Meanwhile, DARPA had awarded a $2.24 million contract to TMC to construct a more powerful scalable system, the Mega-Connection Machine, the prototype of which was to achieve 100 gigaops in 1992, scalable to 1 teraops. The system would be introduced in 1992 and marketed as the CM5.[67]

At the end of 1989 the first operational iWarp chips were produced, with expected performance up to 20 gigaflops. The system was flexible enough to sustain both fine-grain systolic computational models and also coarse-grain message-passing models. The key to the system was an extremely high-bandwidth communications unit that was integrated onto the iWarp cell, along with the computation unit and local memory. This arrangement permitted flexible communications schemes; these modular processing nodes could be connected either in a one-dimensional sequential system (for systolic operation) or a two-dimensional grid pattern (for message-passing operation). It was also a remarkably compact system; each three-inch by five-inch cell could achieve 20 megaflops by itself and boasted up to 6 megabytes.[68]

Other systems included MIT's J-Machine, a database machine that was to scale up to 1K processors in a three-dimensional grid in 1991; Encore's upgrade from the Multimax, called the Gigamax, with its heterogeneous architecture; and the AT&T Aspen, an evolutionary upgrade of the DADO-DSP architecture developed jointly by Columbia University and AT&T with some support from SC.[69] At its peak, the system achieved more than 4 gigaflops. In an experiment conducted with the Navy, a thirty-one-node Aspen BT-100 was sent to sea on board a submarine to perform target motion analysis. It performed over six times faster than other approaches, and sold the navy on the value of high performance computing for such tasks.[70]

But the most impressive system of all was the Touchstone series developed by Intel under contract from DARPA. The success of this program was a surprise, not the least to Intel. As one company official put it,

We started out with what was really a toy in order to learn something about writing parallel programs, and then we developed a machine that was actually competitive with minis and mainframes. We now have a machine that does compete with supercomputers, with the goal to build what we are starting to call ultracomputers.

The Touchstone project was an excellent example of Squires' "gray code" approach. It was planned as a series of prototypes, each of which would add a new, more exotic component to study the impact on performance. The first machine, Iota (completed March 1989), took Intel's iPSC processor and added input-output (I/O) capability to form a parallel processor. The next machine, Gamma (completed December 1989), replaced the iPSC with Intel's state-of-the-art "supercomputer-on-a-chip," the i860. Delta (1991) then increased the number of nodes to sixty-four and added a high-performance communications system based on the high-speed routers invented by Professor Charles Seitz of the California Institute of Technology (CalTech). The final step was to be the Sigma, which would be scaled up to 2,000 processors with an even better communications network and chip packaging technology. This was expected to achieve 500 gigaflops by 1992.[71]

At this point, however, the research community interceded: it loved the Delta and wanted a scaled-up version of that particular machine. DARPA resisted; the purpose of the Delta project was to test Seitz's routers, not build a production machine. Intel pleaded to be allowed to put the Delta into production, and a consortium of users (including CalTech) raised money to fund it. Only then did DARPA agree to put in $9.3 million. The result was a state-of-the-art machine that performed quite well. In May 1991 a 128-node version achieved a top speed of 8.6 gigaflops, breaking the record set by the CM2 two months before. An experimental version being constructed at the same time at Caltech, with 520 i860s, was expected to do even better. The system was ultimately marketed under the name Paragon.[72]

All of these efforts were SC programs, but Squires was already looking ahead to the greater funding and ramped-up effort of the federal HPC program. He anticipated that HPC would allow DARPA to achieve the sort of advances that he had once envisioned for SC. For example, the extra funding would make it possible to achieve computer speeds at least an order of magnitude faster than the systems that were possible under SC, thus bringing about TeraOPS, even ten-TeraOPS, systems. It would also allow DARPA to develop the system modules, packaging technology, scalable mass storage and I/O interfaces, and ultra-compact computing modules for embedding into defense systems. In software technology and networking, too, he saw considerable gains from an accelerated effort.[73] Yet the federal program was by no means assured. It would require almost two years of intense effort before the federal High Performance Computing and Communications initiative was approved by the president and enacted into law.

The Politics of Change

Before that came to pass, two major upheavals visited DARPA, burying the failed trajectory that had once been SC and merging the agency's computer research program to the new trajectory of HPC. First came the firing of Craig Fields. Always a bold and iconoclastic thinker, Fields had seized the opportunity presented by the DARPA directorship to exploit his years of experience with SC. He promoted its philosophy if not its label and he sought ways to leverage DARPA resources. One such way was the High Definition Display Technology program, then funding several related projects that were attempting to move the traditional cathode-ray tube of television and computer screens to higher levels of clarity and compactness. To his mind, this represented a classic DARPA investment in a dual-use technology, with both military and civilian applications. Some critics, however, saw it as government interference in the commercial marketplace, particularly in the high-stakes race between Japan and the United States to bring to market high-density television (HDTV).

When Fields used DARPA funding to support an American firm then flirting with Japanese investors, competitors cried foul. Their complaints resonated at the White House of President George Bush, whose administration shared the widespread belief within the business community that the federal government should eschew "national industrial policy."[74] Though many observers believed that Japan was besting the United States in many commercial arenas by pursuing just such policies, President Bush and his aides and supporters believed that government agencies should not be picking winners and losers. An unrepentant Craig Fields was relieved as DARPA director in April 1990. Within months, a martyr to the free-market economy, he left Washington to become the new president of the Microelectronics and Computer Technology Corporation (MCC).[75]

Fields's successor, Victor Reis, shared an admiration for SC and a belief in DARPA's tradition of high-risk and high-gain. A Princeton Ph.D. and a veteran of Lincoln Laboratory, Reis had come to DARPA in 1983 after a two-year stint in the Office of Science and Technology Policy in the Executive Office of the President. He sailed less close to the wind than Fields, displaying more appreciation for the political imperatives of doing business in Washington. Under his calming leadership, another major transformation befell DARPA. In contrast to the disruption surrounding Fields's departure, this change proved to be amicable and peaceful. It was nevertheless just as decisive for the future of SC.

In a major reorganization, the Information Science and Technology Office (ISTO), successor to the IPTO that J. C. R. Licklider had founded almost three decades earlier, was divided into two new entities (see figure 9.4). In June 1991 Barry Boehm assumed direction of the new Software and Intelligent Systems Technology Office (SISTO), while Steve Squires was elevated to director of the new Computer Systems Technology Office (CSTO).[76] The last remaining veteran of the original SC finally took command of his own ship. Though he still had to answer to Victor Reis, Squires had every reason to believe that he now commanded the vanguard of DARPA computer research. And he meant to sail in harm's way.

Seldom does an institutional reorganization so clearly mirror the deeper currents running beneath the surface. Since the earliest days of SC, Kahn, Squires, and other cognoscenti of the program had insisted that its component parts had to connect. Developments would move up (and later down) the SC pyramid. Infrastructure fed architecture, architecture fed AI, and AI fed applications. Faster machines meant nothing if they did not enable AI. Processing speed meant nothing if it did not support symbolic logic. The achievement of machine intelligence required that all levels of the pyramid connect to that goal.

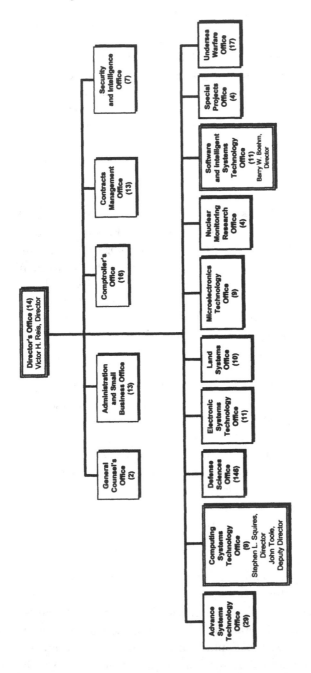

Total Positions: 178
(Some People are Counted More than Once)

Figure 9.4
DARPA organizational chart, July 1991. Source: DARPA telephone directory.

The weak link throughout the program was software. In short, AI, especially expert systems, could not realize the generic capabilities that had been envisioned in 1983. Custom-designed programs ran successfully on a variety of platforms and found their way into defense applications and commercial products. But the rate of advance lagged far behind the dramatic increases in computer speed experienced in the architecture program and in the commercial market as well.

Squires tried valiantly to address this reverse salient by pumping architecture funds into software development. He insisted for a while that new architectures had to run real applications, not just bench tests, to win certification. He recruited William Scherlis in 1986 to help him work the problem. He paid far more than lip service to this intractable problem.

Somewhere in the budget crisis of 1986–1987, however, Squires appears to have given up. Suspicious, perhaps, that AI-simpatico Saul Amarel had favored software in his distribution of reductions, Squires appears to have begun a long-term course adjustment that brought him by 1991 to a single-minded, even obsessive, commitment to speed as an end in itself. Already fixed on the narrower goal of achieving a TeraOPS machine by 1996, Squires had reversed ends and means in his own mind, had given up the goal of machine intelligence, and had put down the burden of trying to match his machines to the prescribed task. In the grand challenges program that was HPC he found tasks to match his machines.

The momentous consequences of this transformation were compounded by the particular architecture to which Squires had hitched his wagon. His greatest achievement was getting massively parallel processing (MPP) "over the wall." When others had doubted that this technology had a future, Squires spotted Thinking Machines and Intel's Touchstone project. He poured SC money into these developments. He purchased the resulting machines and placed them in the hands of DARPA researchers. And he applied his own brilliant management schemes to avoid the trap of point solutions. His architecture program made MPP work and it helped convince a skeptical computer world that this technology had a future.[77]

It did not, however, discover how to program the machines. Some problems, including many of the grand challenges, were inherently parallel. They lent themselves to solution on parallel architectures. But most problems, including most of those in the symbolic logic of AI, were inherently sequential. They could not easily be mapped into parallel architecture. Simply putting parallel machines in the hands of users did not guarantee that they could, or would, be used. Squires was attempting to make a specific technology generic; the technology did not cooperate.

When the venerable ISTO broke into two parts in 1991, one software and one hardware, it signaled the failure of SC to connect the components of its own agenda. Thereafter SISTO would follow its own research trajectory in search of some Alexander to cut the Gordian knot of custom-designed applications. CSTO, under Squires, set off on a different course, in search of the holy grail of TeraOPS. Left behind was the failed trajectory of SC, the integrated, connected vision of a technology base generating intelligent machines.

Disconnecting the Last Veteran

Ironically, Squires ran CSTO aground shortly after jettisoning SC. The program sank from public view with barely a ripple. Squires, in contrast, crashed with a splash, in a public spectacle that wounded him personally without seriously deflecting the program he had floated. His fall from grace occurred even while HPC, the project that he helped to fashion from the wreckage of SC, continued its ascent.

The publication of the HPC strategy in 1987 had been a call to arms; the announcement of the plan in September 1989 brought about full-scale mobilization, both within and outside of the federal government. Scientists, engineers, and computer professionals applauded the plan; computer industry leaders, including John Armstrong of IBM and John Rollwagen of Cray Research, lent their support.[78] In Congress the Office of Technology Assessment issued a call for federal funding of high performance computing the same month that FCCSET's report came out, fueling congressional interest. New bills were introduced in support of the program, and a spate of hearings was held in 1990 and early 1991. Squires himself made a rare appearance on Capitol Hill, testifying before a House subcommittee on 3 October 1989; Robert Kahn and Jack Schwartz also appeared before other committees.

Senator Gore became an early champion of high performance computing, holding his benchmark hearings on the topic in 1988 and introducing the first legislation in the Senate later that year.[79] At the Gore hearings, Kahn had told the senator that "we should consider you an honorary member of the technical community. You have been very helpful."[80] As it happened, Gore went on to play a stronger role than even

Kahn might have imagined, a role that reshaped HPC and contributed to the downfall of Squires.

The selling of HPC encountered more resistance at the White House than it did in Congress. The FCCSET Executive Committee met repeatedly in 1990 to coordinate a campaign to gain President Bush's blessing. Presidential science adviser Alan Bromley supported the plan and held regular meetings on the subject. At the first such meeting on 8 January 1990, Bromley found consensus among the agencies in favor of the program. They agreed to promote the program on the basis of the need of the country for high-performance computing and network technology instead of on the basis of competitiveness, which would sound too much like industrial policy.[81] The agencies announced publicly their support for the program, the DoD doing so in a speech by the Director of International Economic and Energy Affairs on January 11.[82]

Yet the White House itself continued to resist. There was concern about the budget deficit, but more importantly about setting industrial policy. Several of the president's close advisers, particularly Richard Darman, the Budget Director, and Michael Boskin, the Chairman of the Council of Economic Advisors, were particularly opposed to any interference in the functioning of the free market. (Boskin was quoted as saying, "potato chips, semiconductor chips, what is the difference? They are all chips.")[83] The Office of Management and Budget quickly turned down NASA's request for an additional $20 million, representing its share of the program costs for year one. At a meeting of the FCCSET Executive Committee on January 22, Ron York of OSTP admitted that the chances that the White House would embrace the program as a "presidential initiative" were slim. The agencies would have to launch the program themselves with their own budgeted funds, while trying somehow to win over the administration.[84]

DARPA got an early start on the program. In the fall of 1990 Congress voted funds to DARPA for its portion of the program. During FY 1991 over $17 million was applied to forty-one separate projects to "fast start" HPC. This funding formally established the program within DARPA,[85] contributing to the decision to divide ISTO into separate offices on software and hardware.[86]

At last, largely due to the efforts of science adviser Bromley, the White House acquiesced in the program, and included the high performance computing initiative as part of its budget submission for FY 1992. The request asked for $638 million dollars for HPC activities, an increase of $149 million over the previous year (including $49 million extra for

DARPA), but did not endorse the full five-year program proposed by OSTP.[87] Congressional action was delayed by a dispute over turf. A bill favored by the Senate Committee on Energy and Natural Resources gave the lead for the entire program—and the funding—to the Department of Energy, while the Committee on Commerce, Science, and Transportation endorsed the organization of the program as proposed by OSTP, but offered funding only to NSF, NASA, and NIST. At last a compromise was reached favoring OSTP's plan, and the High Performance Computing Act of 1991 (known as the "Gore Act" after its leading sponsor) was signed into law on 9 December 1991.[88]

In spite of Fields's martyrdom on the altar of national industrial policy, Squires had every reason to believe at the end of 1991 that the trajectory to a TeraOPS machine loomed unobstructed before him. That impression could well have been reinforced the following year when Senator Gore was elected vice president in the new administration of President Clinton. Only the elevation of Jeff Bingaman could have been more welcome to the technical community in general and the computer community in particular.

But Gore's agenda, and Clinton's, were not identical with Squires's. There was a populist flavor to their technological enthusiasm, a wish to spread the blessings of technology to all Americans, not just the elite few who might run their grand challenges on Squires's TeraOPS machines. They soon began to shift the focus of HPC away from number-crunching and toward connecting. Their passion was communications—the Internet. The HPC Act of 1991, which had carried no funding, was followed by a High Performance Computing and Communication Initiative in 1993. This program emphasized communication. It was the basis of Vice President Gore's subsequent boast that he had played a major role in developing the Internet. It fed President Clinton's challenge to put a computer in every classroom in America. It significantly redirected the research trajectory that had appeared to be open before Steve Squires in 1991.

This was not Squires's only political problem. In the spring of 1993 complaints surfaced that DARPA's high-performance computing program had favored Thinking Machines and the Intel Touchstone series over other supercomputer vendors. Together these two suppliers controlled about two-thirds of the commercial market for massively parallel processing machines, a position that critics attributed to DARPA favoritism.[89] Congress called for investigations by the Government Accounting Office (GAO) and the Office of Technology Assessment. The GAO report, issued in May 1993, was particularly damning. It accused DARPA of "overemphasizing hardware" at a time when "software remains too primitive to make massively parallel processing systems useful." It also confirmed that DARPA had "not been involved in any major procurement of new machines made by rivals to Intel and Thinking Machines."[90] This may not have been national industrial policy, but it sure looked like picking winners and losers. Even in the new Clinton administration, such charges were taken seriously. Squires was sacked the following August. One year later, Thinking Machines filed for bankruptcy.[91]

In the turbulence surrounding the criticisms of DARPA and the departure of Squires, SC quietly ended on 30 June 1993. Only the accountants noticed. Squires was the last DARPA veteran of the halcyon days when the Strategic Computing Initiative had promised machine intelligence. Institutional memory had faded with the turnover of personnel and the changing face of research priorities. Having long since struck the banner of SC, DARPA quietly interred the program in its silent graveyard of failed initiatives. DARPA directors never invoke it in the litany of agency successes paraded before Congress at budget time. Only in private reunions of SC veterans is the program resurrected and extolled as a valiant campaign in the crusade for world computer supremacy (see figure 9.5).

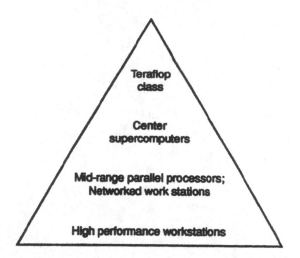

Figure 9.5
Pyramid of High Performance Computing environments. As Steve Squires departed and SC ended, the program's defining icon was transferred to High Performance Computing. *Source:* "From Desktop to Teraflop: Exploring the U.S. Lead in High Performance Computing," August 1993, page viii.

10

Conclusion

DARPA spent $1,000,417,775.68 on SC from 1983 to 1993.[1] Why did it make such an investment? How did it manage its portfolio? What did it buy? None of these questions submit to easy answers. Nor is there a single criterion on which an answer might be based. A common thread, however, runs through the history of SC and through the development of all large-scale technological systems: it is connection. Integration of the SC vision, integration of the SC process, and integration of the SC technology challenged the program's principals from Robert Kahn's first conceptualization in the early 1980s through Stephen Squires's fall from grace in 1993. Those two men understood more clearly than most that SC would rise or fall on the back of systems integration. Following that conceptual thread provides the best map of SC's history and the best measure of its achievement.

As in the story that has gone before, the term "connect" is used here to describe several different kinds of integration. The central meaning, however, is constant. Large-scale technological systems and the R&D programs that produce them require integration. Component development is crucial; connecting the components is more crucial. The system is only as strong as its weakest link. Machine intelligence could not be achieved unless all levels of the pyramid worked and connected.

These concluding remarks seek to appraise SC from the vantage point of connection. They are organized in three parts: why, how, and what? The result, one hopes, is an appreciation of SC's accomplishments, a delineation of its failures, and something of a framework for thinking about all large-scale technological systems. It may well be that R&D differ from the systems they produce; this study argues they do not.

Why?

SC sprang from the convergence of four historical moments. It is a futile exercise in counterfactual history to guess whether it would have happened had any of those moments been different or absent. What one can conclude with some confidence is that the character of the program and the historical trajectory it pursued took shape from this particular convergence. Traces of those four shaping forces were still evident when the program closed its books.

Robert Kahn introduced one factor: the technological paradigm shared by most members of the DARPA computer community. It saw VLSI, architecture, and AI as all being ripe for significant advances in performance. If those advances could be integrated, connected, they might produce intelligent machines. If the United States did not seize this moment, other nations would.

Ronald Reagan assumed the presidency in 1981 with another historical trajectory in train. He wanted to unleash free enterprise in the United States and challenge the Soviet Union more aggressively around the world. He was prepared to spend his popular appeal and political capital to increase defense spending and most especially to invigorate defense R&D. He could provide the dollars to fund Kahn's vision.

The Japanese government entered the picture when it announced its Fifth Generation computer program. Motivated in large measure by the quest for machine translation, and targeted specifically on AI, the program threatened to leap-frog the United States and assume world leadership in high-end computer technology. To many in Congress, this threat proved more alarming than Ronald Reagan's evil empire. To the technical community, the Fifth Generation was both a threat and a lever with which more funding could be pried loose from government.

The fourth ingredient of the historical soup that spawned SC was the dynamic relationship between Robert Kahn and his boss, Robert Cooper. Personally, professionally, and philosophically alike in many ways, they nonetheless harbored distinctly different views of how to organize, sell, and manage the computer research initiative that Kahn proposed. For the most part their differing views complemented one another and provided a powerful combination for launching the program. In some ways, however, their differing views of what SC was and what it should accomplish bifurcated the program from the outset and introduced a conceptual tension that their successors were powerless to overcome. Their collaboration worked; they got the money. Because they never fully integrated their differing visions, however, the program never connected in a single plan of how to proceed.

How?

At least eight different attempts were made to impose a management scheme on SC. Kahn's and Cooper's were among the earliest, but by no means the most influential. None was entirely successful; none lasted long. The catalogue of what they attempted, however, provides something of a primer on the management of large-scale R&D programs. All attempted, in their own way, to connect the parts of the program and to move them forward in concert toward the common goal of machine intelligence.

DARPA's high-risk, high-gain tradition served as the default management scheme. It surely guided SC throughout its history. In practice, however, it proved to be more of a philosophy than a technique or a doctrine. It encouraged a certain kind of research investment without explaining how any given project might connect either with other DARPA programs or with the technological field under development. It provided institutional justification for an undertaking as ambitious as SC, but it offered no clear guidance on how to proceed.

Neither did Kahn. The godfather of SC ran the program out of his hip pocket. No one doubted that he saw the program whole and understood how the various parts would connect. And no one doubted that he had specific strategies in mind, such as his investment matrix. But he also harbored a strong belief that research was contingent and unpredictable. He would not, or could not, deliver to Robert Cooper the detailed, specific plan that the DARPA director needed to sell the program. Instead, he wanted to follow his nose, to go where the research led, to navigate by technology push.

Cooper insisted on demand pull. His clients—Congress, the White House, the military services—were mission driven. They wanted products that would help them achieve their goals. Cooper insisted that the SC plan specify the payoff. Otherwise, he believed, it would not sell. At his urging, the Autonomous Land Vehicle, the Pilot's Associate, and Battle Management replaced Kahn's "technology base" as the focus of SC. Instead of developments rising under their own power through Kahn's pyramid, they would be drawn by applications along Cooper's time lines.

Lynn Conway joined DARPA to articulate and execute the plan. A "process person," she was to impose on SC a management scheme that would integrate Kahn's technology push with Cooper's demand pull. Her first synthesis, the SC plan of October 1983, bode well for her prospects. It included both Kahn's pyramid and Cooper's time line without provoking any visible public concern about their fundamental incompatibility. Her real gift for executing the ideas of others suggested that the differences between Cooper and Kahn might be finessed in practice. Soon, however, Conway's style and ambitions ran afoul of Kahn's, and hopes for a compromise management scheme evaporated. Kahn, Cooper, and Conway all left DARPA within months of each other in 1985 without bestowing on SC a strategy to connect the various parts of the program.

Ronald Ohlander had to manage the technology components for which connection was most important and most difficult. As the program manager for AI, he oversaw developments such as expert systems, the key software in the achievement of machine intelligence. Internally, these systems had to integrate capabilities such as blackboarding, chaining, and reasoning with uncertainty. Externally, they had to run on the new architectures being developed by SC and also execute the specific tasks assigned to the ALV, the Pilot's Associate, and Battle Management. It is little wonder, then, that Ohlander thought more than most about connecting. He produced the next attempt at a management scheme for SC. His "new generation" plan assigned integration to a single contractor.

In this case, one of the AI contractors would assume responsibility for integrating the components provided by researchers in the technology base. This AI team would then work with the applications contractors to install the new capability. Though some critics believed that all the researchers in the program should take responsibility for integration, Ohlander's scheme was nonetheless a plausible solution to the problem. But he left DARPA before it was fully developed, and his successors never pursued it consistently enough to give it a chance.

Saul Amarel, one of the critics of Ohlander's scheme, sought to effect integration at the DARPA office level. He reunited the Information Processing Techniques Office and the Engineering Applications Office, which had been sundered in the Kahn-Conway struggle. The new institutional entity, the Information Science and Technology Office, revealed in its title Amarel's appreciation for the problem facing SC. But Amarel refused to treat SC as a separate DARPA agenda. Rather he scrambled the programs being supported with SC funds and laid out a new set of categories significantly different from those articulated in the SC pyramid. This did not mean that the component projects could not be successfully connected with one another and integrated into useful applications. It did cut some of the old lines of connection, however, and force the creation of new ones. In the process, it killed off whatever chance of success Ohlander's "new generation" scheme might have had. Unfortunately, Amarel spent much of his time and energy during his two years at DARPA trying to restore the funding lost to the budget crises of 1986 and 1987. His plans for integration starved for lack of attention and resources, without demonstrating whether or not they might have worked.

Jack Schwartz brought an entirely new management philosophy to the enterprise. Scornful of the conceit that a technology as complex as machine intelligence could be orchestrated at all, he dismissed the SC plan as a swimmer fighting the current. In its place he proposed to ride the wave. Not so different from Kahn's determination to follow the promising developments, Schwartz's plan nonetheless withdrew support from those research projects that Thomas Hughes would call reverse salients. Schwartz was not prepared to pump more money into expert systems, for example, when they showed so little promise of achieving their goals. But without expert systems, Cooper's applications would not work. Indeed, Kahn's machine intelligence would not work either, at least not in the form that he had originally suggested. One can only guess what he might have done in this circumstance, whether he would have tried to bring expert systems up on line with the rest of this broad technological front or whether he would have revised his objectives to match what the technology could produce. In any case, Schwartz rode the wave of architecture and contributed to a redirection of the SC trajectory.

While all these management schemes were being invented, tried, and abandoned, Stephen Squires slowly developed the most sophisticated and arguably the most successful plan of all. Not a single plan, it was rather an amalgam of management principles that he had extracted from his own experience and from the study of others. At the heart was the maturity model and gray coding. The maturity model, reminiscent of Kahn's matrix, allocated research funds in increasing increments as technological programs displayed more complex capability. Good ideas could qualify easily for small budgets at the outset. If the projects succeeded and demonstrated scalability, they could qualify for ever-increasing levels of support until they finally demonstrated proof of concept and transitioned into commercial or military applications.

Gray coding focused on incremental improvement in existing capabilities by parameter variation that changed one component of a system at a time. This pinpointed the performance of the single changed component and avoided the wholesale assault on point solutions that came with trying to invent a new end-product from scratch. To these and his other management techniques, Squires added the concept of a technology food chain. The technology food chain restated one of the cardinal principles of SC embraced by both Kahn and Cooper at the outset: the program must produce real-world products that migrate from paper studies in the laboratory to real implementations in the hands of users.

In the long journey from Kahn to Squires, SC had returned home to understand itself for the first time. Squires had been at DARPA for ten years when SC ended, just as Kahn had been at DARPA for ten years when he wrote the first SC proposal. Both developed their management styles from years of experience and contemplation. Both demonstrated an ability to manage such programs successfully. But neither was able or willing to articulate their management scheme in a form that might be taken up by others. Kahn managed by example, Squires by fiat. Both created a culture that depended on their presence. Neither left behind an institutional culture of the kind that the legendary J. C. R. Licklider had bequeathed to IPTO.

In sum, SC was managed erratically. The turnover in personnel was itself enough to disrupt the careful orchestration that its ambitious agenda required. Similarly disruptive was the fundamental disconnect between the Kahn and Cooper visions, both of which were embedded in the SC plan. Some technologies were pushed, as in Schwartz's wave model, while others were pulled, as in Squires's TeraOPS obsession. Some researchers waited in vain to be connected with developments in the technology base that never materialized. Other researchers produced technologies whose applications were canceled. When Carnegie Mellon put NavLab III on the road, it connected to a Sparc laptop that was not part of SC at all. If the related technologies that were marshalled in 1983 to make a coordinated assault on machine intelligence took any ground, it was not because of orchestration by SC. The component parts of SC connected with one another haphazardly and incompletely.

What, then, might be said of agency? Did these managers shape the course of technological development? They surely did at a certain gross level. They got the money, and they distributed it to the SC researchers. They nurtured what they perceived to be promising lines of development, and they starved what they perceived to be dead ends. But collectively they displayed no consistent doctrine for distinguishing between promising and discouraging trajectories, and for that reason, in part, they produced a camel instead of a horse. In the end, their program looked more like the one envisioned by Kahn than the one that Cooper had hoped for. They could not make it deliver an ALV or a PA, or perhaps they would not. In any event, they did not. Thus, for all of their agency, their story appears to be one driven by the technology. If they were unable to socially construct this technology, to maintain agency over technological choice, does it then follow that some technological imperative shaped the SC trajectory, diverting it in the end from machine intelligence to high performance computing?

Institutionally, SC is best understood as an analog of the development programs for the Polaris and Atlas ballistic missiles. An elaborate structure was created to sell the program, but in practice the plan bore little resemblance to day-to-day operations. Conceptually, SC is best understood by mixing Thomas Hughes's framework of large-scale technological systems with Giovanni Dosi's notions of research trajectories. Its experience does not quite map on Hughes's model because the managers could not or would not bring their reverse salients on line. It does not quite map on Dosi because the managers regularly dealt with more trajectories and more variables than Dosi anticipates in his analyses. In essence, the managers of SC were trying to research and develop a complex technological system. They succeeded in developing some components; they failed to connect them in a system. The overall program history suggests that at this level of basic or fundamental research it is best to aim for a broad range of capabilities within the technology base and leave integration to others.

What?

The results of SC may be seen as technologically determined. The program set out to develop machine intelligence. When that proved impossible, it achieved what it could—high performance computing. In other words, the program went where the technology allowed it to go.

In another sense, however, SC's results need to be mapped on the expectations that people had for it. The program was different things for different people. For each community, the trajectory it followed wants comparison with the trajectory they anticipated. For those who never believed that SC was an AI program, its failure to achieve machine intelligence is neither surprising nor remarkable.

One group, especially in Congress, believed that it represented the United States response to the challenge of Japan's Fifth Generation. If so, it was a great success. While the Fifth Generation program contributed significantly to Japan's national infrastructure in computer technology, it did not vault that country past the United States. Indeed, the U.S. lead over Japan in AI and software engineering increased over the decade of SC.[2] Credit for this achievement must be shared by many programs and activities outside SC: VHSIC, MCC, Sematech, the chip and computer manufacturers, Microsoft, even the entertainment industry, whose development of computer graphics has set the world standard in the last twenty years.[3] SC played an important role, but even some SC supporters have noted that the Japanese were in any event headed on the wrong trajectory even before the United States mobilized itself to meet their challenge.

Those who saw SC as a cat's paw for the Strategic Defense Initiative (SDI) find themselves left with ironies inside ironies. No evidence has come to light suggesting that the relationship they feared existed in fact. Indeed, Robert Cooper and Richard DeLauer both distrusted SDI and contributed directly to its removal from DARPA. Cooper had to give up $400 million and his Gallium Arsenide program to unburden DARPA of SDI. Nor does SC appear to have done much work directly focused on strategic defense, at least not until Steve Squires offered his consulting services in return for some of the funding lost in the budget battles of the mid-1980s. The real distortion of the SC research agenda came when President Reagan insulated SDI from the Gramm-Rudman-Hollings cuts, letting those reductions fall disproportionately on programs like SC. By that time, however, the Computer Professionals for Social Responsibility had receded from public prominence and shifted their focus to other concerns. SC proved unable to produce the autonomous weapons they had once feared, let alone the Doomsday machine that they imagined Star Wars might become.

Robert Kahn looked upon SC as a way to build the national computer research infrastructure. He believed that the country had surplus talent in computer research that had been starved since the budget cuts of the mid-1970s. He wanted to put funds and equipment in their hands and nurture the development of the next generation of computer researchers—the Conrad Bocks of the world. He funded the MOSIS program even before SC began, because he wanted researchers in universities to have the chance to design and test their own chips just as industry researchers could. Few segments of SC can claim to have been more successful than infrastructure. Though DARPA indulged the researchers' call for LISP machines at just the time when the field was turning to work stations, it quickly recovered and gave people the hardware they needed. MOSIS moved with the state of the art and made its implementation services available to all bona fide applicants. Though the program itself did not fit the DARPA model, the agency quietly sustained it and thereby accelerated the development of both chips and researchers.

The great lacuna in the infrastructure program was networking. No technology is more consonant with the SC objectives; no technology resonates so fully with the SC requirement to connect. Kahn cut his teeth on networking and viewed it as a central component of his plan. But Barry Leiner chose to keep networking out of SC. By the time it finally joined the program in the late 1980s, leadership in government support of networking had passed to the National Science Foundation, working in the context of the Federal Coordinating Council for Science, Engineering, and Technology. Though DARPA's networking research remained important throughout SC, it did not enjoy the prominence it had during the development of ARPANET. Indeed, DARPA did not include communication in its initial vision of high-performance computing. It had become obsessed with building nodes on the information superhighway, not with connecting them.

Many believed that SC was a supercomputer program. The "big iron" of mainframe supercomputers was approaching a "presumptive anomaly," a point beyond which inflexible physical constraints would preclude further advances in speed. Multiprocessing appear to be the most promising answer, especially massively parallel processing. Previous attempts to subdue the bewildering complexity of such architectures had failed for want of practical connections between the processors and memory. New schemes were in the air at the beginning of SC. Squires nurtured all that showed promise, especially Intel's Touchstone and Danny Hillis's Connection Machine, the quintessential SC artifact. Squires got this technology over the wall. Real machines transitioned into practical applications and demonstrated that massively parallel architectures could work.

In many ways, this appears to be the most visible success of SC, the technological development that might be compared with earlier DARPA legends such as ARPANET, time-sharing, and graphics, but it was also fraught with problems. The massively parallel machines have proved devilishly difficult to program for most applications and remain limited to certain problem domains. Thinking Machines went belly-up as a hardware manufacturer and reinvented itself as a software services firm. Squires lost command of ISTO for playing favorites and responding poorly to visions other than his own. His cherished TeraOPS goal has been reached, in part because the government-wide program in high performance computing adopted some of the SC principles that he helped to develop. But too many hands stirred the soup of HPCC to single out TeraOPS as an exclusively SC achievement.

AI provided grist for the mills of both its supporters and its critics. Perception decides. To some, the glass is half full; to others, half empty. At the coarse-grained level, AI fell into another AI winter. The creation of the Software and Intelligent Systems Technology Office in 1991 signaled a turn away from AI in the direction of software engineering. The separation of this office from the Computer Systems Technology Office signaled a new disconnect between architecture and AI. During SC, AI had proved unable to exploit the powerful machines developed in SC's architectures program to achieve Kahn's generic capability in machine intelligence.

On the fine-grained level, AI, including many developments from the SC program, is ubiquitous in modern life. It inhabits everything from automobiles and consumer electronics to medical devices and instruments of the fine arts. Ironically, AI now performs miracles unimagined when SC began, though it can't do what SC promised.

When the specific SC projects are examined, still more irony emerges. The two most promising technologies, expert systems and vision, proved the greatest disappointments. The many component capabilities of expert systems, such as blackboards, logic chaining, and reasoning with uncertainty, could not be connected in a single, generic shell that could coordinate their capabilities. Vision could not produce an algorithm that would consistently connect the millions of data points the hardware could manipulate into a coherent picture of what it was seeing. In contrast, speech recognition and natural language programs met or exceeded the ambitious goals of the 1983 SC plan. In these realms of AI, applications have transitioned out of the laboratory and into military services and the commercial marketplace.

Applications provide a similarly mixed record. Battle Management succeeded. It gave the navy two expert systems that perform useful planning and analysis functions. PA delivered some useful components, but it was canceled short of becoming R2D2. The ALV was also canceled, but NavLab continues to function as a research vehicle as this book goes to press. None of the NavLabs can do what the SC plan promised that ALV would do, but they have far greater capability than any autonomous vehicle in existence when SC began. Some of the applications added as SC proceeded have had remarkable success. The DART logistics loading program, for example, is said to have contributed enough to U.S. prosecution of the 1991 Gulf War to justify all the expenditures made on AI during SC.

In some ways the varying records of the SC applications shed light on the program models advanced by Kahn and Cooper at the outset. Cooper believed that the applications would pull technology development; Kahn believed that the evolving technology base would reveal what applications were possible. Kahn's appraisal looks more realistic in retrospect. It is clear that expert systems enjoyed significant success in planning applications. This made possible applications ranging from Naval Battle Management to DART. Vision did not make comparable progress, thus precluding achievement of the ambitious goals set for the ALV. Once again, the program went where the technology allowed. Some reverse salients resisted efforts to orchestrate advance of the entire field in concert. If one component in a system did not connect, the system did not connect.

Finally, SC tried to transition technology out of the laboratory and into the hands of users. Specifically, it tried to bring university researchers into collaboration with industry producers. It supported start-up companies such as Teknowledge, IntelliCorp, and Thinking Machines that were themselves rooted in academia. It promoted the transition of university research such as the iWarp chip into commercial products. And it funded collaboration among university laboratories and industrial giants such as Martin-Marietta, General Electric, and Westinghouse. Its support was not enough to save Teknowledge and IntelliCorp, however, which reverted to their previous niches.

Some believe that the heady artificiality of DARPA support contributed to the downfall of Thinking Machines by isolating the company for too long from market forces. Large companies such as Intel were better able to take advantage of collaboration with DARPA without suffering unduly if specific projects fell short of expectations. Perhaps most importantly, some of the collaborations between academia and industry proved to be long-standing; Carnegie Mellon and Martin Marietta still cooperate on ALV. Some collaborations led to transition without direct pressure from DARPA; the speech recognition program at MIT has found commercial partners to replace SC funding. And some projects, such as MOSIS, simply could not be commercialized; they continued to receive support from DARPA as a public good.

In the final analysis, SC failed for want of connection. The vision of machine intelligence provided a powerful organizing principle for a

suite of related research programs. It did not, however, solve the coordination problem. How could parts of the system be integrated when their
very existence was contingent? Neither Kahn's pyramid nor Cooper's
time line could make this happen. Stephen Squires came closest to
achieving it, but his answer was to take what the technology offered, to
abandon the SC goal, to follow the siren song of TeraOPS.

The abandonment of SC is one of the most remarkable aspects of the
story. By the late 1980s it had become apparent that the program was
not going to achieve either Cooper's applications or Kahn's "broad base
of machine intelligence technology." Steve Squires flirted with SC2, a
plan for a renewed assault on machine intelligence. Instead, DARPA
elected simply to sweep SC under the carpet and redirect computer research toward the "grand challenges" of high performance computing.
Numerical processing replaced logical processing as the defining goal.
Even while funding continued to flow into the SC budget category, SC
as a program receded from view. The General Accounting Office report
of 1993 discerned this slight of hand, but by then it was a fait accompli;
Congress failed to call for an accounting.

The problem of accountability looms large. Government agencies
such as DARPA often justify their proposals for expensive, ambitious R&
D projects with hyperbole and optimism. Cooper's applications promised point solutions that Kahn distrusted, but without them, there would
have been no assault on the windmills of Kahn's imagination—the
promise of machine intelligence. Little wonder, then, that agencies grow
cynical about the political process, that they promise what they must and
obfuscate when they fail. DARPA may fail on 85 percent of its attempts,
giving it ample cause and opportunity to practice obfuscation. The price
paid for this game of hyperbole and cover-up is that programs such as
SC escape the critical review that $1 billion in public treasure surely call
for. This study cannot begin to provide the technical and managerial
appraisal such an undertaking warrants. But this history suggests that
Congress should mandate such post mortems when it buys into projects
on such a grand scale.

The marvel of SC, however, is not so much that DARPA hid the results
of SC, or even that the program failed, but rather that DARPA tried at
all. For an agency pursuing a high-risk, high-gain philosophy with perhaps an 85 percent failure rate, the odds against SC were overwhelming.
Indeed, one cannot escape the suspicion that SC was always a triumph
of packaging over substance. To get $1 billion in new money from Con-

gress, one has to promise a lot. And one has to hope that no one asks where the bodies are buried at the end of the day. On balance SC probably spent its funding as effectively as any government research program, more effectively than most. And it can claim credit for some important advances, even though they were components, not a system. SC was a pot of money used to nourish the technology base, not a coordinated assault on machine intelligence.

The history of SC therefore remains silent on the issue of agency in technological development. Cooper was right that one can design and mange complex technological systems such as the Apollo spacecraft. The key is that all the components are in hand or within reach. When reverse salients appear, they can be brought back on line by application of sufficient resources. It is a problem in systems integration, connection.

Research and development works differently. The components want more than connection. They want invention. They must be created. No rational plan can hope to connect them in advance. SC promised to do that and failed. It is hard to see how it might have succeeded.

Notes

Preface

1. "Strategic Computing," the subject of this book, began as an "Initiative" in 1983. Soon, however, the term "Initiative" disappeared from DARPA's discussion of the program. "Initiative" is a term of art in Washington, often applied to programs that are funded with "new money," as opposed to funds redirected from other programs. Many administrators connected with the Strategic Computing Initiative found it expedient to remove from its title this reminder that it was essentially an add-on to the normal DARPA agenda. They would therefore refer to the Strategic Computing Program, or simply Strategic Computing. Those terms will be used here, abbreviated as SCP or SC.

Introduction

1. In their research, the authors never found an accounting of the cost for SC. After termination of the authors' contract, Roland submitted a Freedom of Information Act (FOIA) request asking "how much money was appropriated or transferred to [the Strategic Computing Program, 1983–1993] and from what sources, and also . . . how the funding was spent." (Roland to DARPA FOIA Officer, 19 May 1998) Roland received in reply a letter from A. H. Passarella, Director, Department of Defense Directorate for Freedom of Information and Security Review dated 22 July 1998. To it were attached two undated documents, one labeled 1A10E and the other ST-10. These were DARPA account codes that we had encountered in our research.

The first document (four pages) itemized expenditures of $41,817,979.00 in FY 1983–1986. The second document (forty-four pages) itemized expenditures of $958,599,796.68 in FY 1984–1993. The documents do not say whether they represent obligations or expenditures. Nor do they indicate the appropriations or transfers whence the funds came to DARPA. We are left to conclude that by DARPA's official accounts, the program spent $1,000,417,775.68 over eleven years. Of that amount, only one expenditure, $601,984 for Columbia's Non-Von computer project, was made before FY 1984.

The authors were told, without any supporting documentation, that so-called "black money" passed in and out of the SC budget. These are funds for secret programs. Congress's total appropriation to agencies such as the DoD are public. Within these allocations, however, are embedded funds for secret programs which are not identified in the public documents. Agencies of DoD are "taxed" a certain amount of their nominal allocations to generate funds to be redirected to secret programs. Worried that the actual funding for SC might be considerably more or less that the unclassified budget, the authors asked Elias M. Kallis, former director of program management for DARPA's Information Processing Techniques Office. He estimated that the public figures were about right, since the black money flowing into SC about equaled the black money flowing out.

2. Pamela McCorduck, *Machines Who Think: A Personal Inquiry into the History and Prospects of Artificial Intelligence* (New York: W. H. Freeman, 1979).

3. J. C. R. Licklider, "Man-computer Symbiosis," *RIE Transactions on Human Factors in Electronics* (March 1960): 4–11.

4. Simon Nora and Alain Minc, *Information and Society* (Cambridge, MA: MIT Press, 1980), 107. They described this thick fabric as "connecting research centers, universities, and private firms in a web of contracts and information sharing involving an intense exchange of workers. This is, to define it briefly, the 'American model.' It requires great flexibility, great permeability between institutions, and insignificant social cleavages."

5. Thomas P. Hughes, *Networks of Power: Electrification in Western Society, 1880–1930* (Baltimore: Johns Hopkins University Press, 1983).

6. Richard A. Jenkins, *Supercomputers of Today and Tomorrow: The Parallel Processing Revolution* (Blue Ridge Summit, PA: Tab Books, 1986).

7. Merritt Roe Smith and Leo Marx, eds., *Does Technology Drive History?: The Dilemma of Technological Determinism* (Cambridge, MA: MIT Press, 1994).

8. Wiebe E. Bijker, Thomas Parke Hughes, and T. J. Pinch, eds., *The Social Construction of Technological Systems: New Directions in the Sociology and History of Technology* (Cambridge, MA: MIT Press, 1987).

9. Thomas S. Kuhn, *The Structure of Scientific Revolutions*, 2d ed., enl. (Chicago: University of Chicago Press, 1970); Edward W. Constant II, *The Origins of the Turbojet Revolution* (Baltimore: Johns Hopkins University Press, 1980).

10. Hugh G. J. Aitken, *Syntony and Spark: The Origins of Radio* (New York: Wiley, 1976); *The Continuous Wave: Technology and American Radio, 1900–1932* (Princeton: Princeton University Press, 1985).

11. Donald A. MacKenzie, *Inventing Accuracy: An Historical Sociology of Nuclear Missile Guidance* (Cambridge, MA: MIT Press, 1990).

12. Michael Callon, "Society in the Making: The Study of Technology as a Tool for Sociological Analysis," in *The Social Construction of New Technological Systems*, ed. by Bijker, Hughes, and Pinch, 83–103, esp. 93–97; Bruno Latour, *Science in Action* (Cambridge, MA: Harvard University Press, 1987).

13. See most recently Arthur L. Norberg and Judy E. O'Neill, *Transforming Computer Technology: Information Processing for the Pentagon, 1962–1986* (Baltimore: Johns Hopkins University Press, 1996).

14. J. P. Mullins, *The Defense Matrix: National Preparedness in the Military-Industrial Complex* (Avant Books, 1986); F. A. Long and J. Reppy, *Genesis of New Weapons: Decisionmaking and Military R&D* (New York: Pergamon, 1980); Jacques Gansler, *Affording Defense* (Cambridge, MA: MIT Press, 1989); Deborah Shapley and Rustum Roy, *Lost at the Frontier: U.S. Science and Technology Policy Adrift* (Philadelphia: ISI Press, 1985).

15. Burton I. Edelson and Robert L. Stern, *The Operations of DARPA and Its Utility as a Model for a Civilian DARPA* (Washington: Johns Hopkins Foreign Policy Institute, 1989).

16. Lillian Hoddeson and Gordon Baym, *Critical Assembly: A Technical History of Los Alamos during the Oppenheimer Years, 1943–1945* (Cambridge: Cambridge University Press, 1993); Alex Roland, *Model Research: The National Advisory Committee for Aeronautics, 1915–1958*, 2 vols. (Washington: NASA, 1985).

17. Stuart W. Leslie, *Boss Kettering* (New York: Columbia University Press, 1983); Ronald R. Kline, *Steinmetz: Engineer and Socialist* (Baltimore: Johns Hopkins University Press, 1992); George Wise, *Willis R. Whitney, General Electric, and the Origins of U.S. Industrial Research* (New York: Columbia University Press, 1985).

18. Merritt Roe Smith, ed., *Military Enterprise and Technological Change: Perspectives on the American Experience* (Cambridge, MA: MIT Press, 1985).

19. Paul Edwards, *The Closed World: Computers and the Politics of Discourse in Cold War America* (Cambridge, MA: MIT Press, 1996).

20. Alex Roland, *The Military-Industrial Complex* (Washington: American Historical Association, 2001).

21. Joseph Schumpeter, *The Theory of Economic Development* (Cambridge, MA: Harvard University Press, 1934 [1912]); and *Capitalism, Socialism and Democracy* (Cambridge, MA: Harvard University Press, 1943).

22. John Jewkes, David Sawers, and Richard Stillerman, *The Sources of Invention*, 2d ed. (New York: Norton, 1969).

23. R. Nelson and S. Winter, "In Search of a Useful Theory of Innovation," *Research Policy* 6 (1977): 36–76; and *An Evolutionary Theory of Economic Change* (Cambridge, MA: Harvard University Press, 1982). See also Joel Mokyr, *The Lever of Riches: Technological Creativity and Economic Progress* (New York: Oxford University Press, 1990), and compare it with George Basalla, *The Evolution of Technology* (Cambridge: Cambridge University Press, 1988).

24. Giovanni Dosi, "Technological Paradigms and Technological Trajectories," *Research Policy* 11 (1982): 147–162. For a critique of this line of argumentation,

see Donald MacKenzie, "Economic and Sociological Explanation of Technical Change," in *Technological Change and Company Strategies: Economic and Sociological Perspectives,* ed. by Rod Coombs, Paolo Saviotti, and Vivien Walsh (London: Harcourt Brace Jovanovich, 1992), 25–48, esp. 30–35. MacKenzie, a social constructivist, argues that technological trajectories can be self-fulfilling prophecies.

25. Paul A. David, "Clio and the Economics of QWERTY," *American Economics Review* 75 (1985): 332–337; W. Brian Arthur, "Competing Technologies, Increasing Returns, and Lock-in by Historical Events," *The Economic Journal* 9 (March 1989): 116–131; W. Brian Arthur, Yu. M. Ermoliev, and Yu. M. Kaniovski, "Path-dependent Processes and the Emergence of Macro-structure," *European Journal of Operational Research* 30 (1987): 294–303.

26. Partha Dasgupta and Paul A. David, "Toward a New Economics of Science," *Research Policy* 23 (1994): 487–521.

27. Good starting points are Martin-Campbell-Kelly and William Aspray, *Computer: A History of the Information Machine* (New York: Basic Books, 1996); and Paul E. Ceruzzi, *A History of Modern Computing* (Cambridge, MA: MIT Press, 1998). On government support of computer development, see Kenneth Flamm, *Targeting the Computer: Government Support and International Competition* (Washington: Brookings Institution, 1987); and *Creating the Computer: Government, Industry, and High Technology* (Washington: Brookings Institution, 1988); and National Research Council, *Funding a Revolution: Government Support for Computing Research* (Washington: National Academy Press, 1999).

Chapter 1

1. Jeffrey Rothfeder, *Minds over Matter: A New Look at Artificial Intelligence* (New York: Simon & Schuster, 1985), 182. See also Andrew Jenks, "Reston Networking," *Washington Technology* (6 December 1990), which calls Kahn "the godfather of networked computing."

2. Interview of Charles Buffalano by Alex Roland and Philip Shiman, Greenbelt, MD, 2 August 1993.

3. Quoted in M. Mitchell Waldrop, *Man-made Minds: The Promise of Artificial Intelligence* ([New York]: Walker &. Co, 1987), 164.

4. Eric Hoffer, *The True Believer: Thoughts on the Nature of Mass Movements* (New York: Time, 1963).

5. DARPA was originally the Advanced Research Projects Agency. It was created with that title in 1958, a branch of the DoD. In 1972, its title was changed to DARPA; in 1993 it was changed back to ARPA, but only briefly. Because it had the title DARPA throughout the period being studied here, it will be so called throughout this text, unless reference is being made to the earlier or later periods.

6. Keith Uncapher, for one, had suggested something like it to Kahn in the mid-1970s. Interview of Keith Uncapher by Philip Shiman, 8 August 1994, Los Angeles, CA.

7. "Corporation for National Research Initiatives: Building a Strategic Advantage in America," brochure (Reston, VA: CNRI, n.d.). See also John Markoff, "Creating a Giant Computer Highway," *New York Times,* 2 September 1990, Business 1,6; and "A Talk with Robert Kahn; Part 1: National Research Initiatives," *The CPSR [Computer Professionals for Social Responsibility] Newsletter* 5 (spring 1987): 8–13; Fred V. Guterl, "Bob Kahn Wants to Wire the Nation," *Business Month* (April 1987): 30–32.

8. Paul Edwards, *The Closed World: Computers and the Politics of Discourse in Cold War America* (Cambridge, MA: MIT Press, 1996).

9. Interview of Robert Kahn by William Aspray, Reston, VA, 22 March 1989, 1. [Hereafter Kahn-Aspray interview.]

10. Robert Buderi, *The Invention That Changed the World: How a Small Group of Radar Pioneers Won the Second World War and Launched a Technological Revolution* (New York: Simon and Schuster, 1996), 354–379.

11. Interview of Robert Kahn by Judy O'Neill, Reston, VA, 24 April 1990, 8. [Hereafter Kahn-O'Neill interview.]

12. Katie Hafner and Matthew Lyon, *Where Wizards Stay up Late: The Origins of the Internet* (New York: Touchstone, 1998), 89.

13. R. Moreau, *The Computer Comes of Age: The People, the Hardware, and the Software* (Cambridge, MA: MIT Press, 1986), 75–111.

14. Kent C. Redmond and Thomas M. Smith, *From Whirlwind to MITRE: The R&D Story of the SAGE Air Defense Computer* (Cambridge, MA: MIT Press, 2000).

15. Emerson W. Pugh, *Building IBM: Shaping an Industry and Its Technology* (Cambridge, MA: MIT Press, 1995).

16. Buderi, *The Invention That Changed the World*, 392.

17. Ibid., 398.

18. Ibid., 398–399.

19. Arthur L. Norberg and Judy E. O'Neill, *Transforming Computer Technology: Information Processing for the Pentagon, 1962–1986* (Baltimore: Johns Hopkins University Press, 1996), 155–185.

20. Janet Abbate, *Inventing the Internet* (Cambridge, MA: MIT Press, 1999), 48–49 and ff. Unless otherwise noted, the following account of ARPA network research is derived from this pioneering study and from Norberg and O'Neill, *Transforming Computer Technology*.

21. Norberg and O'Neill, *Transforming Computer Technology*, 34.

22. Ibid., 162–167.

23. Packet switching is attributed to Donald Davies of Britain's National Physical Laboratory, who first developed a small-scale prototype of how it might work. His ideas, however, were neither widely known nor broadly applicable when Roberts and Kahn entered the field. The term "packet switching" finally prevailed over the more accurate but less felicitous "adaptive message block switching," which had been proposed by Paul Baran of the RAND Corporation. Kahn-O'Neill interview, 13–16.

24. Robert Kahn comments on early draft manuscript, 12 June 1995.

25. Kahn-Aspray interview, 5.

26. Kahn-O'Neill interview, 5.

27. As Janet Abbate has made clear, this is a wonderful example of social construction of technology. The network initially allowed flexible interpretation of what it was and what it would do. A kind of closure was reached only when users settled on e-mail as the primary use and relegated distributed computing to a clearly secondary function.

28. BBN found itself at cross purposes with DARPA over commercialization of networking. BBN wanted to retain as proprietary information the codes that Kahn and his colleagues had developed. DARPA insisted that they be made available to others on the net. Abbate, *Inventing the Internet*, 70–71.

29. Anthony F. C. Wallace, *Rockdale: The Growth of an American Village in the Early Industrial Revolution* (New York: Norton, 1980), 211–219. I am indebted to David Hounshell for this comparison.

30. Simon Nora and Alain Minc, *The Computerization of Society: A Report to the President of France* (Cambridge, MA: MIT Press, 1980), 107.

31. Herman H. Goldstine, *The Computer from Pascal to von Neumann* (Princeton: Princeton University Press, 1972), 121–235.

32. Edwards, *The Closed World*.

33. I am indebted to W. Mitchell Waldrop for background information on Licklider drawn from *The Dream Machine: J. C. R. Licklider and the Revolution That Made Computing Personal* (New York: Viking, 2001).

34. Norberg and O'Neill, *Transforming Computer Technology*, 27; Edwards, *The Closed World*, 212, 216, 226.

35. J. C. R. Licklider, "Man-Computer Symbiosis," *IRE Transactions on Human Factors in Electronics* 1 (1960), 4–5; quoted in Abbate, *Inventing the Internet*, 43.

36. On the ARPA-DARPA style, see Norberg and O'Neill, *Transforming Computer Technology*, 1–67; Abbate, *Inventing the Internet*, 54–60, 69–78; John

Sedgwick, "The Men from DARPA," *Playboy* 38 (August 1991): 108, 122, 154–156.

37. Hafner and Lyon, *Where Wizards Stay Up Late*, 38; see also Barry M. Leiner, et al., "A Brief History of the Internet," at *http://www.isoc.org/internet-history/brief.html,* 13 March 2000, 2.

38. Newell interview with Norberg, Pittsburgh, PA, 10–12 June 1991.

39. Ibid., 28–30, 47.

40. Kenneth Flamm, *Targeting the Computer: Government Support and International Competition* (Washington: Brookings Institution, 1987), 56. See also H. P. Newquist, *The Brain Makers: Genius, Ego, and Greed in the Quest for Machines That Think* (Indianapolis, IN: SAMS Publishing, 1994), 64. The hacker culture at MIT also prompted some to call MAC "Man against Machines" and "Maniacs and Clowns." Daniel Crevier, *AI: The Tumultuous History of the Search for Artificial Intelligence* (New York: Basic Books, 1993), 65, 69. Others thought it referred to "Minsky against Corbato," key individuals in AI and time-sharing respectively at MIT; interview with Michael Dertouzos by Alex Roland, Cambridge, MA, 12 April 1995.

41. Kenneth Flamm noted in 1987 that "criticism in recent years of the highly concentrated, non-competitive, and informal nature of computer research funding by DARPA has prompted the agency to make some changes." *Targeting the Computer,* 55.

42. The best explication of the controversy over elitist versus egalitarian distribution of federal support for science appears in Daniel J. Kevles, *The Physicists: The History of a Scientific Community in Modern America,* 2d ed. (Cambridge, MA: Harvard University Press, 1987).

43. For example, Daniel Crevier maintains that Marvin Minsky and Seymour Papert of MIT "brought neural-network research in the United States to a virtual halt" with their influential book *Perceptrons* in 1969. Crevier, *AI,* 106.

44. These tendencies in the early IPTO years were moderated somewhat by reforms in the 1970s and 1980s. See pp. 28–32, 111–114 below. They have not, however, been entirely eliminated.

45. Some, such as Gordon Bell and the members of Computer Professionals for Social Responsibility, question some technical decisions, but seldom do they cast aspersions on the integrity of DARPA and its personnel.

46. Indeed, the services moved ARPANET far beyond its initial goals of connecting computer research centers. It soon added operational military centers to the network and contributed to the overload that made a successor network necessary. Aspray-Kahn interview.

47. Ibid., 14–15.

48. Heilmeier interview, quote in Sidney G. Reed, Richard H. Van Atta, and Seymour H. Deitchman, *ARPA Technical Accomplishments: An Historical Review of Selected ARPA Projects*, vol. I (Alexandria, VA: Institute for Defense Analyses, 1990), chapter 18, 10.

49. Quoted in William H. Chafe, *The Unfinished Journey: America since World War II* (New York: Oxford University Press, 1986), 451.

50. "Defense Authorization Act of 1970," (Public Law 91–121) 19 November 1969.

51. Larry Roberts, "Expanding AI Research," in *Expert Systems and Artificial Intelligence: Applications and Management*, ed. by Thomas C. Bartee (Indianapolis, IN: Howard W. Sams & Co., 1988), 229–230. I am indebted to David Hounshell for this citation.

52. J. Merton England, *A Patron for Pure Science: The National Science Foundation's Formative Years, 1945–1957* (Washington: National Science Foundation, 1983).

53. Interview of George H. Heilmeier by Arthur Norberg, 27 March 1991, Livingston, NJ.

54. *ARPA Technical Accomplishments*, chapter 18, 10–11.

55. Licklider departed in August 1975, just one year and eight months into his second tour and seven months after Heilmeier's arrival.

56. Norberg and O'Neill, *Transforming Computer Technology*, 21. The terminology is Kenneth Flamm's translation of government language. See Flamm, *Targeting the Computer*, 52n.

57. Flamm, *Targeting the Computer*, 54–55.

58. Total federal funding for mathematics and computer science research stagnated from the late 1960s through the mid-1970s, not returning to 1967 levels until 1983, the year that SC began. Flamm, *Targeting the Computer*, 53, 46.

59. Kenneth Flamm, *Creating the Computer: Government, Industry, and High Technology* (Washington: Brookings Institution, 1988), 253.

60. Flamm, *Targeting the Computer*, 44–56.

61. Robert Kahn, comments on early draft, 13 June 1995.

62. U.S. Congress, Senate, Committee on Appropriations, *Department of Defense Appropriations for Fiscal Year 1982*, part 4—Procurement/R.D.T. & E., 97th Cong., 1st sess., 1981, 24.

63. This was also the year in which total federal support for mathematics and computer science regained the level it had enjoyed (in constant dollars) in 1967, just before the peak years of American involvement in Vietnam. Flamm, *Targeting the Computer*, 245.

64. Kahn referred to these as "linchpins," technologies that would tie together multiple developments. Interview of Robert Kahn by Alex Roland and Philip Shiman, Reston, VA, 2 August 1993.

65. Goldstine, *The Computer,* 191–197, 204–205. This notion was based on von Neumann's work with the ENIAC, at the University of Pennsylvania, a parallel machine.

66. Edward W. Constant, II, *The Origins of the Turbojet Revolution* (Baltimore: Johns Hopkins University Press, 1980).

67. J. B. Dennis and D. P. Misunas, "A Preliminary Architecture for a Basic Dataflow Processor," *Proceedings 2nd Symposium on Computer Architecture* (New York: IEEE, 1975), 126–132.

68. J. Schwartz, "The Limits of Artificial Intelligence," *The Encyclopedia of Artificial Intelligence,* ed. by Stuart C. Shapiro (New York: Wiley, 1992), 490.

69. These are the specifications for the IBM 3033, introduced in 1977. They are taken from Bryan Ellickson, "Gauging the Information Revolution," RAND Note N-3351-SF, 7. Ellickson gives figures of 22.2 MIPS and 16,384 KB of memory. These have been converted and rounded off at .832 arithmetic functions per MIP and 8 bits per byte. Compare Ray Kurzweil's estimates in *The Age of Spiritual Machines: When Computers Exceed Human Intelligence* (New York: Penguin Books, 2000), 103.

70. Robert Kahn comments on earlier draft manuscript, 13 June 1995.

71. Nora and Minc, *The Computerization of Society.*

72. See the discussion on VHSIC, pp. 51, 126 below.

Chapter 2

1. Interview of Paul Losleben by Alex Roland and Philip Shiman, Palo Alto, CA, 29 July 1994, 14.

2. "Biographical Sketch of Robert S. Cooper," U.S. Congress, Senate, Committee on Armed Services, Hearing, "Nominations of Robert S. Cooper, James P. Wade, Jr., Everett Pyatt, Charles G. Untermeyer, Donald C. Latham, and Robert W. Helm," 98th Cong., 2[d] sess., 31 July 1984, 3; Cooper was being nominated to be Assistant Secretary, Research and Technology, DoD. See also John A. Adam, "The Right Kind of Accidental Career," *IEEE Spectrum* (July 1990): 44–45.

3. Interview of Robert Cooper by Alex Roland and Philip Shiman, Greenbelt, MD, 12 May 1994, 4. Says Cooper: "I had two choices, go to war or join the ROTC and I chose the ROTC. Many of my friends who didn't went to Korea . . . and came home in coffins. So, it was a reasonable choice."

4. "Biographical Sketch of Robert S. Cooper," Hearings of the Senate Armed Services Committee, 12 April 1984.

5. Adams, "Accidental Career," 45.

6. Basil H. Liddell Hart, *Strategy: The Indirect Approach,* 2d rev. ed. (New York: Praeger, 1968).

7. U.S. Congress, Senate, Committee on Appropriations, *Department of Defense Appropriations for Fiscal Year 1982,* 97ᵗʰ Cong., 1ˢᵗ Sess., part 4—Procurement/ R.D.T. & E., 23–24; Eric J. Lerner, "Technology and the Military: DoD's Darpa at 25," *IEEE Spectrum* 20 (August 1983): 70.

8. *Budget of the United States Government, Fiscal Year 1997, Historical Tables* (Washington: GPO, 1996), 52.

9. Ibid., 265, 267.

10. Robert Cooper offered another comparison in the spring of 1982 when he defended his first DARPA budget before Congress. Of the $258 billion budget proposed for DoD, only $24.26 billion, or 17.7 percent, was for the technology base. In turn, only 17.6 percent of that amount, $757 million, was being requested by DARPA. In sum, DARPA was asking for only 0.3 percent. U.S. Congress, House, Committee on Armed Services, *Hearings on Military Posture and H.R. 5968, Department of Defense Authorization for Appropriations for Fiscal Year 1983,* 97ᵗʰ Cong., 2ᵈ sess., part 5 of 7 parts, 1982, 1313.

11. Jeff Bingaman and Bobby R. Inman, "Broadening Horizons for Defense R&D," *Issues in Science and Technology* (Fall 1992): 80–85, at 81. Another unnamed source called DARPA "a shining example of government efficiency"; quoted in Willie Schatz and John W. Verity, "DARPA's big Push in AI," *Datamation* (February 1984): 48–50, at 50.

12. Jeffrey Rothfeder, *Minds over Matter: A New Look at Artificial Intelligence* (New York: Simon & Schuster, 1985), 184, 185.

13. Lerner, "Technology and the Military," 71.

14. Robert A. Divine, *The Sputnik Challenge* (New York: Oxford University Press, 1993), 84–89, 104–112, 162–166.

15. *Project Horizon: A U.S. Army Study for the Establishment of a Lunar Military Outpost,* 2 vols. ([Washington: United States Army,] 8 June 1959).

16. Sidney G. Reed, Richard H. Van Atta, and Seymour H. Deitchman, *ARPA Technical Accomplishments: An Historical Review of Selected ARPA Projects,* 2 vols. (Alexandria, VA: Institute for Defense Analyses, 1990), vol. 1, part 3.

17. Interview of Victor Reis, CBI.

18. Lerner, "Technology and the Military," 70.

19. Ibid.

20. Walter A. McDougall, . . . *the Heavens and the Earth: A Political History of the Space Age* (New York: Basic Books, 1985).

21. Kenneth Flamm, *Creating the Computer: Government, Industry, and High Technology* (Washington: Brookings Institution, 1988), 29–79.

22. McDougall, *Heavens and the Earth*, 113.

23. Robert R. Fossum, "DARPA Instruction No. 2," 31 October 1980.

24. True to his word, Cooper stayed exactly four years, making him the longest-serving director in DARPA-ARPA history.

25. *Budget of the United States Government, Fiscal Year 1997, Historical Tables* (Washington: GPO, 1996), 50–52, 149, 159; U.S. Bureau of the Census, *Historical Statistics of the United States from Colonial Times to 1957* (Washington: GPO, 1961), 719.

26. Andreas Fürst, "Military Buildup without a Strategy: The Defense Policy of the Reagan Administration and the Response by Congress," in *The Reagan Administration: A Reconstruction of American Strength?*, ed. by Helga Haftendorn and Jakob Schissler (Berlin: Walter de Gruyter, 1988), 133.

27. Peter Schweizer, *Victory: The Reagan Administration's Secret Strategy That Hastened the Collapse of the Soviet Union* (New York: Atlantic Monthly Press, 1994), xvi, xix. Schweizer's controversial thesis, that the policies of the Reagan and Bush administrations helped drive the Soviet Union to collapse, lacks documentation. But there is ample evidence in this book that the Reagan administration was pursuing such a policy. For a different view, see Frances Fitzgerald, *Way Out There in the Blue: Reagan, Star Wars, and the End of the Cold War* (New York: Simon & Schuster, 2000).

28. Robert R. Fossum, testimony on 31 March 1981, in U.S. Congress, House, Committee on Appropriations, *Department of Defense Appropriations for Fiscal Year 1982, Hearings*, 97th Cong., 1st sess., part 4—Procurement/R.D.T. & E., 21.

29. Cooper interview with Roland/Shiman, 16.

30. The Republicans who controlled Congress did not intend this legislation to arrest the defense build up, but it nonetheless had that effect. See Robert D. Reischauer, "Reducing the Deficit: Past Efforts and Future Challenges," The 1996 Frank M. Engle Lecture of The American College, Bryn Mawr, Pennsylvania, at http://www.amercoll.edu/WhatsNew/SpecialEvents/Engle/engle96.htm, on 17 January 2001, esp. 6–8.

31. Howard E. McCurdy, *Inside NASA: High Technology and Organizational Change in the U.S. Space Program* (Baltimore: Johns Hopkins University Press, 1993), 135, 140. As early as 1968, 83 percent of NASA's contracting was with private industry; Arnold S. Levine, *Managing NASA in the Apollo Era* (Washington: NASA, 1982), 71. On DARPA's percentage, see, for example, Robert Cooper's testimony in U.S. Congress, House, Committee on Appropriations, *Department of Defense Appropriations for 1985, Hearings*, 98th Cong., 2d Sess., part 5, 494, which reports that 51.3 percent of DARPA's FY 1984 budget went to industry.

32. National Science Foundation, Division of Science Resources Studies, *Federal Funds for Research and Development, Detailed Historical Tables: Fiscal Years 1955–1990* (Washington: NSF, n.d.), 115–116.

33. Cooper interview with Roland/Shiman, 20.

34. David H. Schaefer, "History of the MPP," in *The Massively Parallel Processor,* ed. by Jerry L. Porter (Cambridge, MA: MIT Press, 1985), 1–5.

35. Cooper interview with Roland/Shiman, 26.

36. Ibid.; Ohlander interview, Roland/Shiman, 21 October 1994.

37. Interview of Michael Dertouzos by Alex Roland, Cambridge, MA, 12 April 1995. Dertouzos believed that the Japanese had patterned their Fifth Generation program on MIT's multiprocessor research. He feared that they might prove more effective than the United States in transforming these ideas into marketable goods and services.

38. Michael Dertouzos interview by Arthur Norberg, Cambridge, MA, 20 April 1989.

39. Cooper interview with Roland/Shiman, 26.

40. Kenneth Flamm, *Targeting the Computer: Government Support and International Competition* (Washington: Brookings Institution, 1987), 116–117; Eric Bloch, "SRC: The Semiconductor Industry Draws on University Resources," *IEEE Spectrum* (November 1983): 56–57.

41. Quoted in Tobias Naegele, "Ten Years and $1 Billion Later, What Did We Get from VHSIC?" *Electronics* (June 1989): 97–103, at 98.

42. Ray Connolly, "40% Boost Sought in VHSIC Budget," *Electronics* (19 May 1981): 40, 41. The program got under way in 1979 with a projected budget of $198 million. "DoD Will Request Proposals Next Month for VHSIC," *Electronics* 52 (15 March 1979): 41. It ultimately spent more than $1 billion.

43. Ray Connolly, "Pentagon Moves to Expand VHSIC," *Electronics* (5 May 1982): 96, 98, at 98.

44. M. Mitchell Waldrop, *Man-Made Minds: The Promise of Artificial Intelligence* (n.p.: Walker and Co., 1987), 172–175; John Walsh, "MCC Moves Out of the Idea Stage," *Science* 220 (17 June 1983): 1256–1257.

45. "Uncle Sam's Reply to the Rising Sun," *New Scientist* (15 November 1984): 19.

46. James D. Meindl, "A University-Industry-Government Paradigm for Joint Research," *IEEE Spectrum* 20 (November 1983): 57–58.

47. John Walsh, "New R&D Centers Will Test University Ties," *Science* 227 (11 January 1985): 150–152.

48. Flamm, *Targeting the Computer*, 117.

49. Harvey Newquist, *The Brain Makers* (Indianapolis, IN: SAMS Publishing, 1994), 251. Robert Cooper confirmed this interpretation in 1984 congressional testimony. See U.S. Congress, House, Committee on Science and Technology, Subcommittee on Investigations and Oversight and Subcommittee on Science, Research and Technology, *Japanese Technological Advances and Possible United States Responses Using Research Joint Ventures, Hearings*, 98th Cong., 1st Sess., 29–30 June 1983, 127.

50. Flamm, *Targeting the Computer*, 114.

51. Jerry Werner and Jack Bremer, "Hard Lessons of Cooperative Research," *Issues in Science and Technology* (Spring 1991): 44–49, at 44.

52. Walsh, "MCC Moves Out of the Idea Stage," 1257.

53. Robert Kahn comments on earlier draft, 13 June 1995.

54. Robert E. Kahn, "A Defense Program in Supercomputation from Microelectronics to Artificial Intelligence for the 1990s," appendix to IPTO, DARPA, "FY84 President's Budget Review," 1 September 1982.

55. Ibid., 2.

56. Cooper emphasized the same point in his congressional testimony supporting the program. See, for example, his testimony on 14 June 1983 in U.S. Congress, House, Committee on Science and Technology, Subcommittee on Energy Development and Applications and Subcommittee on Energy Research and Production, *Computers and their Role in Energy Research: Current Status and Future Needs*, 98th Cong., 1st Sess., 58; and his testimony on 29 June 1983 in U.S. Congress, House, Committee on Science and Technology, Subcommittee on Investigations and Oversight and Subcommittee on Science, Research and Technology, *Japanese Technological Advances and Possible United States Responses Using Research Joint Ventures*, 98th Cong., 1st Sess, 133.

57. Kahn, "Defense Program in Supercomputation." This statement referred specifically to applications in category II, but the whole document suggests that this was a larger, general goal for the program.

58. Ibid., 1. Kahn began by asserting that recent experience in the Middle East and the Falklands War had demonstrated "the crucial importance of computers in the battlefield," but he did not elaborate.

59. Ibid.

60. Kahn interview, 27 July 1995.

61. Cooper interview, 12 May 1994, 34.

62. Kahn says "it was not so much a dispute as a failure to come to a meeting of the minds." Robert Kahn comments on earlier draft, 13 June 1995.

63. Of the term "act of faith," Kahn says "not quite—it was a carrot." Ibid.

64. Cooper testimony of 14 June 1983 in *Computers and Their Role in Energy Research*, 58.

65. Interview of Edward Feigenbaum by William Aspray, Palo Alto, CA, 3 March 1989, 16–17.

66. Michael L. Dertouzos to LCS [Laboratory for Computer Science] Faculty and Research Associates, 24 April 1980, with attached seating chart, in MIT Archives, collection 268, box 19, Folder 7.

67. Michael L. Dertouzos to [Distribution] List, 13 October 1982, ibid., folder 4.

68. Interview of Michael Dertouzos by Arthur Norberg, Cambridge, MA, 20 April 1989, 8.

69. Interview of Allen Newell by Arthur Norberg, Pittsburgh, PA, 10–12 June 1991, 78, 103.

70. Peter Denning, past president of the Association for Computing Machinery and editor of *Communications,* reportedly believed that SC was based on a desire to rebuild the academic computer infrastructure. Jacky, "The Strategic Computing Program," 197. Lynn Conway, former director of SC, also ascribed to Kahn "an interest in supporting the community in general." Interview of Lynn Conway by Alex Roland and Philip Shiman, Ann Arbor, MI, 7 March 1994, 5.

71. Interview of Charles Buffalano by Alex Roland and Philip Shiman, Greenbelt, MD, 2 August 1983.

72. Richard S. Westfall, "Galileo and the Telescope: Science and Patronage," *Isis* 76 (1985): 11–30.

73. Levine, *Managing NASA in the Apollo Era.*

74. Both techniques, it must be noted, had more value to program personnel in selling these programs to Congress than they did for management. See Harvey Sapolsky, *The Polaris System Development: Bureaucratic and Programmatic Success in Government* (Cambridge, MA: Harvard University Press, 1972); and John Lonnquest, "The Face of Atlas: General Bernard Schriever and the Development of the Atlas Intercontinental Ballistic Missile, 1953–1960," Ph.D. dissertation, Duke University, 1996.

75. Kahn interview, 2 August 1993, 17–22.

76. "Technological Paradigms and Technological Trajectories," *Research Policy* 11 (1982): 147–162.

77. Richard R. Nelson and Sidney G. Winter, "In Search of a Useful Theory of Innovation," *Research Policy* 6 (1977): 36–76.

78. See, for example, Paul A. David, "Clio and the Economics of QWERTY," *The American Economic Review* 75 (May 1985): 332–37.

79. Cooper interview, 12 May 1994, 54–55. "Initiative" was a Washington term of art. It distinguished the labeled activity from ongoing programs, thus justifying significant "new money," as opposed to an increase in existing appropriations.

80. The two were soon to be institutionally divorced. See pp. 87–89 below.

81. Jonathan Jacky, "The Strategic Computing Program," in *Computers in Battle— Will They Work?* ed. by David Bellin and Gary Chapman (New York: Harcourt Brace Jovanovich, 1987), 171–208, at 183–186.

82. P. 55.

83. Ibid.

84. Ibid., 59. Italics in original.

85. Ibid., 21.

86. See note 76 above.

87. Interview of Paul Losleben by Alex Roland and Philip Shiman, Palo Alto, CA, 29 July 1994, 2

88. John Walsh, "Xerox Scientist Joins DoD Supercomputer Program," *Science* (24 June 1983): 1359.

89. C. Mead and L. Conway, *Introduction to VLSI Systems* (Reading, MA: Addison-Wesley, 1980).

90. Conway interview, 7 March 1994, 4–6.

91. Richard H. Van Atta, et al., *DARPA Technical Accomplishments,* vol. II, *An Historical Review of Selected DARPA Projects* ([Washington]: Institute for Defense Analyses, 1991), 18–1 to 18–12.

92. Cooper interview, 12 May 1994, 67.

93. Interview of Robert Kahn by William G. Aspray, Reston, VA, 22 March 1989, 22–23.

94. Conway interview, 7 March 1994, 14. Robert Kahn recalls that IPTO Deputy Director Duane Adams visited him in the hospital and reported that Cooper had asked Conway to "crispen up" the document, to "make it punchier." On his next visit, Adams reported that Cooper had said Kahn's document was terrible and that Conway was to redo it from scratch. Kahn interview, 2 August 1993, 21.

95. G. H. Greenleaf, vice president, Advanced Technology Systems, BDM, to Director, DARPA, 8 September 1983; ARPA order 4959, 27 September 1983. The order was not activated until 11 October 1983, after Congress had approved the SC program.

96. Ibid., 22–23.

97. P. 58.

98. The plan provides the following definitions: "'Rules' represent the codification of an expert system process and a 'firing' indicates the examination, interpretation, and response to one rule in a particular context," p. 22.

99. R2D2 was the anthropomorphic robot that rode in the back seat of Luke Skywalker's Rebel star fighter in *Star Wars,* the 1977 George Lukas film that launched the enormously popular Star Wars trilogy.

100. This was a measurement used in evaluating planning programs. Alex Lancaster comments on earlier manuscript, oral presentation, Washington, DC, 12 June 1995.

101. P. 27.

102. P. 28.

103. In the pyramid, navigation and planning and reasoning are also included, but they do not receive the same emphasis in the text.

104. P. 60.

105. *Strategic Computing,* 61.

106. "Strategic Computing and Survivability," 8 July 1983, 10.

Chapter 3

1. John Logsdon, *The Decision To Go to the Moon: Project Apollo and the National Interest* (Chicago: University of Chicago Press, 1970).

2. An earlier, undated paper authored by Cooper and Kahn, "Strategic Computing and Survivability," was cleared for open publication in May 1983. Genevieve T. Motyka, "Memorandum to Information Processing Techniques Office," 26 May 1983; copy in possession of authors. The paper appears to have been written originally for the series of presentations to industry that Cooper and Kahn made the previous year.

3. Peter Denning, past president of the Association of Computing Machinery and editor in chief of its journal, *Communications of the ACM,* said: "DARPA wanted a program to rebuild the academic computer scientific infrastructure. It had to create a program that would fly in the political real world. By playing on feelings about Japan and Germany catching up with us, they could sell the research part. By adopting visions of new computer-based defensive systems, they could obtain DoD support." Quoted in Jonathan Jacky, "The Strategic Computing Program," in *Computers in Battle—Will They Work?,* ed. by David Bellin and Gary Chapman (New York: Harcourt Brace Jovanovich, 1987), 171–208, at 197.

4. I am indebted to Keith Uncapher for pointing out the critical importance of new money.

5. Interview with Jonathan Jacky by Alex Roland and Philip Shiman, Seattle, Washington, 1 August 1994.

6. See, for example, Robert S. Cooper and Robert E. Kahn, "SCS [Strategic Computing and Survivability]: Toward Supersmart Computers for the Military," *IEEE Spectrum* 20 (November 1983): 53–55.

7. Willie Schatz and John W. Verity, "DARPA's Big Push in AI," *Datamation* (February 1984): 48–50, at 48. See also Marjorie Sun, "The Pentagon's Ambitious Computer Plan," *Science* (16 December 1983): 1312–1315.

8. Dwight B. Davis, "Assessing the Strategic Computing Initiative," *High Technology* 5 (April 1985): 41–49, at 47.

9. See pp. 111–114 below.

10. Joseph Weizenbaum, "Facing Reality: Computer Scientists Aid War Effort," *Technology Review* (January 1987): 22–23; Jonathan B. Tucker, "The Strategic Computing Initiative: A Double-edged Sword," *Science for the People* 17 (1985): 21–25; Willie Schatz and John W. Verity, "Weighing DARPA's AI Plans," *Datamation* (1 August 1984): 34, 39, 42–43.

11. CPSR, "History," http://www.cpsr.org/cpsr/history.html, June 1998; CPSR, "A Year-by-Year Review," http://www.cpsr.org/cpsr/timeline.html, 2 June 1998; Jonathan Jacky, "The Strategic Computing Program," in *Computers in Battle—Will They Work?*, ed. by Bellin and Chapman, 194; M. Mitchell Waldrop, *Man-made Minds: The Promise of Artificial Intelligence* (n.p.: Walker & Co., 1987), 181.

12. Jeff Johnson, "Conflicts and Arguments in CPSR," http://www.cpsr.org/cpsr/conflicts.html, 2 June 1998.

13. Terry Winograd, "Terry Winograd's Thoughts on CPSR Mission," http://www.cpsr.org/cpsr/winnog.html, 2 June 1998.

14. Jacky, "The Strategic Computing Program," 172–173; Jacky interview. The characterization was not limited to enemies of the program within CPSR; Jacky refers to a *Newsweek* story entitled "Birth of the Killer Robots."

15. MIT computer scientist Joseph Weizenbaum called the ALV "a sort of traveling free fire zone" in "Computers in Uniform: A Good Fit?" *Science for the People* (Cambridge, MA: Science for the People, 1985): 26–29, quotation on 28.

16. Such concerns were not restricted to SCI. Social critics such as Jacques Ellul and Lewis Mumford were warning in the 1960s and 1970s of the insidious power of technology to take over human affairs. See Jacques Ellul, *The Technological Society* (New York: Vintage Books, 1964); and Lewis Mumford, *The Myth of the Machine*, 2 vols. (New York: Harcourt Brace Jovanovich, 1967–1970). Langdon Winner examined the controversy in *Autonomous Technology: Technics Out of Control as a Theme in Political History* (Cambridge, MA: MIT Press, 1977).

17. William J. Perry, the future secretary of defense, and Duane A. Adams, the future deputy director and acting director of DARPA, disagreed. See their analysis, "Ballistic Missile Defense," in "Report of the Defense Science Board Task Force on Military Applications of New-Generation Computing Technologies" (Washington: Office of the Under Secretary of Defense for Research and Engineering, December 1984), app. G.

18. U.S. Senate, Committee on Foreign Relations, *Strategic Defense and Anti-Satellite Weapons,* Hearing, 98th Cong., 2[d] sess., 25 April 1984, 68–74; quote at 74.

19. Severo M. Ornstein, Brian C. Smith, and Lucy A. Suchman, "Strategic Computing: An Assessment," *Communications of the ACM* 28 (February 1985): 134–136; Robert S. Cooper, "Strategic Computing Initiative: A Response," ibid. (March 1985): 236–237; James D. Gawn, David B. Benson, and Severo M. Ornstein, "Strategic Computing Initiative: Rejoinders," ibid. (August 1985): 793–794, 877.

20. Cooper testimony, U.S. Congress, House, Committee on Science and Technology, Subcommittees on Energy Development and Applications and Energy Research and Production, *Computers and Their Role in Energy Research: Current Status and Future Needs, Hearings,* 98th Cong., 1st Sess., June 14 and 15, 1983, 90.

21. Quoted in Davis, "Assessing the Strategic Computing Initiative," 48–49.

22. Cooper interview, 61.

23. Ibid., 60–61.

24. Those trajectories converged down the road, but the relationship was never what the critics feared at the outset. See p. 264 below.

25. On Thinking Machines, see chapter 5 below; on Teknowledge and Intellicorp, see chapter 6 below.

26. Lynn Conway interview by Alex Roland and Philip Shiman, Ann Arbor, MI, 7 March 1994, 11.

27. "Report of the Defense Science Board Task Force on Military Applications of New-Generation Technologies" (Washington: Office of the Under Secretary of Defense for Research and Engineering, December 1984).

28. Kent C. Redmond and Thomas M. Smith, *From Whirlwind to MITRE: The R&D Story of the SAGE Air Defense Computer* (Cambridge, MA: MIT Press, 2000), 55–57, 4141–4427. Frederick P. Brooks, Jr., *The Mythical Man-Month: Essays on Software Engineering* (Reading, MA: Addison-Wesley, 1975).

29. Robert E. Everett to Joshua Lederberg, 5 March 1984, in appendix L, "Ballistic Missile Defense."

30. Frederick P. Brooks, Jr., to Joshua Lederberg, 22 May 1984, ibid. Italics in original.

31. Ibid., appendix J.

32. Edward Feigenbaum and Pamela McCorduck, *The Fifth Generation: Artificial Intelligence and Japan's Computer Challenge to the World* (Reading, MA: Addison-Wesley, 1983).

33. U.S. Congress, House, Committee on Science, Space, and Technology, *Japanese Technological Advances and Possible U.S. Responses Using Research Joint Ventures. Hearings,* 98th Cong., 1st sess., 29–30 June 1983, 116–143, at 119.

34. In a 1983 revised edition of their book, Feigenbaum and McCorduck singled out SC as being a particularly appropriate response to the Japanese initiative.

35. Battista might have extended his list to include radios, tape recorders, sound equipment, watches, motorcycles, cameras, optical equipment, pianos, bicycles, ski equipment, ship building, and more. See Ezra F. Vogel, *Japan as Number One: Lessons for America* (Cambridge, MA: Harvard University Press, 1979). On America's reluctance to appreciate the problem, see 225–231 and also Vogel's *Japan as Number One Revisited* (Pasir Panjang, Singapore: Institute for Southeast Asian Studies, 1986).

36. U.S. Congress, House, Committee on Armed Services, *Defense Department Authorization and Oversight,* Hearings on H.R. 2287, part 5, "Research, Development, Test, and Evaluation," 98/1, 1983, 994–997.

37. Kenneth Flamm, *Targeting the Computer: Government Support and International Competition* (Washington: Brookings Institution, 1987), 114–115.

38. John Walsh, "MCC Moves Out of the Idea Stage," *Science* 220 (17 June 1983): 1256–1257; Mark A Fischetti, "MCC: An Industry Response to the Japanese Challenge," *IEEE Spectrum* 20 (November 1983): 55–56; "Uncle Sam's Reply to the Rising Sun," *New Scientist* (15 November 1984): 19.

39. Fred Warshofsky, *The Chip War: The Battle for the World of Tomorrow* (New York: Scribner, 1989).

40. Cooper interview; Kahn interview; Squires interview.

41. Jack Schwartz, a mathematician at New York University who would subsequently play a major role in SC, declared the Fifth Generation "a foolish project misconceived from the beginning." Quoted in Karen Fitzgerald and Paul Wallich, "Next-generation Race Bogs Down," *IEEE Spectrum* (June 1987): 28–33, quote on 28.

42. Cooper interview, 33–35. DARPA Program Manager Paul Losleben later said that contrary to the "party line," without "the Fifth Generation Project, Strategic Computing would never have happened." Paul Losleben interview with Alex Roland and Philip Shiman, Stanford, CA, 29 July 1994, 15.

43. Schatz and Verity, "DARPA's Big Push in AI."

44. Philip J. Klass, "DARPA to Develop Techniques for Computer Intelligence," *Aviation Week and Space Technology* (1 July 1983): 14–15. One iteration of the SC plan had called it "A Defense Program in Supercomputation"; see p. 53 above.

45. Willie Schatz, "DARPA Goes Parallel," *Datamation* (1 September 1985): 36–37.

46. Alan Borning interview by Alex Roland and Philip Shiman, Seattle, WA, 1 August 1994. William Scherlis called SC "a pile of money"; telephone interview with Alex Roland, 1 February 1994. Cooper interview.

47. "From my point of view, Strategic Computing is already a success. We've got the money." David Mizell, talk at a conference on SC organized by the Silicon Valley Research Group, 1985, quoted in Tom Athanasiou, "Artificial Intelligence, Wishful Thinking and War," in *Cyborg Worlds: The Military Information Society,* ed. by Les Levidow and Kevin Roberts (London: Free Association Books, 1989), 122.

48. See J. Merton England, *A Patron for Pure Science: The National Science Foundation's Formative Years, 1945–1957* (Washington: National Science Foundation, 1983).

49. Wiebe E. Bijker, Thomas P. Hughes, and Trevor J. Pinch, eds., *The Social Construction of Technological Systems: New Directions in the Sociology and History of Technology* (Cambridge, MA: MIT Press, 1987).

50. Donald MacKenzie, "Economic and Sociological Explanation of Technical Change," in Rod Coombs, Paolo Saviotti, and Vivien Walsh, *Technological Change and Company Strategies: Economic and Sociological Perspectives* (London: Harcourt Brace Jovanovich, 1992), 25.

51. Joint interview with Stephen Squires and John Toole, conducted by Alex Roland and Philip Shiman, Arlington, VA, 29 November 1993; Squires speaking.

52. Kahn sketched such a matrix in an interview with the authors at the Reston, VA, offices of CNRI on 29 November 1994. Subsequently, however, he resisted entreaties to elaborate on what a complete matrix might look like and how he might use it.

53. Cooper interview, 67.

54. Ibid., 62, 64.

55. Robert Kahn interview with Alex Roland and Philip Shiman, Reston, VA, 29 November 1994.

56. Conway interview, 12.

57. Cooper interview, 68; Conway interview, 15.

58. Kahn interview, 29 November 94.

59. U.S. Congress, House, Committee on Science and Technology, Subcommittees on Energy Development and Applications and Science, Research, and Technology, *Federal Supercomputing Programs and Policies, Hearing,* 99th Cong., 1st Sess., 10 June 1985 (Washington: GPO, 1986), 44.

60. Kahn interview, 29 November 1994.

61. Interview with Charles Buffalano.

62. Losleben interview, 9. The Hubble telescope was subsequently repaired and went on to give widely acclaimed service. True to form, Lynn Conway told *Science* reporter M. Mitchell Waldrop that SC was "like the Space Telescope." Waldrop, "The Fifth Generation: Taking Stock," *Science* (30 November 1984): 1063.

63. James E. Webb, *Space Age Management: The Large-Scale Approach* (New York: McGraw-Hill, 1969).

64. Arnold S. Levine, *Managing NASA in the Apollo Era: An Administrative History of the U.S. Civilian Space Program, 1958-1969* (Washington: NASA, 1983).

65. Roger E. Bilstein, *Stages to Saturn: A Technological History of the Apollo/Saturn Launch Vehicles* (Washington: NASA, 1980).

66. It was especially important in their view because it "transitioned" new technology to industry.

67. Interview of Keith Uncapher by Philip Shiman, Los Angeles, CA, 8 Aug. 1994, 29.

68. In comments on an earlier draft, Robert Kahn wrote here "You mean *did* turn into. . . ." Comments, 13 June 1995.

69. Natural language understanding seeks to program computers to understand normal human communication, in English or other languages. Speech recognition seeks to program computers to recognize and translate the spoken word into natural language. See chapter 6 below.

70. The SC annual report for 1984 listed twenty-four principals, 11; Conway interview, 29.

71. *Strategic Computing First Annual Report* (n.p.: DARPA, February 1985), 9–10.

72. "Strategic Computing and Survivability Research: Final Report," BDM/W–537–TR, BDM Corporation, Arlington, VA, 28 May 1985; prepared under ARPA Order 4959, contract no. MDA903–84–C–0059, II–15, II–30.

73. Losleben interview.

74. Robert Kahn comments on earlier draft, 13 June 1995.

75. Kahn interview.

76. On reverse technological salients in large-scale technological systems, see Thomas P. Hughes, *Networks of Power: Electrification in Western Society, 1880–1930* (Baltimore: Johns Hopkins University Press, 1983), 79–80 et passim.

77. Cooper testimony in U.S. Congress, House, Committee on Armed Services, *Defense Department Authorization and Oversight*, Hearings, part 5, "Research, Development, Test, and Evaluation—Title II," 98/1, 1983, 995.

78. Cooper interview, 61. In testimony before Congress, Cooper put the best light on this loss, calling it "one of the most spectacular transitions of technology in DARPA history." U.S. Congress, House, Committee on Armed Services, *Defense Department Authorization and Oversight*, Hearings on H.R. 5167, part 4, "Research, Development, Test, and Evaluation—Title II, 98/2, 1984, 918.

79. U.S. Congress, House, Committee on Appropriations, *Department of Defense Appropriations for 1985*, Hearings, 98/2, part 5, 1984, 79, 515.

80. Cooper later regretted transferring GaAs. He did it in part because the existing program was committed to an expensive joint venture with Rockwell International to build a large pilot line. By transferring the program to SDI, Cooper unloaded a heavy obligation on DARPA's future funds.

81. This campaign may be traced in part in the index to the legislative history of public law 98–369 (Div. B, Title VII), in *Congressional Information Service/Index Legislative Histories,* 98th Cong., 2d sess., January-December 1994, 207–212.

82. Public Law 98–369, 18 July 1984, 98 stat. 1175, Subtitle A—"Amendments to the Federal Property and Administrative Services Act of 1949," sec. 303 (a).

83. U.S. Congress, House, Committee on Appropriations, *Department of Defense Appropriations for 1985*, Hearings, 98/2, part 5, 1984, 508.

84. Public law 98–72, 11 August 1983, 97 stat. 403. Again, provisions were included to waive this requirement for cause.

85. Losleben interview, 7–9.

86. M. Mitchell Waldrop, "The Fifth Generation: Taking Stock," 1063.

87. Stephen Squires, meeting of SC history steering committee, Washington, DC, 19 October 1994.

88. Losleben interview, 9.

Chapter 4

1. David B. Gibson and Everett M. Rogers, *R&D Collaboration on Trial: The Microelectronics and Computer Technology Corporation* (Boston: Harvard Business School Press, 1994), 4.

2. Interview of Robert Cooper by Alex Roland and Philip Shiman, Greenbelt, MD, 12 May 1994.

3. *DARPA Strategic Computing*, 3rd Annual Report, 1987, 26, 32–33.

4. MOSIS has two translations. It derives from MOS implementation system, meaning a program to manufacture designs of MOS integrated circuit chips. MOS in turn generally means metal oxide semiconductor, a descriptive term listing the three materials that are layered on these chips to produce the desired electrical effects. Silicon is the most frequently used semiconductor in these chips, so the S in MOS was sometimes taken to stand for silicon. But MOSIS also implements GaAs chips, so the more appropriate rendering of MOSIS is metal oxide semiconductor implementation system.

5. Some switches achieve the yes/no capability by creating and reacting to differences in voltage.

6. Hans Queisser, *Conquest of the Microchip* (Cambridge, MA: Harvard University Press, 1990), 101; *Encyclopedia of Computer Science*, ed. by Anthony Ralston, Edwin D. Reilly, and David Hemmendinger (New York: Grove Dictionaries, 2000), 691–693.

7. Ibid., 691, 879. Somewhat different values are offered in Richard A. Jenkins, *Supercomputers of Today and Tomorrow: The Parallel Processing Revolution* (Blue Ridge Summit, PA: TAB Books, 1986), 62: SSI = 2 to 64; MSI = 64 to 2,000; LSI = 2,000 to 64,000; VLSI = 64,000 to 2,000,000; ULSI = 2,000,000 to 64,000,000.

8. Gibson and Rogers, *R&D Collaboration*, 8.

9. Ibid., 38–43.

10. Glenn R. Fong, "The Potential for Industrial Policy: Lessons from the Very High Speed Integrated Circuit Program," *Journal of Policy Analysis and Management* 5 (1986): 264–291.

11. Ivan E. Sutherland, Carver A. Mead, and Thomas E. Everhart, "Basic Limitations in Microcircuit Fabrication Technology," RAND report R-1956-ARPA, November 1976, 2.

12. Richard H. Van Atta, Sidney Reed, and Seymour J. Deitchman, *DARPA Technical Accomplishments*, Volume II: *An Historical Review of Selected DARPA Projects*, Institute for Defense Analyses, IDA paper P-2429, April 1991, chapter 17; Arthur L. Norberg and Judy E. O'Neill, *Transforming Computer Technology: Information Processing for the Pentagon, 1962–1986* (Baltimore: Johns Hopkins University Press), 271–275.

13. See Ross Knox Bassett, "New Technology, New People, New Organizations: The Rise of the MOS Transistor, 1945–1975," unpublished Ph.D. dissertation, Princeton University, 1998, especially chapters 6–8 on IBM and Intel.

14. Charles L. Seitz, "Concurrent VLSI Architectures," *IEEE Transactions on Computers* 33, no. 12 (December 1984): 1250–1251.

15. Carver Mead and Lynn Conway, *Introduction to VLSI Systems* (Reading, MA: Addison-Wesley, 1979); see especially ix–xi.

16. Danny Cohen and George Lewicki, "MOSIS—The ARPA Silicon Broker," *CalTech Conference on VLSI* (January 1981), 29.

17. XEROX PARC had been experimenting with just such procedures for its own purposes since the mid-1970s. "VLSI Implementation—MOSIS," 7 June 1990, draft, copy obtained from Information Sciences Institute of the University of Southern California, courtesy of Ronald Ohlander, 6. [Hereafter "VLSI Implementation."] Unless otherwise indicated, the account of the MOSIS program presented here is based on this document.

18. Uncapher interview with Shiman, 8 August 1994; ARPA order 4242.

19. The contracts were MDA903 80 C 0523, under ARPA order 2223, 1 October 1979; and MDA903 81 C 0335, under ARPA order 4012, 29 April 1980.

20. Van Atta, Reed, Deitchman, *Technical Accomplishments*, Vol. II, 17–9, 17–11, 17–12.

21. One document refers to "the MOSIS philosophy." See "MOSIS," (n.p., n.d.), part 1, 7.

22. G. Lewicki, "MOSIS," photocopies of overhead transparencies, May 1988, DARPA Cold Room, no. 106047.

23. Christine Tomavich, "MOSIS—A Gateway to Silicon," *IEEE Circuits and Devices Magazine* (March 1988): 22.

24. E-mail, Kathy Fry of ISI to John Toole of DARPA, 25 February 1988, in Toole Collection, Colladny Brief.

25. See the 1990 price schedule in César Piña, "MOSIS Semi Annual Report" for March 1990, presented at Spring 1990 DARPA Contractor's Meeting, Salt Lake City, Utah, Cold Room no. 107304.

26. "ARPA Order Decision Summary," ARPA order 5009, Amendment no. 1, 6 February 1984; ibid., Amendment no. 64, 26 August 1987. This contract also had a second ARPA order, 4848, which was modified to add SC projects in response to "Memorandum for the Director, Program Management; Subj: Amendment to ARPA Order 4848—USC/ISI," 25 Aug. 1983. This request was made before the program was officially approved.

27. G. Lewicki, "MOSIS," May 1988, photocopy of presentation transparencies, Cold Files, no. 106047, 14. Note that the numbers in the slide do not add up; the numbers given above have been recomputed.

28. Modification P00068 to DARPA contract no. MDA903–81–C–0335, 7 May 1987, 2, in ISTO Files, "ISTO/Denise." These numbers, as with most DARPA figures, especially on working documents, must be taken with a grain of salt. DARPA funding to ISI reportedly began in 1972, but the institute does not show up on DARPA's list of "Contracts and Contract Mod Info for FO > $500K and FY between 1969–1984" (18 December 1995) until November of 1981. Through

FY 1984, DARPA paid ISI more than $39 million for work in the previous three fiscal years. In all, over ten years, ISI received more that $75 million in SC funding and an indeterminate amount of other DARPA funding. (A. H. Passarella to Alex Roland, 22 July 1998, enclosures marked "1A10E" and "ST-10.") From the available records it is impossible to determine the exact source and amount of ISI funding from DARPA in any given year.

29. "VLSI Implementation—MOSIS," 3. In 1990, César Piña reported the 1989 numbers as 1887 and more than 2000; see César A. Piña, "MOSIS Semi Annual Report," prepared for Spring 1990 DARPA Contractor's Meeting, Salt Lake City, UT, March 1990.

30. Mead and Conway, *Introduction to VLSI Systems,* pp. vii et passim.

31. See p. 122 above.

32. James D. Meindl, "Microelectronic Circuit Elements," *Scientific American* 237 (September 1997): 77–78, 81.

33. Van Atta, et al., *DARPA Technical Accomplishments,* vol. II, table 18–1.

34. Edward C. Constant, *The Origins of the Turbojet Revolution* (Baltimore: Johns Hopkins University Press, 1980).

35. César Piña, interview with Alex Roland and Philip Shiman, Marina del Rey, CA, 21 October 1994. As it turns out, ingenious technological advances have allowed feature sizes to shrink more than expected. They are currently predicted to reach .05 µ in 2012. *Encyclopedia of Computer Science,* 898.

36. Ibid., 901.

37. In 1985 a trade journal's annual poll found thirty-nine silicon foundries and eight GaAs foundries. See "Survey of Semiconductor Foundries," *VLSI Design* (August 1985): 90–118.

38. ISI, *Annual Technical Report,* 1985, 94, 102.

39. In 1986 part of the MOSIS system was transferred to NSA to allow that agency to establish a classified implementation facility. This eliminated the need for MOSIS to be subjected to the security restrictions required to handle highly classified designs. "RMOSIS (MOSIS for Government)," FY 1988 Project Review (Microsystems Design and Prototyping), from the files of John Toole, "FY 1988 Project Reviews."

40. "Study of the Potential Future Market for MOSIS," Penran Corporation, 5 Cleland Place, Menlo Park, CA, 25 April 1987, 3; copy in possession of the authors, received from Paul Losleben.

41. "VLSI Implementation—MOSIS, Edition of June 7, 1990," draft, 29; in possession of the authors.

42. Keith Uncapher, oral communication, Washington, DC, 13 June 1995.

43. The following account is derived largely from "MOSIS Success Stories," in Saul Amarel, *Information Science and Technology Office Program Review* (January 1987).

44. John Hennessy, "VLSI RISC Processors," *VLSI Design Systems* 6 (October 1985): 22–24, 28, 32; John L. Hennessy, "VLSI Processor Architecture," *IEEE Transactions on Computers* C-33 (December 1984): 1221–1246; J. L. Hennessy and D. A. Patterson, *Computer Architecture: A Quantitative Approach* (San Mateo, CA: Morgan Kaufman, 1990).

45. See Charles L. Seitz, "Concurrent VLSI Architectures," *IEEE Transactions on Computers* C-33 (December 1984): 1247–1265. Indeed, one reason for pursuing VLSI was its potential usefulness in designing parallel architectures. Mead and Conway, *Introduction to VLSI Systems,* 264. See also chapter 5 below.

46. *Encyclopedia of Computer Science,* 1345–1346.

47. See chapter 5 below.

48. For example, in 1988, George Lewicki reported that over a twelve-month period, the program had cost a total of $9.2 million. DARPA paid about 63 percent of that, the rest coming from NSA, NSF, and other sources. Of the total, fabrication costs made up 74.3 percent, the balance going to salaries and overhead at ISI-USC. G. Lewicki, "MOSIS," May 1988.

49. *Strategic Computing: New-Generation Computing Technology: A Strategic Plan for its Development and Application to Critical Problems in Defense* (Washington: DARPA, 28 October 1983), iii.

50. John Toole interview with Alex Roland and Philip Shiman, Arlington, VA, 11 July 1994.

51. "Study of the Potential Future Market for MOSIS."

52. Oral presentation at "DSSA's Role in Development and Infrastructure: What's There and What's Missing?," quarterly workshop of the ARPA program in Domain Specific Software Applications [DSSA], Seattle, WA, 3 August 1994.

53. Ronald Ohlander, "Infrastructure Resulting from the Strategic Computing Initiative," unpublished paper prepared for this study, 11 June 1995, 1–2. Unless otherwise noted, the following account of SC's machine acquisition program is drawn from this paper.

54. Daniel Crevier, *AI: The Tumultuous History of the Search for Artificial Intelligence* (New York: Basic Books, 1993), 59–62.

55. H. P. Newquist, *The Brain Makers: Genius, Ego, and Greed in the Quest for Machines That Think* (Indianapolis, IN: SAMS Publishing, 1994).

56. Crevier, *AI,* 209.

57. By the end of 1984, ISI had negotiated discounts of 15–30 percent from Symbolics, 57 percent from LMI, 40 percent from TI, and 30–50 percent from Xerox. (E-mail, Robert Balzar to Ronald Ohlander, 21 December 1984.) And

it won comparably advantageous maintenance agreements. This service was provided by ISI for a fee of 2 percent on total program cost, that is, $169,000 out of the $8.4 million expended through March 1986. (E-mail, Dougherty to Sears, 27 March 1986).

58. A draft of the sole-source justification to purchase thirty Symbolics 3600 systems in March 1984 said "DARPA has chosen to work in a Common Lisp framework as a standard across the Strategic Computing and Survivability Program. (E-mail from [Gary] McReal of ISI to multiple addressees at DARPA, 20 March 1984.) Keith Uncapher is not sure that the $8.5 million was actually spent; he recalls that it was reduced to $4 million. Oral comments, Washington, DC, 13 June 1995.

59. Norberg and O'Neill, *Transforming Computer Technology*, 40.

60. Ohlander, "Infrastructure," 4.

61. Keith W. Uncapher to Robert Kahn, "Re: Proposal for 'Strategic Computing and Survivability Development Computers,' " ISI proposal number: 83–ISI–7; contract: MDA903–81–C–0335, 5 August 1983.

62. E-mail from Balzer to Ohlander, 8 January 1985; Newquist, *The Brain Makers*, 343–346.

63. E-mail from Davey to Ohlander, 4 December 1985.

64. Newquist, *The Brain Makers*, 334, 421.

65. E-mail, Stephen Dougherty of ISI to Allen Sears of DARPA, 13 March 1986.

66. E-mail Stephen Dougherty of ISI to Allen Sears of DARPA, 27 March 1986. Indirect cost recovery was at the off-campus rate.

67. Interview of Hisham Massoud by Alex Roland, Durham, NC, 21 August 1995.

68. See, for example, Paul A. David, "Clio and the Economics of QWERTY," *American Economic Review* 75 (1985): 332–337.

69. See, for example, the standardization of American screw thread dimensions in Bruce Sinclair, *Philadelphia's Philosopher Mechanics: A History of the Franklin Institute, 1824–1865* (Baltimore: Johns Hopkins University Press, 1974).

70. Barry Leiner, comments at Strategic Computing Study Advisory Committee meeting, Arlington, VA, 24 January 1995.

71. DARPA, "FY86 APPORTIONMENT/FY87 POM REVIEW: Information Processing Techniques Office," April 1985, 1.

Chapter 5

1. W. Daniel Hillis, *The Connection Machine* (Cambridge: MIT Press, 1985), 3–5. The term "von Neumann bottleneck" was coined by J. Backus in his Turing

Lecture, "Can Programming Be Liberated from the Von Neumann Style?" *Communications of the ACM* 8 (1978): 613–641.

2. Hillis, "The Connection Machine," AI Lab memorandum 646, MIT, 1981.

3. Hillis, "The Connection Machine: A Computer Architecture Based on Cellular Automata," *Physica* 10D (1984): 213–228. See also [MIT AI Lab], proposal for the Connection Machine, [November 1981], DARPA Technical Library, DynCorp, TDL no. 102460 (hereafter cited as MIT, Connection Machine proposal).

4. For the history of parallel processing see R. W. Hockney and C. R. Jesshope, *Parallel Computers: Architecture, Programming, and Algorithms* (Bristol, UK: Alger Hilger Ltd., 1981), 2–24; and Richard A. Jenkins, *Supercomputers of Today and Tomorrow: The Parallel Processing Revolution* (Blue Ridge Summit, PA: TAB Books, 1986), 36–60.

5. Jenkins, *Supercomputers*, 80–84; Hockney & Jesshope, *Parallel Computers*, 20.

6. Hillis, *Connection Machine*, 53–54.

7. MIT, Connection Machine proposal, 22–24.

8. Aaron Zitner, "Sinking Machines," *The Boston Globe*, 6 September 1994.

9. Ibid.

10. "The Connection Machine," in Richard H. Van Atta, Sidney G. Reed, and Seymour J. Deitchman, *DARPA Technical Accomplishments*, Volume II: *An Historical Review of Selected DARPA Projects*, IDA paper p-2429, Institute for Defense Analyses, April 1991, chapter 17-C:1.

11. Marvin Minsky, "A Framework for Representing Knowledge," MIT-AI Laboratory memo 306, June 1974, abridged in Patrick Winston, ed., *The Psychology of Computer Vision* (New York: McGraw-Hill, 1975), 211–277; Roger C. Schank and Robert P. Abelson, *Scripts, Plans, Goals, and Understanding: An Inquiry into Human Knowledge Structures* (Hillsdale, NJ: L. Erlbaum Associates, 1977).

12. Harvey Newquist, *The Brain Makers* (Indianapolis: Sams Publishing [Prentice Hall], 1984), 160–163.

13. Hillis, *The Connection Machine*, 27, 31–48; MIT, Connection Machine proposal, 1–4.

14. Hillis, "The Connection Machine" (draft), in *Proceedings of the Islamadora Workshop on Large Scale Knowledge Base and Reasoning Systems, February 25–27, 1985*, ed. Michael L. Brodie, 267. Unpublished document for workshop participants, found in CSTO reference collection, document 105137.

15. MIT, Connection Machine proposal, 21.

16. AI Lab proposal, 22–24; "VLSI Architecture and Design," 21 June 1982, IPTO program summary filed in AO 2095.

17. Newquist, *The Brain Makers*, 163–173, 189–196, 335.

18. Zitner, "Sinking Machines."

19. "VLSI Architecture and Design," 21 June 1982, IPTO program summary filed in AO 2095.

20. As with VLSI and MOSIS, the project was already supported by IPTO. SC was simply another pot of money into which it might dip.

21. Jenkins, *Supercomputers*, 49–54.

22. Ibid., 51–53.

23. Ivan E. Sutherland and Carver A. Mead, "Microelectronics and Computer Science," *Scientific American* 237 (Sept. 1977): 212–216.

24. Edward W. Constant, *The Origins of the Turbojet Revolution* (Baltimore: Johns Hopkins University Press, 1980).

25. Jenkins, *Supercomputers*, 51–53, 69–70.

26. Kahn, "Strategic Computing and Survivability" [DARPA, May 1983], 56, 61.

27. Hockney and Jesshope, *Parallel Computers*, 353. See also Jenkins, *Supercomputers*, 14.

28. Kahn, "Strategic Computing and Survivability," 55, 59–66.

29. Creve Maples (Lawrence Berkeley Laboratory), as quoted in Jenkins, *Supercomputers of Today and Tomorrow*, 78.

30. Kenneth Flamm, *Targeting the Computer: Government Support and International Competition* (Washington, D.C.: The Brookings Institution, 1987), 49–51.

31. Interviews of Stephen Squires by Alex Roland and/or Philip Shiman, Arlington, VA, 17 June 1993, 12 July 1994, 21 December 1994. The remainder of this chapter relies heavily on these interviews with Steve Squires, who went on to be director of ISTO. Squires did not allow taping of his interviews, so the account here is based entirely on notes taken during the interviews. It is important to note that not all of the evidence offered by Squires can be verified with reliable documentation. This chapter therefore represents, to a certain extent, the architecture story according to Squires. He is the best witness and at the same time the most biased witness to these events.

32. Squires interview, 12 July 1994.

33. Squires interview, 17 June 1993.

34. Ibid.

35. Squires interviews, 17 June 1993, 12 July 1994.

36. Squires interview, 17 June 1994.

37. Jenkins, *Supercomputers of Today and Tomorrow*, 59.

38. Kahn interview, 27 July 1995; Arthur L. Norberg and Judy E. O'Neill, *Transforming Computer Technology: Information Processing for the Pentagon, 1962–1986* (Baltimore: Johns Hopkins University Press), 388–392; Jenkins, *Supercomputers*, 56–60; Sidney G. Reed, Richard H. Van Atta, and Seymour J. Deitchman, *DARPA Technical Accomplishments*, volume I: *An Historical Review of Selected DARPA Projects*. IDA paper P–2192, prepared under contract MDA 903–84–C–0031 (Alexandria, Va.: Institute for Defense Analyses, February 1990), chapter 18.

39. Norberg, 390–391.

40. The gray code was a technique originally devised for the conversion of analog to binary digital measurements in certain mechanical applications. The code was so arranged that for each successive digital reading only one bit would be changed in each binary number, thus keeping them in their proper sequential order. Anthony Ralston and Edwin D. Reilly, eds., *Encyclopedia of Computer Science*, 3rd ed. (New York: Van Nostrand Reinhold, 1993), 182.

41. Squires interviews, 17 June 1993, 17 June 1994, 21 December 1994.

42. SCI plan, October 1983, 42–49.

43. Ibid.

44. The name "butterfly switch" stems from the butterfly-shaped layout of the communications network, which was so designed to perform fast Fourier transforms efficiently.

45. See AO files 4906 (BBN) and 5755 (BBN); Stephen H. Kaisler, "Parallel Architectures Support Program," in National Security Industrial Association, Software Committee, *A National Symposium on DARPA's Strategic Computing Program* (n.p., n.d.), 306. Another twenty-three machines, with 2–10 nodes each, had been installed for various communications functions between June 1983 and February 1986; ibid.

46. AO file 4864 (CMU); Ellen P. Douglas, Alan R. Houser, and C. Roy Taylor, eds., *Final Report of Supercomputer Research, 15 November 1983–31 May 1988*, CMU–CS–89–157, prepared under contract N00039–85–C–0134, School of Computer Science, Carnegie Mellon University, Pittsburgh, Pa., June 1989, chapters 2 and 4.

47. AO file 4864 (CMU); Douglas et al., *Final Report*, chapter 3.

48. AO files 2095 (MIT), 4920 (MIT).

49. AO file 4465 (Columbia U.). The development plan for the Non-Von project is a good example of Squire's "gray code" and progressive development strategy. The initial design laid out by Columbia—Non-Von 1—was to be prototyped using custom VLSI chips, each of which would contain a single 1-bit processor. While that work was proceeding, Columbia researchers would design and simulate the next generation of the machine, the Non-Von 3 (Non-Von 2, as previ-

ously designed, "was an interesting architectural exercise," according to the contract, but it was not going to be implemented.) If Non-Von 1 went well, the researchers would implement Non-Von 3, again with custom VLSI, this time with four or eight 8-bit processors per chip.

Meanwhile, the researchers would concurrently begin designing Non-Von 4, exploring a new approach that would permit a number of different instructions to be executed simultaneously. This was called multiple-instruction-multiple-data, or MIMD (the earlier approaches were SIMD—single-instruction-multiple-data—because they permitted only one instruction to be executed at a time, albeit on several items of data concurrently). Columbia was not authorized to construct a prototype of Non-Von 4, however, at least not on the current contract.

50. AO file 4974 (Texas Instruments), CBI.

51. Squires interviews, 17 June 1993, 12 July 1994.

52. "Multiprocessor Computer System Architecture for the Strategic Computing Program," *Commerce Business Daily*, Issue PSA-8516, 3 February 1984, 32; Squires interview, 12 July 1994.

53. Defense Advanced Research Projects Agency, *Strategic Computing, First Annual Report*, February 1985, 8.

54. Squires interviews, 17 June 1993, 12 July 1994.

55. Squires defined the various elements of his strategy in interviews with the authors, and in a draft outline, "Strategic Computing Architectures," December 1994, document provided by the author.

56. Thomas P. Hughes, *Networks of Power: Electrification in Western Society, 1890–1930* (Baltimore: Johns Hopkins University Press, 1983). Squires cited this book in an interview with the authors at DARPA headquarters on 11 November 1992, though his description of the book at that time demonstrated no familiarity with the book's argument. It appears likely that Squires read the book years earlier, absorbed some of its concepts, and later forgot that these ideas came from Hughes.

57. This represents an elaboration of Kahn's grid concept for pegging levels of support. See pp. 95–96 above.

58. Ibid. For the layout of the maturity model see solicitation N00039-84-R-0605(Q), issued by the Department of the Navy, Naval Electronic Systems Command, 14 September 1984 (copy found in file AO 4465 [Columbia U.], section C). The model is also described in Paul B. Schneck, Donald Austin, and Stephen L. Squires, "Parallel Processor Programs in the Federal Government," *Computer* 18 (June 1985): 43–56.

59. Squires interview, 2 December 1994; see also Squires's marginal note, 16 August 1985, in ESL Inc. proposal no. Q4557, "Statement of Work for Multipro-

cessor System Architectures, Medium and Full Scale Signal Processors," 15 August 1985, in response to RFP–N00039–84–R–0605(Q), Naval Electronic Systems Command, in ISTO office files 1310.

60. Solicitation N00039–84–R–0605(Q), sections C, L, and M.

61. Squires interview, 21 December 1994. Squires' "wave" strategy is well described in Gary Anthes, "Parallel Processing," *Federal Computer Week* 1 (30 November 1987): 25–28, which was largely based on an interview with Squires.

62. Squires interview, 12 July 1994.

63. Ibid.

64. Ibid.

65. AO file 5058. Unfortunately, the original documentation, including the MRAO, ARPA Order, and contract, are all missing from the file.

66. Solicitation N00039–84–R–0605(Q), issued by the Department of the Navy, Naval Electronic Systems Command, 14 September 1984 (copy found in file AO 4465 [Columbia U.].

67. Gary Anthes, "A Sampler of DARPA-Funded Parallel Processors," *Federal Computer Week* 1 (30 November 1987): 27.

68. The university efforts that had already been put under contract prior to the solicitation were also incorporated in the program, mostly as stage-2 projects.

69. AO files 5450 (various), CBI; 5475 (Various), CBI; 5809 (ESL); 5876 (BBN), ARPA; 5944 (IBM), CBI; 5946 (Kestrel Inst.), ARPA; 9133 (Harris Corp.), CBI.

70. This figure is derived from ARPA order decision summary, 18 March 1985, AO file 5450. The yearly breakdown is:

FY 1985: $8,059,000
FY 1986: $36,937,000
FY 1987: $39,278,000

As of September 1984, $20 million had been budgeted for the architectures program for FY 1985 and $45 million for FY 1986, but by the following spring this had clearly been reduced. See "FY86 Presidents Budget Review," Information Processing Techniques Office, DARPA, September 1984.

71. "Strategic Computing Architecture Projects," undated briefing slide [ca. 1986]; see also "Budget Planning—ISTO, July 1986," document supplied by Saul Amarel.

Architectures Program, 1986

Full Prototypes:

Warp (Carnegie Mellon University)

MOSAIC (ESL)

The Connection Machine (Thinking Machines Corporation)

Logic programming accelerator (University of California-Berkeley)

Cross-Omega Connection Machine (General Electric)

DADO/DSP (AT&T)

Monarch (BBN)

Ultramax (Encore)

Simulations/Small Prototypes:

Reconfigurable systolic arrays (Hughes Corp.)

Data flow symbolic multiprocessor (Georgia Institute of Technology)

MASC (Systems Development Corporation)

FAIM (Schlumberger)

Tree machines [DADO and Non-Von] (Columbia University)

Data flow emulation facility (MIT)

RP3 (IBM)

Massive memory machine (Princeton University)

Performance Modeling:

Evaluation of parallel architectures (University of Texas)

Measurement of performance (National Bureau of Standards)

Scalable parallel algorithms (Stanford)

Generic Software/Tools:

Software workbench [Mach] (Carnegie Mellon University)

QLISP for parallel processing (Stanford)

Multiprocessor Ada (Incremental Systems Inc.)

Parallel software tools (Kestrel Institute)

72. See ARPA order file 5809 (ESL); see also *Strategic Computing: The MOSAIC Technology, Answers to Questions,* ESL proposal Q4557, 16 September 1985, in response to RFP N00039–84–R–0605(Q), ITO office files 1310.

73. Duane Adams to Lou Kallis, e-mail, 19 August 1987, filed in ARPA order file 4864, MRAO amendment 9; Douglas, Houser, and Taylor, eds. *Final Report on Supercomputer Research, 15 November 1983 to 31 May 1988.*

74. The chips were tested by running the game "Life" on them.

75. Squires interview, 12 July 1994; AO file 5058, CBI.

76. Hillis, "Parallel Computers for AI Databases," in *The Islamadora Workshop on Large Scale Knowledge Base and Reasoning Systems, February 25–27, 1985: Presenta-*

tions and Discussions, 159; document published for the workshop participants and found in CSTO reference collection, no. 105451.

77. Hillis, *The Connection Machine.*

78. Sheryl L. Handler, "High-Performance Computing in the 1990s," in *Frontiers of Supercomputing II: A National Reassessment,* ed. Karyn R. Ames and Alan Brenner, Los Alamos Series in Basic and Applied Sciences (Berkeley: University of California Press, 1994), 523.

79. Hillis, "The Connection Machine," *Scientific American* 256 (June 1987): 108–115, at 111.

80. MRAO, amendment, 2 Jan. 1986, AO file 5058, CBI.

81. Squires interview, 12 July 1994.

82. MRAO 5943/2, 5 February 1988, AO files, CBI.

83. Squires interviews, 12 July and 21 December 1994; Anthes, "A Sampler," 27.

84. See statement of work, contract N00039–86–C–0159, filed with AO 5876 (ARPA/DynCorp).

85. Ibid.

86. Christopher B. Hitchcock (BBN) to Dr. Tice DeYoung (SPAWAR), 6 April 1989, filed in ibid.

87. Tice DeYoung to Lt. Col. Mark Pullen (DARPA/ISTO), Memorandum, "Request for Additional Funds to Cover Cost Growth in SPAWAR Contract Number N00039–86–C–0159, with BBN," 10 April 1989, and MRAO 5876/3, 26 April 1989, all filed in ibid; Ben Barker, "Lessons Learned," in *Frontiers of Supercomputing II,* ed. Ames and Brenner, 455, 466–467.

88. Barker, "Lessons Learned."

89. Squires interview, 21 December 1983.

90. "The person who had the ability to choose the name liked Egyptian gods," an NSA official explained. "There is no other meaning to the name." Larry Tarbell, "Project THOTH: An NSA Adventure in Supercomputing, 1984–1988," in *Frontiers of Supercomputing II,* ed. by Ames and Brenner, 481.

91. This account of project THOTH is taken from Tarbell, "Project THOTH," 481–488.

92. Ibid., 486.

93. Ibid., 487.

94. Zitner, "Sinking Machines," 21, 24–25.

95. Ibid.; Elizabeth Corcoran, "Thinking Machines Rather Than Markets . . ." *Washington Post News Archives*, 28 August 1994.

96. John Markoff, "In Supercomputers, Bigger and Faster Means Trouble," *New York Times*, 7 August 1994, 5.

Chapter 6

1. The following account, unless otherwise indicated, relies primarily on Stuart Russell and Peter Norvig, *Artificial Intelligence: A Modern Approach* (Englewood Cliffs, NJ: Prentice Hall, 1995); Daniel Crevier, *AI: The Tumultuous History of the Search for Artificial Intelligence* (New York: Basic Books, 1993); and M. Mitchell Waldrop, *Man-made Minds: The Promise of Artificial Intelligence* (New York: Walker, 1987).

2. A. M. Turing, "On Computable Numbers, with an Application to the *Entscheidungsproblem*," *Proceedings of the London Mathematics Society*, 2d ser., 42 (1936): 230–265; "Computing Machinery and Intelligence," *Mind* 59 (1950): 433–460.

3. J. David Bolter, *Turing's Man* (Chapel Hill: University of North Carolina Press, 1984). In 1990 Dr. Hugh Loebner joined with the Cambridge Center for Behavioral Studies in offering a $100,000 prize for "the first computer whose responses were indistinguishable from a human's"; an annual prize is offered for the machine that is the most human. See John Schwartz, "Competition Asks: Is It Live or Is It the Computer Talking?" *Washington Post* (13 December 1993): A3; the "Home Page of The Loebner Prize," http://pascal.com.org/~loebner/loebner-prize.htmlx, 11 August 1998; and http://www.cs.flinders.edu.au/research/AI/LoebnerPrize/, 11 August 1998.

4. Herbert A. Simon, *Models of My Life* (New York: Basic Books, 1991), 206–207; cited in Crevier, *AI*, 46.

5. W. S. McCulloch and W. Pitts, "A Logical Calculus of the Ideas Immanent in Nervous Activity," *Bulletin of Mathematical Biophysics* 5 (1943): 115–137.

6. Russell and Norvig, *Artificial Intelligence*, 5.

7. Russell and Norvig put the Turing test in the "acting humanly" category. Ibid. 5. The interdisciplinary roots of machines intelligence are explored in Paul N. Edwards, *The Closed World: Computers and the Politics of Discourse in Cold War America* (Cambridge, MA: MIT Press, 1996). See also Stuart C. Shapiro, "Artificial Intelligence," in *Encyclopedia of Computer Science*, ed. by Anthony Ralston and Edwin D. Reilly (New York: Van Nostrand and Reinhold, 1993), 87–90.

8. Hubert L. Dreyfus, *What Computers Still Can't Do: A Critique of Artificial Intelligence* (Cambridge, MA: MIT Press, 1992), 57. This is a revised and updated version of the 1972 original. John R. Searle, a Berkeley philosopher, has been equally prominent in challenging AI, most notably in a classic paper on the "Chinese room," which questioned the validity of the Turing test; John R. Searle,

"Minds, Brains and Programs," *Behavioral and Brain Sciences* 3 (1980): 417–458. For their efforts, both philosophers received the 1994 Simon Newcomb award of the American Association for Artificial Intelligence. For a different view, see Bruce Mazlish, "The Fourth Discontinuity," *Technology and Culture* 8 (January 1967): 1–15, and *The Fourth Discontinuity: The Co-Evolution of Humans and Machines* (New Haven: Yale University Press, 1993), which takes an evolutionary approach to arrive at a prediction of "combots," computer-robots that will take on many—but not all—human characteristics.

9. Dreyfus, *What the Computer* Still *Can't Do*, 151, 303.

10. Russell and Norvig, *Artificial Intelligence*, 13.

11. Noam Chomsky, *Syntactic Structures* (The Hague: Mouton, 1957).

12. Quoted in Russell and Norvig, *Artificial Intelligence*, 20.

13. Pamela McCorduck, *Machines Who Think* (San Francisco: W.H. Freeman, 1979), 188.

14. See pp. 141–143 above.

15. Among the notable achievements of this work were Patrick Winston, "Learning Structural Descriptions from Examples," Ph.D. dissertation, MIT, 1970; and Terry Winograd, "Understanding Natural Language," *Cognitive Psychology* 3 (1972). Winston would go on to direct the MIT AI Laboratory; Winograd joined the faculty at Stanford, where he was an active member of Computer Professionals for Social Responsibility and a critic of the SC Program.

16. NP-completeness theory was just appearing in the early 1970s when Winston, Winograd, and others were working on their block problems. See S. A. Cook, "The Complexity of Theorem-Proving Procedures," *Proceedings of the Third ACM Symposium on Theory of Computing* (1971): 151–158; and R. M. Karp, "Reducibility among Combinatorial Problems," in *Complexity of Computer Computations*, ed. by R. E. Miller and J. W. Thatcher (New York: Plenum Press, 1972), 85–104. The complexity of problems and of the algorithms needed to solve them may be measured by the size of the problem, as measured by some parameter N, such as the number of rows in a matrix or the length of a list to be searched. If the number of steps or operations required to solve the problem is an exponential function of N, then the problem is "intractable." This means that it may be solved for small values of N, but not for large; that is, the problem does not scale. The blocks on the table top do not scale into railroads yards and ship ports. Russell and Norvig, *Artificial Intelligence*, 12, 21; "Computational Complexity," in Anthony Ralston, Edwin D. Reilly, and David Hemmendinger, eds., *Encyclopedia of Computer Science*, 4th ed. (New York: Nature Publishing Group, 2000), 260–265; "NP-Complete Problems," ibid., 1252–1256.

17. Cited in Russell and Norvig, *Artificial Intelligence*, 21; see also 691–693.

18. Stuart C. Shapiro, "Artificial Intelligence," in *Encyclopedia of Artificial Intelligence*, ed. by Stuart C. Shapiro, 2d ed. (2 vols.; New York: Wiley, 1992), 56.

19. MIT Archives, Collection AC 268, Laboratory for Computer Science, box 13, folder 4.

20. Norberg and O'Neill, *Transforming Computer Technology*, 224–237.

21. Edward A. Feigenbaum, "A Personal View of Expert Systems: Looking Back and Looking Ahead," Knowledge Systems Laboratory report no. KSL 92–41 (Stanford, CA: Stanford University, April 1992), 3.

22. This treatment of Feigenbaum is based on his own accounts in "Expert Systems," *The Encyclopedia of Computer Science*, ed. by Ralston and Reilly, 536–540; and "A Personal View of Expert Systems: Looking Back and Looking Ahead," Stanford University Knowledge Systems Laboratory report no. KSL 92–41, April 1992. The latter was the author's acceptance speech for the first Feigenbaum Medal, presented at the World Congress on Expert Systems, Orlando, FL, December 1991.

23. Russell and Norvig, *Artificial Intelligence*, 22

24. E. Feigenbaum, B. G. Buchanan, and J. Lederberg, "On Generality and Problem Solving: A Case Study Using the DENDRAL Program," in *Machine Intelligence 6*, ed. by B. Meltzer and D. Michie (New York: American Elsevier, 1971), 165–190.

25. Feigenbaum, "Expert Systems," 2.

26. Feigenbaum, "A Personal View," 14.

27. F. Hayes-Roth, "The Knowledge-based Expert System: A Tutorial," *Computer* 9 (1984): 11–28. See also F. Hayes-Roth, "Ruled-based Systems," *Communications of the ACM* 28 (September 1985): 921–932.

28. Crevier, *AI*, 197–203. See also P. Harmon and D. King, *Expert Systems: Artificial Intelligence in Business* (New York: Wiley, 1985).

29. Barbara Hayes-Roth, "Blackboard Systems," in *Encyclopedia of Artificial Intelligence*, ed. by Shapiro, 1:73–79.

30. Tool is the more general term. A shell is a tool tailored for a certain domain or task and embodying a certain kind of inferencing procedure. Mark Richer, "An Evaluation of Expert System Development Tools," in *AI Tools and Techniques*, ed. by Mark H. Richer (Norwood, NJ: Ablex Publishing, 1989), 68.

31. *Strategic Computing: New-Generation Computing Technology: A Strategic Plan for its Development and Applications to Critical Defense Problems* (Washington: DARPA, 28 October 1983), 30.

32. Robert E. Kahn, "A Defense Program in Supercomputation from Microelectronics to Artificial Intelligence for the 1990s," 1 September 1982, 6.

33. See pp. 53–56 above.

34. [Robert Kahn,] "Strategic Computing & Survivability: A Program to Develop

Super Intelligent Computers with Application to Critical Defense Problems: Program Description" (Arlington, VA: DARPA [May 1983], 54.

35. See the section on "Generic Systems" in BDM, "Strategic Computing and Survivability Research: Final Technical Report," BDM/W–537–TR, 28 May 1985; and the section on "Generic AI Systems" in Craig I. Fields and Brian G. Kushner, "The DARPA Strategic Computing Initiative," *Spring Compcon 86* (Digest of Papers) (San Francisco, CA: IEEE Computer Society, 1986), 352–358.

36. Saul Amarel interview with Alex Roland and Philip Shiman, New Brunswick, NJ, 16 December 1994.

37. *Strategic Computing* (28 October 1983), i–ii.

38. GASOIL in 1986 would take one person-year to produce 2,500 rules; Crevier, *AI*, 200.

39. Feigenbaum, "Personal View," 9; Paul E. Ceruzzi, *A History of Modern Computing* (Cambridge, MA: MIT Press, 1999), 73.

40. *Strategic Computing* (28 October 1983), 22, and app. II.1.4. The program plan stated that "'rules' represent the codification of an expert system process and a 'firing' indicates the examination, interpretation, and response to one rule in a particular context. In current systems, the firing of one rule can require the execution of tens of thousands of instructions, and as contexts become more complex the number of instructions for rule-firing increases." Five times real time mean that if it takes two seconds to make a statement, it will take the computer ten seconds to understand it. Interview of Victor Zue by Alex Roland, Cambridge, MA, 16 June 1995.

41. Ronald Ohlander interview by A. L. Norberg, Marina del Rey, CA, 25 September 1989, 1–24.

42. Commerce Business Daily (21 Feb. 1984).

43. Memorandum, Ronald B. Ohlander for Deputy Director, Research, and Director, Information Processing Techniques Office, on "Competitive Evaluation of Bidders for Research in Expert Systems Technology under Strategic Computing Program," n.d., from e-mail files of Ronald Ohlander; Memorandum, Robert E. Kahn to Commander, Rome Air Development Center, on "Selection of Contractors for Expert Systems Technology Research," n.d., from ibid.

44. Ohlander interview.

45. Conrad Bock interviews with Alex Roland and Philip Shiman, Palo Alto, CA, 28 July and 24 October 1994.

46. Douglas Hofstadter, *Gödel, Escher, Bach: An Eternal Golden Braid* (New York: Basic Books, 1979).

47. Edward A. Feigenbaum and Pamela McCorduck, *The Fifth Generation: Artifi-*

cial Intelligence and Japan's Challenge to the World (Reading, MA: Addison-Wesley Publishing, 1983).

48. E-mail, Conrad Bock to Alex Roland, 19 August 1998.

49. Newquist, *The Brain Makers*, 185.

50. Harvey Newquist maintains they were selling entirely service, but their offering had some aspects of a product; you develop it once and then sell it over and over again on line. The problem was not that they were tied down to a service, but that there simply was not enough demand for their product. See Newquist, *The Brain Makers*, 263–264, et passim, on which the following account is largely based.

51. B. G. Buchanan and E. H. Shortliffe, eds., *Rule-Based Expert Systems: The MYCIN Experiments of the Stanford Heuristic Programming Project* (Reading, MA: Addison-Wesley, 1984).

52. Richard Fikes interview with Alex Roland and Philip Shiman, Palo Alto, CA, 26 July 1994.

53. Newquist, *The Brain Makers*, 265. KEE stood for "Knowledge Engineering Environment," an IntelliCorp trademark. Richer, "An Evaluation of Expert System Development Tools," in *AI Tools and Techniques*, ed. by Richer, 82–89.

54. By this time Feigenbaum had resigned from the boards of both companies. They were, after all, now competitors. He continued, however, as a consultant to both, and even returned to IntelliCorp as chairman in 1988. Newquist, *The Brain Makers*, 266.

55. Allen Sears, MRAO, 24 July 1984, for AO 5292; Richard Fikes, "New Generation Knowledge System Development Tools," e-mail Fikes to Sears and Ohlander, 26 October 85, in Ohlander collection; ARPA order 6130, copy at Meridian Corp.

56. Thomas Gross interview by Alex Roland and Philip Shiman, Pittsburgh, PA, 9 March 1994; David McKeown interview, ibid., 8 March 1994.

57. ARPA order no. 6130, 20 April 1987; copy in possession of the authors obtained through Freedom of Information Act request, H.O. McIntyre to Alex Roland, 26 April 1998, ref. 98–F–2207.

58. Newquist, *The Brain Makers*, 219; Frederick Hayes-Roth comments on an earlier draft of this chapter, in e-mail, Ronald Ohlander to Roland and Shiman, 4 April 1996.

59. Paul Carroll, *Big Blues: The Unmaking of IBM* (New York: Crown, 1994).

60. Frederick Hayes-Roth interview by Alex Roland and Philip Shiman, Palo Alto, CA, 29 July 1994, 9.

61. Frederick Hayes-Roth, James E. Davison, Lee D. Erman, and Jay S. Lark, "Framework for Developing Intelligent Systems," *IEEE Expert* (June 1991): 40.

62. Hayes-Roth interview, 13.

63. F. Hayes-Roth, "Rule-based Systems," in *Encyclopedia of Artificial Intelligence,* ed. by Stuart C. Shapiro and David Eckroth, 2 vols. (New York: Wiley, 1990 [1987]), 963–973, quote on 971. Newquist, *Brain Makers,* 266.

64. Bruce Bullock, Teknowledge Federal Systems, "ABE: An Environment for Integrating Intelligent Systems," twenty-four slides prepared in 1984 and published in NSIA Software Committee National Symposium on "DARPA's Strategic Computing Initiative," Arlington, VA, 7–8 May 1986; Frederick Hayes-Roth, James E. Davidson, Lee D. Erman, and Jay S. Lark, "Framework for Developing Intelligent Systems: The ABE Systems Engineering Environment," *IEEE Expert* (June 1991): 30–40. See also, Lee Erman e-mail to Allen Sears, 14 April 1986.

65. Hayes-Roth interview, 14.

66. Hayes-Roth contends that Teknowledge's major goal in the 1980s was "to make AI technology into a modular component that could interoperate with more conventional technologies." E-mail, Ohlander to Roland and Shiman, 4 April 1996.

67. Memorandum for the Director, Program Management, from Ohlander, "Request for a New ARPA Order" [hereafter MRAO] [27 June 1984]. This request was staffed through the ARPA bureaucracy and signed by Cooper on 13 July. It designated Rome Air Development Center in New York as the actual contracting agent. Rome did not sign a contract with Teknowledge until 8 August 1985, though the company was able to recover expenses incurred back to 1984. Award/Contract F30602–85–C–0135 from U.S. Air Force, Air Force Systems Command, Rome Air Development Center, through DCASMA Van Nuys, Van Nuys, CA, to Teknowledge Federal Systems, 8 August 1985. The amount of the contract finally awarded was $1,763,137.

68. MRAO, 2.

69. Robert Simpson interview by A. L. Norberg, Washington, DC, 14 March 1990, 10.

70. Ohlander interview, 25 Sept. 1989, 3.

71. Hugh G. J. Aitken, *Syntony and Spark: The Origins of Radio* (New York: Wiley, 1976).

72. Bock interviews; Fikes interview, 24. Ohlander estimated that when he was PM he visited every site in his portfolio twice a year; Ohlander interview by A. L. Norberg, Marina del Rey, CA, 25 September 1989, 35.

73. Fikes interview, 26.

74. Ibid., 25–26. Nor, according to Fikes, would PIs inform on each other when meeting privately with the PM. He recalls that Sears would meet individually with PIs late in the evening over wine, another opportunity to glean insight and

information. But Fikes believes that PIs relied on external evaluations to make independent judgments about which programs were progressing and which were stalled.

75. This perception derives from the authors' attendance at a PI meeting on "DSSA's Role in Development and Infrastructure, What's There and What's Missing," Quarterly DSSA [Domain-Specific Software Architecture] Workshop, Seattle, WA, 3–5 August 1994. This meeting, led by PM Kirstie Bellman, hosted nine PIs and about fifty people in all.

76. MRAO, for an amendment (no. 2) to ARPA order 5255, from Sears, dated 14 October 1986. In a handwritten note to ISTO Director Saul Amarel, Sears said that "our goal is to have ABE have enough *performance* to be credible [sic] as a development and demo[nstration] system." [Italics in original.]

77. E. M. Kallis, ARPA order Decision Summary for ARPA order 5292, amendment no. 3, 3 November 1986.

78. See pp. 256–258.

79. Jack Schwartz and Susan Goldman interview by Alex Roland and Philip Shiman, New York, NY, 16 December 1994, 1–2. Fields's title was deputy director of research. This appears to have meant that he was DARPA deputy director for research.

80. "The Dark Ages of AI: A Panel Discussion at AAAI-84," *The AI Magazine* (Fall 1985): 122–134. Drew McDermott introduced the session by noting in part: "Suppose that five years from now the strategic computing initiative collapses miserably as autonomous vehicles fail to roll. The fifth generation turns out not to go anywhere, and the Japanese government immediately gets out of computing. Every startup company fails. . . . And there's a big backlash so that you can't get money for anything connected with AI. Everybody hurriedly changes the names of their research projects to something else. This condition [is] called 'AI Winter' by some." (122) See also Ruth Simon, "The Morning After," *Forbes* (19 Oct. 1987): 164; Stephen J. Andriole, "AI Today, Tomorrow and Perhaps Forever," *Signal* (June 1986): 121–123; Robert J. Drazovich and Brian P. McCune, "Issues in Military AI," in National Academy of Sciences, *Proceedings of the Workshop on AI and Distributed Problem Sovling*, May 16–17, 1985 (Washington: National Academy Press, 1985), 29–40; and Gary R. Martins, "AI: The Technology That Wasn't," *Defense Electronics* (December 1986): 56, 57, 59. Martins said "the AI bandwagon has developed a flat tire. On closer inspection we shall see that all five wheels have come off." (56)

81. See p. 145.

82. Jack Schwartz, "The Limits of Artificial Intelligence," in *Encyclopedia of Artificial Intelligence*, ed. by Shapiro and Eckroth, 488–503, quote at 493.

83. Ibid., 499.

84. Ibid., 492.

85. Fikes interview.

86. Hayes-Roth interview.

87. The successor to the Information Processing Techniques Office (IPTO), created when IPTO and the Engineering Applications Office (EAO) were merged in 1986. See pp. 256–258.

88. Fikes interview.

89. Hayes-Roth, Davison, and Erman, "Framework for Developing Intelligent Systems," 39.

90. Hayes-Roth interview.

91. Newquist, *The Brain Makers*, 379–389.

92. Lee D. Erman, Jay S. Lark, and Frederick Hayes-Roth, "ABE: An Environment for Engineering Intelligent Systems," *IEEE Transactions on Software Engineering* 14 (December 1988): 1758–1770; Frederick Hayes-Roth, et al., ABE: A Cooperative Operating System and Development Environment," in *AI Tools and Techniques*, ed. by Richer, 323–355.

93. Hayes-Roth comments.

94. The final MRAO for this project, amendment 8 to the basic contract, dated 15 September 1989, said that the program's "overall objectives have been tested and, *to the point possible,* validated through ABE's use in both Lockheed's Pilot's Associate Program and Mitre's ALBM/Rage program." [Italics added.]

95. Hayes-Roth interview.

96. Newquist, *The Brain Makers*, 367.

97. DARPA, "Strategic Computing and Survivability" (May 1983), 71.

98. ABE allowed the "development of modules" in "C"; Frederick Hayes-Roth, Lee D. Erman, Scott Fouse, Jay S. Lark, and James Davidson, "ABE: A Cooperative Operating System and Development Environment," in *AI Tools and Techniques*, ed. by Richer, 350. See also Lee D. Erman, Jay S. Lark, and Frederick Hayes-Roth, "ABE: An Environment for Engineering Intelligent Systems," *IEEE Transactions on Software Engineering* 14 (December 1988): 1768; Frederick Hayes-Roth, James E. Davidson, Lee D. Erman, and Jay S. Lark, "Framework for Developing Intelligent Systems: The ABE Systems Engineering Environment," *IEEE Expert* (June 1991): 35; and William S. Mark and Robert L. Simpson, "Knowledge-Based Systems," ibid., 16.

99. Richard Fikes's comments on earlier draft of this chapter, forwarded in e-mail, Ronald Ohlander to Roland and Shiman, 4 April 1996; Richard Fikes, et. al, "OPUS: A New Generation Knowledge Engineering Environment. Phase 1 Final Report of the project in the Knowledge Systems Technology Base of the Defense Advanced Research Projects Agency's Strategic Computing Program," (Mountain View, CA: IntelliCorp, July 1987).

100. Paul Morris interview by Alex Roland and Philip Shiman, Palo Alto, CA, 29 July 1994. Conrad Bock does not recall a lack of focus following Fikes's departure; Bock interview, 28 July 1994, 42.

101. Gary Fine interview, Palo Alto, CA, 29 July 1994. IntelliCorp has spun off a new product called Kappa, a development environment in C language for UNIX-based client-server applications. See the series of Kappa Application Profiles, published by IntelliCorp.

102. Newquist, *The Brain Makers*, 379–398.

103. Fine interview; Hayes-Roth interview.

104. E-mail, Conrad Bock to Alex Roland, 5 August 1998.

105. Bock interview, 22 October 1994.

106. Conrad Bock and James Odell, "A Foundation for Composition," *Journal of Object-Oriented Programming* 7 (October 1994): 10–14.

107. Hayes-Roth, "Rule-based Systems," 972.

108. Frederick Hayes Roth, "Rule-Based Systems," *Encyclopedia of Artificial Intelligence*, 2d ed., 1417–1426, quote at 1426.

109. A good overview is provided in Frederick Hayes-Roth and Neil Jacobstein, "The State of Knowledge-based Systems," *Communications of the ACM* 37 (March 1994): 27–39. They conclude that "the current generation of expert and KBS [knowledge-based system] technologies had no hope of producing a robust and general human-like intelligence." (36) See n. 70 above.

110. Ronald Ohlander, "Integration, Transition, and Performance Evaluation of Generic Artificial Intelligence Technology," undated draft chapter provided to the authors from Ohlander computer files, subsection "continuous speech recognition research approach."

111. Russell and Norvig, *Artificial Intelligence*, 757–759.

112. Jonathan Allen, "Speech Recognition and Synthesis," in *Encyclopedia of Computer Science*, 2d ed., 1264–1265. The authors are grateful to Ralph Grishman for providing helpful suggestions on modification of an earlier version of this section. See Grishman's comments in e-mail, Ronald Ohlander to Shiman and Roland, 29 April 1996.

113. Victor Zue interview by Alex Roland, Cambridge, MA, 16 June 1995. Zue maintains that by the mid-1990s systems recognizing 40,000 words were up and running. As a referee for the original SC specifications, Zue was skeptical and had to be convinced by Raj Reddy to endorse them.

114. Again, the authors are indebted to Ralph Grishman for sharing with them "Message Understanding Conference 6: A Brief History," forwarded in Ohlander e-mail to Shiman and Roland, 29 April 1996. This study chronicles

the evolution of the MUCs, beginning in 1987. Grishman concludes that the sixth MUC (1995) provided "valuable positive testimony on behalf of information extraction, but further improvement in both portability and performance is needed for many applications. . . . We may hope that, once the task specification settles down, the availability of conference-annotated corpora and the chance for glory in further evaluations will encourage more work in this area" (11–12).

Raymond Perrault reports that the MUC conferences were a response to DARPA Program Manager Robert Simpson's recommendation in 1985 that SC researchers in this field "come up with an evaluation methodology, so that systems could be compared and progress could be assessed." E-mail, Perrrault to Ohlander, 3 January 1996, forwarded in Ohlander to Shiman and Roland, 3 January 1996. Perrault says that "the Strategic Computing Initiative had a huge impact on the field of text understanding. Indeed,the success of the field today, as exemplified by several commercial and near-commercial systems and by the Tipster Program, can be traced directly to the SCI."

115. This periodization is based on "Strategic Computing Vision," e-mail, Ronald Ohlander to Shiman and Roland, 22 March 1996, an account to which Martin Fischler and Ram Nevatia also contributed.

116. David Marr, *Vision: A Computational Investigation into the Human Representation and Processing of Visual Information* (San Francisco: W. H. Freeman, 1982).

117. Russell and Norvig, *Artificial Intelligence,* 769–770; Crevier, *AI,* 183–190; Marr, *Vision.*

118. Hayes-Roth and Jacobstein, "The State of Knowledge-based Systems"; interview of Edward Feigenbaum by Alex Roland and Philip Shiman, Arlington, VA, 28 July 1995.

119. Department of Commerce, *Critical Technology Assessment of the U.S. Artificial Intelligence Sector* ([Washington, DC:] DoC, 1994, vii.

120. Michael Schrage, "Senile Dementia for Artificial Intelligentsia?" *Computerworld* (9 May 1994): 39.

121. Hayes-Roth and Jacobstein, "The State of Knowledge-Based Systems," 27. See also F. Hayes-Roth, "The Evolution of Commercial AI Tools: The First Decade," *International Journal on Artificial Intelligence Tools,* 2 (1993): 1–13.

122. Crevier, *AI,* 209.

123. Feigenbaum, "A Personal View of Expert Systems"; and Feigenbaum, Robert Englemore, Thomas Gruber, and Yumi Iwaski, "Large Knowledge Bases for Engineering: The How Things Work Project of the Stanford Knowledge Systems Laboratory," KSL 90–83 (Palo Alto, CA: Stanford University Knowledge Systems Laboratory, November 1990).

124. Hayes-Roth and Jacobstein, "The State of Knowledge-Based Systems," 36.

125. Stephen B. Sloane, "The Use of Artificial Intelligence by the United States Navy: Case Study of a Failure," *AI Magazine* 12 (Spring 1991): 80–92.

126. As Shapiro put it, "as soon as a task is conquered, it no longer falls within the domain of AI. Thus, AI is left with only its failures; its successes become other areas of computer science." Shapiro, "Artificial Intelligence," 88.

Chapter 7

1. Interview of Clinton W. Kelly III, by Alex Roland and Philip Shiman, Durham, NC, 23 May 1994.

2. Ibid.

3. Carl S. Lizza, "History and Retrospective of the Pilot's Associate Program," unpublished document provided by the author, 2–3.

4. Statement of Work, section 2.4, contract N00039–84–C–0467 (Carnegie Mellon University), filed in AO 4864; Ellen P. Douglas, Alan R. Houser, and C. Roy Taylor, eds., *Final Report on Supercomputer Research, 15 November 1983 to 31 May 1988,* School of Computer Science, Carnegie Mellon University, Report CMU–CS–89–157, June 1989, sect. 5:1–10; John P. Flynn and Ted E. Senator, "DARPA Naval Battle Management Applications," *Signal* 40 (June 1986): 59, 64–65.

5. Flynn and Senator, "DARPA Naval Battle Management Applications," 59–60.

6. Ibid., 60.

7. William Broad, *The Universe Below: Discovering the Secrets of the Deep Sea* (New York: Simon & Schuster, 1997).

8. Paul Losleben, interview conducted by Alex Roland and Philip Shiman, Palo Alto, Ca., 29 July 1994.

9. Clinton Kelly, "Autonomous Robot Vehicle Program," *Signal* 38 (August 1984), cited in Strategic Computing Program, *Representative Press Releases, February 1983–January 1986* (n.p.: DARPA, n.d.), 59–61; "Robots Take Their First Steps," *Saudi Gazette,* 8 October 1984.

10. Richard H. Van Atta, Sidney Reed, and Seymour J. Deitchman, *DARPA Technical Accomplishments,* Volume II: *An Historical Review of Selected DARPA Projects* (Alexandria, VA: Institute for Defense Analyses, April 1991), chapter 14:4–11; Arthur L. Norberg and Judy E. O'Neill, *Transforming Computer Technology: Information Processing and the Pentagon, 1962–1986* (Baltimore: Johns Hopkins University Press, 1996), 352–375.

11. Real time in a moving vehicle loses some of its meaning. It still means processing input as fast as it is received. But as the vehicle speeds up, changes in the visual data it receives also accelerate. Recognizing an approaching obstacle

at 20 kph is easier than at 60 kph, even though the computer may work at nominally real time in both cases. The vision works at the same speed, but the image
understanding must work faster.

12. DARPA, *Strategic Computing: New Generation Computing Technology: A Strategic Plan for Its Development and Application to Critical Problems in Defense* (Arlington, VA: DARPA, 28 Oct. 1983), 22–23.

13. Kelly interview, 23 May 1994; Ron Ohlander, interview conducted by Alex Roland and Philip Shiman, Los Angeles, California, 21 October 1994.

14. "Research and Development and Systems Integration of Specific Advanced Computing Technologies which will result in the Development and Demonstration of an Autonomous Land Vehicle," *Commerce Business Daily*, PSA-8511, 27 January 1984, 2.

15. Ibid. Contract DACA76–84–C–0005, filed with ARPA Order 5132, ARPA Order files; Kelly interview, 23 May 1994, Durham, NC.

16. Van Atta, Reed, and Deitchman, *DARPA Technical Accomplishments,* chapter 14:7.

17. Interview of Dr. Larry Davis by Philip Shiman, College Park, MD, 11 January 1995.

18. It did this because of the unfortunate resemblance of the original initials, AI&DS, to the disease just then becoming epidemic.

19. Hughes was funded by DARPA's Engineering Applications Office (ARPA Order no. 5372); ADS subcontracted directly with Martin Marietta.

20. "Image Understanding," *Commerce Business Daily*, 20 January 1984, Issue PSA 8506, 32; *Strategic Computing, First Annual Report*, 8.

21. For these tasks see the statements of work in ARPA Order files no. 5351 (CMU), 5147 (U. Of Rochester), 5352 (CMU), 5359 (MIT), 6365 (U. Of Southern California), 5373 (Hughes), and 5375 (U. Of Massachusetts).

22. Purchase Description for Road-Following, attached to DACA76–85–C–0003, filed in ARPA Order no. 5251; Ohlander interview by Roland & Shiman, 21 October 1994.

23. Takeo Kanade and Charles Thorpe, "CMU Strategic Computing Vision Project Report: 1984 to 1985," CMU–RI–TR–86–2, November 1985, 3–5.

24. For GE, see ARPA Order 5512; for Honeywell, see ARPA Order 5628; for Columbia, see ARPA Order 5577.

25. J. Lowrie, et al., *Autonomous Land Vehicle Annual Report,* December 1985 (Martin Marietta Corporation, MCR–85–627), section XI.

26. Davis interview, 11 January 1995.

27. Lowrie et al., *Autonomous Land Vehicle Annual Report*, section V and VII.

28. See Lowrie et al., *Autonomous Land Vehicle Annual Report*, sections IV–VIII.

29. William E. Isler, "The Autonomous Land Vehicle," in NSIA Software Committee, *A National Symposium on DARPA's Strategic Computing Program*, 7–8 May 1986, proceedings, 107.

30. ETL was then engaged in creating a detailed digital map of the test site for the use of the ALV and the SCVision researchers. *The Autonomous Land Vehicle 1st Quarterly Report*, MCR–84–600, issue 17, Martin Marietta Corporation, May 1986, 6.

31. Takeo Kanade and Charles Thorpe, *Strategic Computing Vision Project Report: 1984–1985*, CMU–RI–TR–86–2, the Robotics Institute, Carnegie Mellon University, November 1985.

32. Kelly interview, 23 May 1994; Roger Schappel, telephone interview conducted by Dr. Philip Shiman, 12 December 1994.

33. *The Autonomous Land Vehicle Workshop*, April 16–17 1985, proceedings; John B. Gilmer, Jr., "Report on Autonomous Land Vehicle Workshop, September 26–27, 1985" (draft), BDM Corp., 25 October 1985; *The Autonomous Land Vehicle 1st Quarterly Report*, 8.

34. John B. Gilmer, Jr., "ALV System Architecture Meeting of 17 July 1985" (draft), BDM Corp., 25 July 1985.

35. John B. Gilmer, Jr., "Report on DARPA ALV Vision System Hardware Architecture Meeting of October 1, 1985," BDM/R–85–0984–TR, 25 October 1985; T. J. Mowbray, "DARPA/MMC ALV Long Range Architecture Study, Review of Computer Architectures Studied," *The Autonomous Land Vehicle 2nd Quarterly Review*, June 1986, conference proceedings, Martin Marietta Corp., 87–155; *The Autonomous Land Vehicle (ALV) Second Quarterly Report*, MCR–84–600, Issue 18, Martin Marietta Corp., September 1986, 54–95.

36. Lowrie et al., *Autonomous Land Vehicle, Annual Report*, X:18.

37. Ibid., II:3.

38. John Aloimonos, Larry S. Davis, and Azriel Rosenfeld, "The Maryland Approach to Image Understanding," *Image Understanding Workshop: Proceedings of a Workshop Held at Cambridge, Massachusetts, April 6–8, 1988*, 2 vols. (McLean, VA: Science Applications International Corporation, 1988), 1:154.

39. Gilmer, "Report on Autonomous Land Vehicle Workshop, September 26–27, 1985."

40. Robert Simpson dismisses these early debates as just "posturing"; telephone interview conducted by Philip Shiman, 13 December 1994.

41. *The Autonomous Land Vehicle 1st Quarterly Report*, 10.

42. Lowrie et al., *Autonomous Land Vehicle Annual Report,* section VII:27; *Autonomous Land Vehicle 1st Quarterly Report,* 36–39.

43. *Autonomous Land Vehicle Phase I Final Report,* II:2–5, V:1–3; Lowrie et al., *Autonomous Land Vehicle Annual Report,* VII:27.

44. *Autonomous Land Vehicle 1st Quarterly Report,* 86.

45. The panning system was activated when vision detected only one road edge in the images it received, an indication that the road was curving away. Vision therefore instructed the camera to turn in the direction of the missing road edge, where the center of the road was expected to be. The problem was that while it was sending those instructions to the camera, it was dutifully passing on the scene model (with the single edge) to the navigator, which recognized that the vehicle was no longer in the middle of the road. While vision turned the camera, navigator turned the vehicle. The camera, the movements of which were based on the last scene model and did not take into account the *current* movement of the vehicle, therefore panned too far. The video image again showed one edge (this time on the opposite side); vision interpreted this to mean that the road was again curving in the opposite direction, and panned the camera that way—and navigator turned the vehicle again. Thus the camera turned repeatedly from side to side and the vehicle repeatedly overcompensated, until the vehicle ran off the road. *The Autonomous Land Vehicle 1st Quarterly Report,* 39–42.

46. *Autonomous Land Vehicle, 1st Quarterly Report,* 36–42; *Autonomous Land Vehicle Phase I Final Report,* MCR–84–600, Issue 24, Martin Marietta Corp., 4 March 1988, IV:1.

47. Lowrie et al., *Autonomous Land Vehicle Annual Report,* IX–6.

48. Gilmer, *Report on Autonomous Land Vehicle Workshop,* September 26–27, 1985.

49. *Autonomous Land Vehicle Workshop, April 16–17, 1985,* proceedings.

50. *Autonomous Land Vehicle 1st Quarterly Report,* 8; *Autonomous Land Vehicle 2nd Quarterly Report,* 4–5; *The Autonomous Land Vehicle (ALV) Program, 3rd Quarterly Report,* MCR–84–600, issue 19, Martin Marietta Corp., November 1986, 60.

51. *Autonomous Land Vehicle 6th Quarterly Report,* 4:3. Martin Marietta did not positively report its connection to the ARPANET until March 1988 (*Autonomous Land Vehicle Phase I Final Report,* 9:2).

52. Gilmer, *ALV System Architecture Meeting of 17 July 1985,* 15.

53. "Vision Provisions for Transition," draft document enclosed in Chuck Thorpe to Ron Ohlander, e-mail, 9 August 1985, copy supplied by Ron Ohlander.

54. Ibid.

55. Kanade and Thorpe, *CMU Strategic Computing Vision Project Report: 1984– 1985*, 3–6, 10.

56. Carnegie Mellon University, "Revised Proposal for Road-Following," 20 September 1984, enclosed in MRAO, 17 October 1984, ARPA order 5351.

57. Simpson interview.

58. Charles Thorpe and William Whittaker, *Proposal for Development of an Integrated ALV System*, Robotics Institute, Carnegie Mellon University, May 1985, enclosed in MRAO, ARPA Order 5351, amendment 2.

59. See chapter 5.

60. Alan Newell, interview conducted by Arthur L. Norberg, Pittsburgh, Pa., 10– 12 June 1991.

61. Thorpe and Whittaker, *Proposal for Development of an Integrated ALV System*, 28–29.

62. MRAO, May 1985, filed in ARPA Order 5351; exact date illegible on document.

63. Gilmer, "ALV System Architecture Meeting of 17 July 1985," 15.

64. Ohlander interview.

65. Ibid.

66. Ron Ohlander, "Strategic Computing Program: Integration, Transition, and Performance Evaluation of Generic Artificial Intelligence Technology" (draft), September 1985, Copy supplied by Dr. Ohlander.

67. Interview with Charles Thorpe, conducted by Philip Shiman, Pittsburgh, Pa., 17 November 1994.

68. Charles Thorpe and Takeo Kanade, *1986 Year End Report for Road Following at Carnegie Mellon*, CMU–RI–TR–87–11, the Robotics Institute, Carnegie Mellon University, May 1987, 17; C. Thorpe, S. Shafer and T. Kanade, "Vision and Navigation for the Carnegie Mellon Navlab," *Second AIAA/NASA/USAF Symposium on Automation, Robotics and Advanced Computing for the National Space Program, March 9–11, 1987*, AIAA–87–1665, 1, 7.

69. Thorpe and Kanade, *1986 Year End Report*, 36–37.

70. *Autonomous Land Vehicle 2nd Quarterly Report*, 5.

71. Ibid., 83–84.

72. Simpson interview.

73. *Autonomous Land Vehicle 2nd Quarterly Report*, 96. Martin did the same at Maryland.

74. *Autonomous Land Vehicle (ALV) Program Sixth-Quarterly Report*, MRC–84–600, issue 22, Martin Marietta Corp., 31 August 1987, III:6.

75. *Autonomous Land Vehicle 2nd Quarterly Report*, 96.

76. *Autonomous Land Vehicle Fourth Quarterly Report*, VII:1–2.

77. Ibid., II:50–52; *Autonomous Land Vehicle Phase I Final Report*, II:9.

78. Karen E. Olin and David Y. Tseng, "Autonomous Cross-Country Navigation," *IEEE Expert* 6 (August 1991):16, 23–25.

79. *The Autonomous Land Vehicle Program Seventh Quarterly Report*, MCR–84–600, Issue 23, Martin Marietta Corp., 30 November 1987, I:6.

80. Takeo Kanade, "Report for ALV Research Review," document provided by the author.

81. *The Autonomous Land Vehicle Program, Phase II, First Quarterly Scientific and Technical Report*, June 30, 1988, MCR–88–600, issue 1, VIII:8–9.

82. See ARPA Order decision summary, ARPA Order 5132, amendment 25; "ISTO by Program Element and Project," 8 January 1988, attached to John Toole to budget analyst, ISTO, memorandum on "Descriptive Summaries," 11 January 1988, document provided by the author.

83. Kelly interview.

84. William Isler to Mary L. Williams, 19 April 1988, and Mark S. Wall to LaVerne Bonar, 3 January 1989, filed in ARPA Order 5132.

85. Erik G. Mettala, "The OSD Tactical Unmanned Ground Vehicle Program," *Image Understanding Workshop: Proceedings of a Workshop Held at San Diego, California, January 26–29, 1992*, 159–171.

86. Charles Thorpe et al., "Toward Autonomous Driving: The CMU Navlab, Part I—Perception," *IEEE Expert* 6 (August 1991): 31–42; Thorpe et al., "Toward Autonomous Driving: The CMU Navlab, Part II—Architecture and Systems," ibid., 44–52; discussion with Charles Thorpe, Pittsburgh, Pa., 17 November 1994.

87. Thorpe to Philip Shiman, private e-mail, November 1994; see also Charles Thorpe et al., "Toward Autonomous Driving: The CMU Navlab, Part II—Architecture and Systems," *IEEE Expert* 6 (August 1991): 50–51.

88. Simpson interview.

Chapter 8

1. *Strategic Computing, Second Annual Report*, DARPA, February 1986, 27–30. Of the $63.2 million spent on SC in 1985, industry received 47.5 percent ($32.4

million), universities 40.7 percent ($27.8 million), and government agencies and research labs 11.8 percent ($8.0 million); ibid., 26.

2. Saul Amarel, interview conducted by Arthur L. Norberg, New Brunswick, N.J., 5 October 1989.

3. Amarel interview, CBI.

4. Arthur L. Norberg and Judy E. O'Neill, *Transforming Computer Technology: Information Processing for the Pentagon, 1962–1986* (Baltimore: Johns Hopkins University Press, 1996), 24–67.

5. Amarel interview, CBI.

6. Ibid; Amarel, "Comments on the Strategic Computing Program at IPTO," La Jolla Institute, September 1985 (copy obtained from the author), 1; Amarel, "AI Research in DARPA's Strategic Computing Initiative," *IEEE Expert* 6 (June 1991): 10.

7. Amarel, "Comments on the Strategic Computing Program," 2–3. Amarel's criticism was not limited to vision, but applied to natural language as well. Ibid., 4.

8. Ibid., 2–3, 5–6; Amarel interview, CBI.

9. Amarel, "Comments on the Strategic Computing Program," 5.

10. Interview with Clinton Kelly by Alex Roland and Philip Shiman, Durham, NC, 23 May 1994.

11. Amarel interview, CBI.

12. Amarel, "Organizational plan for computer science and technology at DARPA," 3 April 1986 (document obtained from the author).

13. Interview of Saul Amarel conducted by Alex Roland and Philip Shiman, Rutgers, NJ, 16 December 1994.

14. This plan formalized an ad hoc group previously assembled by Cooper.

15. Ibid.

16. Ibid.

17. Flyer obtained from Saul Amarel.

18. NSIA Software Committee, *A National Symposium on DARPA's Strategic Computing Program,* Arlington, VA, 7–8 May 1986 (n.p., n.d.).

19. Originally Helena Wizniewski's computational mathematics program constituted a sixth program area in ISTO; but Wizniewski, who was not pleased with the move from EAO to ISTO, complained loudly, and she was allowed to transfer to the Defense Sciences Office that fall.

20. "Automation Technology Briefing, 12 January 1987," in *Information Science and Technology Office, Program Review* [January 1987]; and *Information Processing Research* (briefing documents), February 1986; both documents obtained from Saul Amarel.

21. Amarel, "Information Science and Technology at DARPA," January 1986, unpublished document obtained from the author; "Budget Planning—ISTO, July 1986," unpublished document obtained from Amarel. By the summer of 1987, robotics and software technology constituted separate program areas; see Amarel, memorandum for ISTO staff on ISTO long-range planning contribution, 24 July 1987, document obtained from the author.

22. Figures derived from DARPA telephone directories for various years, the closest things to organization charts that exist.

23. Amarel, ISTO strategic plan, 24 June 1987, unpublished document obtained from the author.

24. Saul Amarel, "Plan Outline for a Computer Aided Productivity (CAP) Program" (draft), 29 May 1987, copy obtained from the author.

25. Unsigned, undated document filed in AO file 2095 (MIT). The document is not entirely clear, but this appears to be what happened.

26. Chronology of AO actions, ARPA Order 5577; see also chronology of AO actions, ARPA Order 5682.

27. Amarel, memorandum to Tegnalia, 7 November 1985, document obtained from the author; "IPTO Budget Issue: Release of Funds on Hold at PM for FY86–87," in "IPTO Budget Review, FY1986–FY1990," 28 January 1986, unpublished document obtained from Amarel. The quote is from the Amarel memorandum cited above.

28. Senator Jeff Bingaman, speech delivered 7 May 1987, in NSIA Software Committee, *A National Symposium on DARPA's Strategic Computing Program*, Arlington, VA, 7 May 1987, 90–91; "IPTO Budget Issue: Release of Outyear S/C Funds," in "IPTO Budget Review, FY1986–FY1990."

29. Bingaman, speech of 7 May 1987, NSIA Software Committee, *National Symposium*, 91–92; DARPA, *Fiscal Year 1987, Research and Development Program (U): Annual Statement*, DARPA, February 1986, 7.

30. "Congress Enacts Strict Anti-Deficit Measure," *Congressional Quarterly Almanac*, 99th Cong., 1st Sess., 1985, 459–468.

31. "Relative Impact of FY86 Budget Reductions," in "Budget Planning—ISTO, July 1986," document obtained from Amarel. Duncan testified before Congress that the total DARPA reduction in 1986 was $117.4 million; U.S. Congress, House, Subcommittee of the Committee on Appropriations, *Department of Defense Appropriations for 1987*, 99th Cong., 2[d] Sess. (Washington: GPO, 1986), 5:262.

32. "Budget Planning—ISTO, July 1986," document obtained from Amarel. Some of the non-SC money, about $10 million, went to the Tactical Technology Office when two program managers transferred there with their basic programs. By the end of August, ISTO's basic budget had rebounded somewhat, to $93.5 million. The total cut still stood at 27 percent, however. See "ISTO Budget Review (28 August 1986)," document obtained from Amarel.

33. Ibid.

34. "IPTO Issue: Congressional Submission and Release of Outyear S/C Funds," in "FY86 Apportionment/FY87 POM Review, Information Processing Techniques Office," April 1985, document obtained from Amarel; "Budget Planning—ISTO, July 1986," document obtained from Amarel. DARPA eventually reported that it had spent $62.9 million on SC in 1985; see Defense Advanced Research Projects Agency, *Strategic Computing, Third Annual Report* (DARPA, February 1987), 26. A significant part of this reduction occurred when DARPA's gallium arsenide program went to SDIO in 1985.

35. Bingaman, speech of 7 May 1986, NSIA Software Committee, *National Symposium*, 91–92; Senate debate of 26 April 1985, *Congressional Record*, 99th Cong., 1st Sess., 1985, 12919–12922.

36. Bingaman, speech of 7 May 1986, in NSIA Software Committee, *National Symposium*, 92–93; senate debate of 10 December 1985 [Legislative day 9 December 1985] on H. J. Res. 465, *Congressional Record*, 99th Cong., 1st Sess., 1985, S17277–17282.

37. U.S. Congress, House, Subcommittee of the Committee on Appropriations, *Department of Defense Appropriations for 1988*, 100th Cong., 1st Sess. (Washington: GPO, 1987): 6:549.

38. Ibid., 517.

39. *SC Fourth Annual Report*, November 1988, 46.

40. These figures are calculated from "IPTO Budget Review, FY1986–FY1990," 28 January 1986, "Budget Planning—ISTO, July 1986," "Strategic Computing Budget Review," 18 August 1986; Information Science and Technology Office, Program Review, January 1987. Copies of these documents were obtained from Amarel.

41. The budget for the architectures program for FY 1986 looked like this:

(January 1986)

Prototype architectures/scalable modules	$19,450,000
Generic software/system development tools	$10,085,000
Performance modeling/algorithms	$10,700,000
Total	*$40,235,000*

(July 1986)

Prototype architectures/scalable modules	$18,712,000
Generic software/system development tools	$ 4,507,000
Performance modeling/algorithms	$ 1,682,000
Total	*$24,901,000*

("Budget Planning—ISTO, July 1986"; "IPTO Budget Review, FY1986–FY1990," 28 January 1986.)

42. Squires interview, 12 July 1994; memorandum, Kelly to the director, program management, 23 October [1986], filed with AO 4864 (Carnegie Mellon University); AO 9133 (Harris Corporation), CBI. Figures for the money offered by SDIO taken from Strategic Defense Initiative organization, work package directive, revised 16 October 1986, copy found in AO file 9173. Of the $7.8 million, three-year contract with IBM, DARPA provided only $2 million; SDIO provided the rest ($5.8 million); AO 5944/2 (IBM), CBI.

43. Testimony of Dr. Robert Duncan, 9 April 1986, in U.S. Congress, House, Subcommittee of the Committee on Appropriations, *Department of Defense Appropriations for 1987*, 99th Cong., 2[d] Sess. (Washington: GPO, 1986), 5:264; "Strategic Computing Budget Review (28 August 1986)," unpublished document supplied by Amarel.

44. Except where otherwise noted, the information from this section is drawn from DARPA, *SC Third Annual Report;* DARPA, *SC Fourth Annual Report;* and "Strategic Computing Plan, 1987, For the Period FY1987–FY1992," manuscript document (undated) in the files of Stephen L. Squires, DARPA.

45. Ellen P. Douglas, Alan R. Houser, and C. Roy Taylor, eds., *Final Report on Supercomputer Research, 15 November 1983 to 31 May 1988*, CMU–CS–89–157, report on contract N00039–85–C–0134, School of Computer Science, Carnegie Mellon University, Pittsburgh, Pa., June 1989, chapter 5:10.

46. Carl Lizza, "History and Retrospective of the Pilot's Associate Program," unpublished paper supplied by the author, 3–7.

47. This program was run by Major David N. Nicholson in the Tactical Technology Office.

48. "Strategic Computing Plan 1987, For the Period FY1987–FY1992," (n.p., n.d.), section 6.2, 1.

49. SCI architectures briefing in Amarel, "Information Science and Technology Office, Program Review" (January 1987).

50. Squires interview, 21 December 1994; "Defense Advanced Research Projects Agency, Information Science and Technology Office," information briefing by Col. A. E. Lancaster, Jr. [ca. summer 1987?], document obtained from Col. John Toole; Gary Anthes, "Parallel Processing," *Federal Computer Week* 1 (30 November 1987): 26.

51. Dmitris N. Chorafas and Heinrich Seinmann, *Supercomputers* (New York: McGraw-Hill, 1990), 122–132; DARPA, *Strategic Computing, Fourth Annual Report* (DARPA, Nov. 1988), 27; United States General Accounting Office, *High Performance Computing: Advanced Research Projects Agency Should Do More to Foster Program Goals*, Report to the Chairman, Committee on Armed Services, House of Representatives, May 1993, 36.

52. "Advanced Computing Systems," *Commerce Business Daily*, issue PSA-9308, 1 April 1987, 1; GAO, "High Performance Computing," 40–41.

53. "Strategic Computing Plan 1987, For the Period FY1987–FY1992," (n.p., n.d.), sections 6, 7.

54. Defense Advanced Research Projects Agency, *Strategic Computing, Fourth Annual Report* (DARPA, November 1988), 41–42.

55. Interview of Jacob Schwartz and Susan Goldman conducted by Alex Roland and Philip Shiman, New York, NY, 16 December 1994.

56. Ibid.

57. Ibid. See also Amarel interview, CBI.

58. ISTO program review, August 1988; Schwartz & Goldman interview.

59. Early in his tenure, on a visit to the West Coast in August 1987, Schwartz requested a paper by Berkeley and Stanford researchers on the application of CAD to VLSI design; see "Trends and Open Problems in CAD for VLSI," paper enclosed with memorandum from John Toole, 14 September 1987, documents obtained from John Toole, ARPA.

60. ISTO Program Review—August 1988, unpublished document obtained from John Toole, ARPA.

61. Schwartz and Goldman interview.

62. Ibid.

63. Jaime Carbonell, interview conducted by Alex Roland and Philip Shiman, Pittsburgh, PA, 8 March 1994.

64. Interview with Edward Feigenbaum, conducted by Alex Roland and Philip Shiman, 28 July 1995, Arlington, Va.

65. ISTO program reviews, January 1987 and August 1988. These figures are only projections; exact figures of final expenditures are not known.

66. Amended FY 1988–1989 biennial budget, RDT&E descriptive summary, project ST-11, and FY 1990–1991 biennial RDT&E descriptive summary, project ST-11, documents obtained from John Toole, ARPA.

67. Schwartz and Goldman interview.

68. See pp. 204–205 above.

69. ISTO program reviews, August 1986 and August 1988.

70. Schwartz was to write an article on the subject; see his "The New Connectionism: Developing Relationships Between Neuroscience and Artificial Intelligence," in *The Artificial Intelligence Debate: False Starts, Real Foundations*, ed. Stephen R. Graubard (Cambridge, MA: MIT Press, 1988), 123–141.

71. ISTO program review, August 1988.

72. Ibid.

73. Robert Simpson, interview conducted by Arthur L. Norberg, Washington, D.C., 14 March 1990.

74. Schwartz and Goldman interview.

75. Raj Reddy, "Foundations and Grand Challenges of Artificial Intelligence," presidential address to the American Association for Artificial Intelligence, 1988, repr. in Stephen J. Andriole and Gerald W. Hopple, eds., *Applied Artificial Intelligence: A Sourcebook* (New York: McGraw-Hill, 1992), 644.

76. Simpson interview, CBI.

77. Ibid.; FY 1990–1991 Biennial RDT&E descriptive summary, project ST-11, document obtained from John Toole.

78. "Report of the Research Briefing Panel on Computer Architecture," in *Research Briefings 1984, for the Office of Science and Technology Policy, the National Science Foundation, and Selected Federal Departments and Agencies* (Washington: National Academy Press, 1984), 1–16.

79. "Advanced Computing Systems," *Commerce Business Daily*, issue PSA-9308, 1 April 1987, 1; GAO, "High Performance Computing," 40–41.

80. Stephen L. Squires, "TeraOPS Computing Technology Initiative Towards Achieving TeraOPS Performance Systems: The TOPS Program," 13 October 1987 (draft), document supplied by the author.

81. Figures from ISTO and SC program and budget reviews.

82. ISTO program review, August 1988; "Computing Infrastructure FY 89 Chart 4," 11 October 1988, attached to memorandum, Mark Pullen to John Toole, 11 October 1988, SC briefing file, document supplied by Col. Toole.

83. Notes of meeting with Craig Fields, 7 October 1988, document supplied by John Toole; see also memorandum, Jack Schwartz to Program Managers, "Upcoming Congressional Review of DARPA programs, including ST-10 and ST-11," 15 September 1988, document obtained from John Toole. Toole recorded that Fields emphasized in the meeting cited above, "Dispell [sic] notion that pgm. will never end!!"

84. FY 1990–1991 biennial RDT&E descriptive summary, project no. ST-10, copy supplied by John Toole; FY 1992–1997 POM RDT&E descriptive summary, program element 0602301E, copied supplied by John Toole, DARPA. The *SC 4th Annual Report*, published in November 1988, indicates the FY 1988 SC allocation as only $115 million—of which $12 million was again earmarked by Congress. The reasons for this decline are unclear. Possibly Congress withdrew funds that DARPA had failed to obligate in a timely manner; from time to time DARPA had more funding authorized than it was able to spend.

85. Notes of meeting with Craig Fields, Washington, DC, 23 January 1995.

86. Schwartz & Goldman interview.

87. Saul Amarel, "AI Research in DARPA's Strategic Computing Initiative," *IEEE Expert* (June 1991): 7–11.

88. Katie Hafner and Matthew Lyon, *Where Wizards Stay up Late: The Origins of the Internet* (New York: Touchstone, 1998), 257–258.

89. Daniel Crevier, *AI: The Tumultuous History of the Search for Artificial Intelligence* (New York: Basic Books, 1993), 210–214.

90. United States, President's Blue Ribbon Commission on Defense Management, *A Quest for Excellence: Final Report to the President* (Washington: The Commission, June 1986), 55–57; U.S. Congress, Senate, Committee on Armed Services, Subcommittee on Defense Acquisition Policy, *Acquisition Findings in the Report of the President's Blue Ribbon Commission on Defense Management*, S. Hearing 99–805, 99th Cong., 2d Sess., 8 April 1986.

91. See testimony of Robert C. Duncan at U.S. Congress, House, Committee on Appropriations, Subcommittee on the Department of Defense, *Department of Defense Appropriations for 1987*, Hearings, 99th Cong., 2d sess., 9 April 1986, 233–273.

92. Its contract was finally terminated in 1990 after "almost five years of failure interrupted by intermittent success," according to an air force participant. Lizza, "Pilot's Associate Program."

93. Ibid., 10–14, 15–16.

94. Ibid., 12.

95. In June 1988, only three months after demo 2, the House Committee on Appropriations recommended deleting funds for the program because of "out-year budget shortfalls and uncertain transition potential." Clint Kelly theorizes, however, that this move was prompted by a faction in the air force that was promoting the concept of unmanned drones, and that considered the PA as a threat to that idea. U.S. Congress, House, Committee on Appropriations, *Department of Defense Appropriations Bill, 1989: Report of the Committee on Appropriations*, 100th Cong., 2d Sess., H. Rept. 100–681, 187.

96. Lizza, "Pilot's Associate Program," 17–20, 23.

97. Testimony of Craig Fields, 3 March 1990, in U.S. Congress, House, Committee on Armed Services, Research and Development Subcommittee, *Hearings on National Defense Authorization Act for Fiscal Year 1991—H.R. 4739 and Oversight of Previously Authorized Programs,* Research and Development Subcommittee Hearings on *Research, Development, Test, and Evaluation,* 101ˢᵗ Cong., 2ᵈ Sess. (Washington: GPO, 1991), 4–60.

Chapter 9

1. Peter Galison, *How Experiments End* (Chicago: University of Chicago Press, 1987).

2. Carl S. Lizza, "History and Retrospective of the Pilot's Associate Program," unpublished paper provided by the author.

3. Herman H. Goldstine, *The Computer from Pascal to von Neumann* (Princeton: Princeton University Press, [1972] 1993); Kenneth Flamm, *Creating the Computer: Government, Industry and High Technology* (Washington: Brookings Institution, 1988), 1–79.

4. Richard A. Jenkins, *Supercomputers of Today and Tomorrow: The Parallel Processing Revolution* (Blue Ridge Summit, PA: TAB Books, 1986), 11.

5. L. T. Davis, "Advanced Computer Projects," in *Frontiers of Supercomputing,* ed. by N. Metropolis, et al. (Berkeley: University of California Press, 1986), 25–33, at 27; IEEE-USA Scientific Supercomputer Subcommittee, *Supercomputer Hardware: A Report,* November 1983, foreword, 3; published in IEEE-USA Committee on Communications and Information Policy, Scientific Supercomputer Subcommittee, *Supercomputing: A Collection of Reports* (IEEE, n.p., n.d.).

6. Larry Smarr, "NSF Supercomputing Program," in *Frontiers of Supercomputing II: A National Reassessment,* ed. Karyn R. Ames and Alan Brenner, Los Alamos Series in Basic and Applied Sciences (Berkeley: University of California Press, 1994): 404.

7. National Science Foundation, *A National Computing Environment for Academic Research* (Washington: NSF, 1983).

8. *Report to the Federal Coordinating Council for Science, Engineering, and Technology Supercomputer Panel on Recommended Government Actions to Retain U.S. Leadership in Supercomputers,* U.S. Department of Energy Report (1983).

9. Peter D. Lax, *Report of the Panel on Large Scale Computing in Science and Engineering,* National Science Foundation Report NSF 82–13 (December 1982).

10. Nancy R. Miller, "Supercomputers and Artificial Intelligence: Recent Federal Initiatives," 6 September 1985, in U.S. Congress, House Committee on Science, Space, and Technology, *Federal Supercomputer Programs and Policies,* 99ᵗʰ Cong., 1ˢᵗ Sess., 10 June 1985, 5–6.

11. *Report of the Federal Coordinating Council for Science, Engineering, and Technology Panel on Advanced Computer Research in the Federal Government* (June 1985), table A-1.

12. Ibid., ix.

13. Eight were sponsored by DARPA as part of SC. Ibid., tables D-1 and D-2.

14. Ibid, xii.

15. Lincoln Faurer, "Supercomputing since 1983," in *Frontiers of Supercomputing II*, 450.

16. Clifford Barney, "Who Will Pay for Supercomputer R&D?" *Electronics* 56 (September 8, 1983): 35–36; N. Metropolis et al., eds. *Frontiers of Supercomputing*.

17. FCCSET was not an obvious savior. Senator Jeff Bingaman described it as "otherwise moribund" when it took up the supercomputing challenge. Jeff Bingaman, "Supercomputing as a National Critical Technologies Effort," in *Frontiers of Supercomputing II*, 7–14, at 9.

18. *Report of the Federal Coordinating Council for Science, Engineering, and Technology Panel on Advanced Computer Research in the Federal Government* (June 1985), 1.

19. Ibid.

20. The White House Science Council, for example, declared forcefully:

> The bottom line is that any country which seeks to control its future must effectively exploit high performance computing. A country which aspires to military leadership must dominate, if not control, high performance computing. A country seeking economic strength in the information age must lead in the development and application of high performance computing in industry and research.

White House Science Council, *Research in Very High Performance Computing* (November 1985), as quoted in Executive Office of the President, Office of Science and Technology Policy, *A Research and Development Strategy for High Performance Computing* (20 November 1987), 8.

21. Saul Amarel, "Status Report on Subcommittee on Computer Research and Development of the Federal Coordinating Council for Science, Engineering and Technology," 20 March 1987, unpublished document provided by the author.

22. See chapter 3 above.

23. Lynn Conway believed that DARPA had to be forced to add communication to high-performance computing in the 1990s. Interview of Lynn Conway by Alex Roland and Philip Shimian, Ann Arbor, MI, 7 March 1994.

24. Federal Coordinating Council for Science, Engineering and Technology,

Committee on High Performance Computing, *Annual Report [for Fiscal Year 1986]* (Washington: OSTP, January 1987), 1–4.

25. OSTP, *Research and Development Strategy*, 10.

26. "Outline of FCCSET Presentation," slides of briefing delivered to Dr. William Graham, 8 June 1987, document supplied by Saul Amarel.

27. Office of Science and Technology Policy, *A Research and Development Strategy for High Performance Computing* (20 November 1987).

28. Ibid., 4.

29. Ibid.

30. Ibid., 26–28.

31. U.S. Senate, Committee on Commerce, Science, and Transportation, *Computer Networks and High Performance Computing*, Hearing before the Subcommittee on Science, Technology, and Space, 100ᵗʰ Cong., 2ᵈ sess., 11 Aug. 1988. The title was an early clue to Senator Gore's priorities; he placed networks, that is, communications, before supercomputing.

32. Stephen L. Squires, "TeraOp Computing Technology Initiative Towards Achieving TeraOp Performance Systems: The TOPS Program" (draft), 13 October 1987, and [Squires,] "DARPA TeraOPS Systems Program Overview," briefing slides dated 2 November 1987, both unpublished documents supplied by the author.

33. Squires, "TeraOPS Computing Technology Initiative."

34. "DARPA TeraOPS Systems Program Overview."

35. Jacob Schwartz to Teraops-related Program Managers (Squires, Mark Pullen, Bill Bandy, and John Toole), Memorandum, "Teraops Program Plan," 4 January 1988, unpublished document supplied by Squires.

36. [Stephen L. Squires,] "DARPA TeraOPS Computing Technology Program," briefing slides dated 28 September 1988, provided by the author.

37. Ibid.

38. "Broad Agency Announcement: Advanced Computing Systems," BAA 88–07, *Commerce Business Daily*, 29 February 1988, 1.

39. Senator Gore introduced S. 2918, "a bill to provide for a national plan and coordinated Federal research program to ensure continued U.S. leadership in high-performance computing," on 19 October 1988. *Congressional Record*, vol. 134, No. 149, 100ᵗʰ Cong., 2ᵈ sess., 19 October 1988, S 16889. The bill died in the Committee on Governmental Affairs.

40. See n. 4 above.

41. Pittsburgh Supercomputing Center, *Projects in Scientific Computing* (Pittsburgh: Pittsburgh Supercomputing Center, 1989).

42. Ibid., 4.

43. Institute of Electrical and Electronics Engineers, United States Activities Board, Committee on Communications and Information Policy, Scientific Supercomputer Subcommittee, *U.S. Supercomputer Vulnerability: A Report* (IEEE, 8 August 1988, reprinted in IEEE, *Supercomputing*), 2–3, 7; David E. Sanger, "A High-Tech Lead in Danger," *New York Times*, 18 December 1988, sect. 3, 6. The IEEE report notes that the Japanese explicitly refused to provide the newer technologies to its Asian competitors as well, 9.

44. Sanger, "A High-Tech Lead," 6.

45. Office of Science and Technology Policy, *The Federal High Performance Computing Program*, 8 September 1989, 43.

46. Ibid., 17–42.

47. National Research Council, *Evolving the High Performance Computing Initiative to Support the Nation's Information Infrastructure* (Washington: National Academy Press, 1995), 49.

48. Ibid., 45–46.

49. Ibid., 8, 13, 23–25, 49–50.

50. Barry Boehm, interview conducted by Philip Shiman, Los Angeles, Ca., 8 August 1994.

51. "DARPA ISTO Strategic Plan," January 1990, unpublished briefing slides provided by Barry Boehm, University of Southern California.

52. Craig Fields invoked this model in a 1991 interview, "The Government's Guiding Hand: An Interview with Ex-DARPA Director Craig Fields," *Technology Review* 94 (March 1991): 35–40, at 40.

53. Ibid.

54. "Operations Support System (OSS): SecNav Program Review, 1 September 1988," unpublished briefing; Ron Crepeau, "Operations Support System (OSS) Program Overview," unpublished briefing slides, 4 January 1989; (Briefing on OSS development), n.d., unpublished slides; and "Report on the Independent Assessment [sic] of the Operations Support System" (draft), 6 February 1989, unpublished document, all provided by Dr. Albert Brandenstein.

55. Barry Boehm, "PE 0602301: Strategic Technology: Plans, Programs, and Payoffs," 30 March 1990, unpublished briefing slides provided by John Toole.

56. Boehm interview.

57. Boehm, "PE 0602301: Strategic Technology."

58. "Current Difficulties and Limitations," slide in briefing by Steve Cross on "Proposed Strategic Computing Initiative in Transportation Planning," in ibid.

59. Cross, "Proposed Strategic Computing Initiative in Transportation Planning," briefing in ibid.; "Strategic Computing in Action: DART," *IEEE Expert* 6 (June 1991): 9; Sheila Galatowitsch, "DARPA: Turning Ideas into Products," *Defense Electronics* 23 (July 1991): 24–25

60. Boehm interview.

61. Rand Waltzman, "Vision Environments Program," February 1990, briefing in Boehm, "PE 0602301: Strategic Technology."

62. This was one of the first speech recognition systems to "transition" to commercial success. See Simson Garfinkel, "Enter the Dragon," *Technology Review* 101 (September/October 1998): 58–64.

63. "Spoken Language: Speech Recognition and Understanding," 14 November 1989, briefing in Boehm, "PE 0602301: Strategic Technology"; Boehm interview.

64. See briefing in Boehm, "PE 0602301: Strategic Technology."

65. One slide in Boehm's 1990 briefing said "He who has and exploits supercomputing will rule the HPC/Microsystems world."

66. In the March program review, one of the stated purposes of the SC architectures program was to "develop basic results to enable the U.S. to maintain dominant position for information science and technology." This was perhaps an unstated goal of the SC program, but was never explicit. See ibid.

67. See briefing in Boehm, "PE 0602301: Strategic Technology."

68. Ibid.; H. T. Kung, "Parallel Processing: Moving into the Mainstream," *Frontiers of Supercomputing II*, 106–110.

69. See pp. 173–175 above in chapter 5.

70. See briefing in Boehm, "PE 0602301: Strategic Technology."

71. Joe Brandenburg, "It's Time to Face Facts," in *Frontiers of Supercomputing II*, 117–119.

72. Stephen Squires, interview conducted by Philip Shiman, Arlington, VA, 21 December 1994; Don Clark, "Intel Claims Computer Speed Record," *San Francisco Chronicle*, 31 May 1991.

73. Briefing in Boehm, "PE 0602301: Strategic Technology."

74. See Otis Graham, *Losing Time: The Industrial Policy Debate* (Cambridge, MA: Harvard University Press, 1992), esp. pp. 231–232.

75. William J. Broad, "Pentagon Wizards of Technology Eye Wider Civilian Role," *New York Times*, 22 October 1991, c1, c11; Michael Schrage, "Will Craig

Fields Take Technology to the Marketplace via Washington?" *Washington Post,* 11 December 1992, D3.

76. The reorganization also spun off a small Electronics Systems Technology Office (ESTO).

77. Squires's goal of a TeraOPS machine was realized late in 1996, when Intel Corporation and Sandia National Laboratory linked together 7,264 high-end Pentium processors to achieve 1.06 TeraFLOPS performance. Rajiv Chandrasekaran, "New $55 Million Machine Claims Crown as Champion of Supercomputing," *Washington Post,* 17 December 1996, p. A01; Intel News Release at http://www.intel.com/pressroom/archive/releases/cn121796.htm, dated 17 December 1996.

78. Armstrong to Dr. D. Allan Bromley, 12 December 1989, and Rollwagen to Bromley, 8 January 1990, copies supplied by Stephen Squires.

79. U.S. Congress, Senate, Committee on Commerce, Science, and Transportation, *Computer Networks and High Performance Computing,* Hearing before the Subcommittee on Science, Technology, and Space, 100th Cong., 2[d] sess., 11 Aug. 1988; S. 2918, "A bill to provide for a national plan and coordinated Federal research program to ensure continued U.S. leadership in high-performance computing," *Congressional Record,* 134, no. 149 (19 October 1988), S 16889; the bill died in the Committee on Governmental Affairs.

80. *Computer Networks and High Performance Computing,* 35.

81. William L. Scherlis to Barry Boehm and Stephen Squires, Memorandum, "FCCSET Executive," 30 January 1990, copy provided by Stephen Squires.

82. David Silverberg, "DoD Prepares National Strategy to Pursue Important Technologies," *Defense News* (22 January 1990): 25; Bromley to Don Atwood et al., Memorandum, "Followup to January 8 Meeting on High Performance Computer," 9 January 1990, copy provided by Squires; I. Lewis Libby to Bromley, 25 January 1990, with attachment, ibid.

83. David Wehrly, "Industry Perspective: Remarks on Policy and Economics for High-Performance Computing," *Frontiers of Supercomputing II,* 542.

84. Scherlis to Boehm and Squires, 30 January 1990.

85. "HPC Efforts for FY91," in U.S. Congress, House, Subcommittee of the Committee on Appropriations, *Department of Defense Appropriations for 1994,* 103rd Cong., 1st Sess., (Washington: GPO, 1993), 5:374–375.

86. This was part of a general reorganization of DARPA. The agency had grown during the 1980s, from 141 officials in 1981 to 174 a decade later. Some offices had grown enormously, including IPTO-ISTO, which went from six technical officers in early 1983 to sixteen by 1991. DARPA Telephone Directories, 23 January 1983, February 1991.

87. U.S. Congress, Senate, Committee on Commerce, Science, and Transportation, *High-Performance Computing Act of 1991*, S. Report 102–57, 102ᵈ Cong., 1ˢᵗ Sess., 16 May 1991 (legislative day, 25 April), 3–4.

88. *High Performance Computing Act of 1991*, Public Law 102–194, 105 Stat. 1594–1604. See also President George Bush, "Remarks on Signing the High-Performance Computing Act of 1991," 9 December 1991, *Public Papers of the President of the United States: George Bush* (Washington: GPO, 1992), 1579–1580.

89. Glenn Zorpette, "Introduction," to *Teraflops Galore*, a special issue of *IEEE Spectrum* (September 1992): 27–33.

90. United States General Accounting Office, *High Performance Computing: Advanced Research Projects Agency Should Do More to Foster Program Goals*, Report to the Chairman, Committee on Armed Service, House of Representatives, GAO/IMTEC–93–24 (Washington: GAO, May 1993), 3–4.

91. Aaron Zatner, "Sinking Machines," *Boston Globe* 6 September 1994, 21–25.

Conclusion

1. See introduction, n. 1.

2. Martin Fransman, *Japan's Computer and Communications Industry: The Evolution of Industrial Giants and Global Competitiveness* (Oxford: Oxford University Press, 1995), 165–167, 446–449.

3. Committee on Innovations in Computing and Communications: Lessons from History, Computer Science and Telecommunications Board, Commission on Physical Sciences, Mathematics, and Applications, National Research Council, *Funding a Revolution: Government Support for Computing Research* (Washington: NRC Press, 1999).

Note on Sources

Documenting the history of modern technological development poses challenging problems and exhilarating opportunities. Documentation abounds, but finding the important documents often proves to be more difficult than in fields with sparser records. Photocopying has both helped and hindered; it multiplies the chances of finding at least one copy of key documents, but it simultaneously swells the pile of paper that must be searched.

This study rests on a solid but incomplete base of primary documents, complemented by extensive interviews and supplemented by secondary literature. Even the interviews posed an embarrassment of riches. We identified more knowledgeable people than we had the time or opportunity to interview, just as we identified more records than we had the time or understanding to master.

In the case of both interviews and documents, we were grateful for what they revealed but painfully aware of what they concealed. Most of our interviewees, especially the principals in the story, harbored their own interpretations of SC. They shaped their conversations with us accordingly. Robert Kahn spoke for many when he insisted that we focus on the technology and discount the people and the politics. For him, and for many of his colleagues, this was a technological story. Most of the participants, with a few exceptions, refused to speak ill of their colleagues; all was harmony and light. Informal conversations made clear that all held strong views about the strengths and weaknesses of the others in the program, but they were generally unwilling to air those views. So too did they guard their judgments about the technical developments. As Stephen Squires told us once, the SC principals made many mistakes, but we would have to find them for ourselves.

The documents displayed the same elusive quality. At one meeting of the history project's advisory board, the participants dismissed the documentary evidence we had offered for one of our claims. They said that the document in question was cooked to meet the expectations of the audience, in this case Congress. We then asked if they could identify any document in the SC program that spoke the truth, that could be accepted at face value. They found this an intriguing question. They could not think of a single such document. All documents, in their view, distorted reality one way or another—always in pursuit of some greater good. One reason for taking this position was the impulse to retain interpretive flexibility. By undermining the reliability of the documents, the participants were enhancing the authority of their recollections. Still, their warning gave us pause, as it should to all historians.

We therefore tried to substantiate our story in four different ways. First, we sought consensus among the participants. Often they disagreed, but when they shared the same recollection, we developed some confidence in it. We still sought documentary verification, but we set the validation bar lower than we might have otherwise. Sometimes such consensus revealed nothing more than shared misconceptions, but even these shed useful light on what actually happened. Second, we sought documentary verification of oral testimony and oral verification of documents. When we got one or the other, we ascribed high, though not absolute, reliability to the information.

Third, we used a technique of comparison by analogy to test the plausibility of our evidence. For example, we attended the 1994 conference of the American Association for Artificial Intelligence, the August 1994 workshop of DARPA's Domain Specific Software Architecture program, and two DARPA technical symposia. By immersing ourselves in the cultural milieu of our story, we hoped to develop some sense for how this community worked and what representations of its history rang true. Finally, we circulated our draft manuscript to participants and observers for their critiques and suggestions. We believe that these reviews allowed us to enhance the accuracy of the factual data and the rigor of the interpretations without compromising our ability to tell the story as we saw it.

We also applied our own experience as DARPA contractors. Having learned first-hand the joys and frustrations of dealing with DARPA, we were able to develop real appreciation for what the agency does and empathy for the contractors who get it done. DARPA, in our experience, can be stimulating and censorious, fast-paced and dilatory, direct and Byzantine, solicitous and imperious, lean and bureaucratic, imaginative and plodding. Knowing this range of possibilities allowed us to imagine more fully and realistically the DARPA that invented, sold, administered, and abandoned SC.

Interviews

We profited enormously by having available the interviews conducted in preparation of the prequel to this story, Arthur L. Norberg and Judy E. O'Neill, *Transforming Computer Technology: Information Processing for the Pentagon, 1962–1986* (Baltimore: Johns Hopkins University Press, 1996). We used the following:

Michael L. Dertouzos	4/20/89	Cambridge, MA	N
Feigenbaum, Edward	3/3/89	Palo Alto, CA	A
Herzfeld, Charles M.	8/6/90	Washington, DC	N
Kahn, Robert	3/22/89	Reston, VA	A
Kahn, Robert	4/24/90	Reston, VA	O
Lukasik, Stephen	10/17/91	Redondo Beach, CA	O
McCarthy, John	3/2/89	Palo Alto, CA	A
Newell, Allen	6/10–12/91	Pittsburgh, PA	N

Nilsson, Nils	3/1/89	Palo Alto, CA	A
Ornstein, Severo	3/6/90	Woodside, CA	O
Reddy, Dabbala Rajagopal	6/12/91	Pittsburgh, PA	N
Roberts, Lawrence	4/4/89	San Mateo, CA	N
Simpson, Robert	3/14/90	Washington, DC	N
Taylor, Robert	2/28/89	Palo Alto, CA	A
Uncapher, Keith	7/10/89	Los Angeles, CA	N
Winograd, Terry	12/11/91	Stanford, CA	N
Winston, Patrick	4/18/90	Cambridge, MA	N
Winston, Patrick	5/2/90	Cambridge, MA	N

Interviewers: N = Arthur Norberg; A = William Aspray; O = Judy O'Neill

We also conducted the following interviews specifically for this project:

Adams, Duane	11/29/94	Arlington, VA	R, S
Amarel, Saul	12/16/94	Rutgers, NJ	R, S
Balzer, Robert	8/8/94	Marina del Rey, CA	S
Bandy, William	7/7/94	Ft. Meade, MD	S
Bock, Conrad	7/28/94	Mountain View, CA	R, S
Bock, Conrad	10/24/94	Mountain View, CA	R, S
Boehm, Barry	8/8/94	Los Angeles, CA	S
Borning, Alan	8/1/94	Seattle, WA	R, S
Brandenstein, Albert	9/12/94	Washington, DC	R, S
Brooks, Frederick	1/29/01	Chapel Hill, NC	R
Buffalano, Charles	8/2/93	Greenbelt, MD	R, S
Buffalano, Charles	5/12/94	Greenbelt, MD	R, S
Carbonnel, Jaime	3/8/94	Pittsburgh, PA	R, S
Conway, Lynn	1/12/94	telephone	S
Conway, Lynn	3/7/94	Ann Arbor, MI	R, S
Cooper, Robert	5/12/94	Greenbelt, MD	R, S
Cross, Stephen	11/17/94	Pittsburgh, PA	S
Davis, Larry S.	1/11/95	College Park, MD	S
Dertouzos, Michael	6/8/95	Cambridge, MA	R
Dertouzos, Michael	4/12/95	Cambridge, MA	R
Duncan, Robert	5/12/94	McLean, VA	R, S
Englemore, Robert	10/24/94	Palo Alto, CA	R, S
Feigenbaum, Edward	7/28/95	Arlington, VA	R, S
Fields, Craig	1/23/95	Washington, DC	R, S

Fikes, Richard	7/28/94	Palo Alto, CA	R, S
Fine, Gary	7/28/94	Mountain View, CA	R, S
Flynn, John P.	12/15/94	Rosslyn, VA	S
Goldman, Susan (see Schwartz, below)			
Gross, Thomas	3/9/94	Pittsburgh, PA	R, S
Hayes-Roth, Richard	7/29/94	Mountain View, CA	R, S
Hennessy, John	10/24/94	Palo Alto, CA	R, S
Jacky, Jonathan	8/1/94	Seattle, WA	R, S
Kahn, Robert	8/2/93	Reston, VA	R, S
Kahn, Robert	11/29/94	Reston, VA	R, S
Kahn, Robert	7/26/95	Reston, VA	R, S
Kahn, Robert	7/27/95	Reston, VA	R, S
Kallis, Lou	9/13/94	Washington, DC	R, S
Kanade, Takeo	11/17/94	Pittsburgh, PA	S
Kelly, Clinton	5/23/94	Durham, NC	R, S
Kelly, Clinton	5/24/94	Durham, NC	R, S
Kelly, Clinton	11/17/94	Pittsburgh, PA	S
Lancaster, Alex	12/3/93	Arlington, VA	S
Lancaster, Alex	12/16/93	Arlington, VA	S
Leiner, Barry	1/24/95	Arlington, VA	R, S
Losleben, Paul	7/29/94	Palo Alto, CA	R, S
Massoud, Hisham	8/21/95	Durham, NC	R
McKeown, David	3/8/94	Pittsburgh, PA	R, S
Moravec, Hans	11/16/94	Pittsburgh, PA	S
Morris, Paul	7/29/94	Mountain View, CA	R, S
Ohlander, Ronald	8/9/94	Marina del Rey, CA	S
Ohlander, Ronald	10/21/94	Los Angeles, CA	R, S
Piña, César	10/21/94	Marina del Rey, CA	R, S
Pullen, J. Mark	7/8/94	Fairfax, VA	S
Reddy, Raj	11/16/94	Pittsburgh, PA	R, S
Schappell, Roger	12/12/94	telephone	S
Scherlis, William	2/1/94	telephone	R
Scherlis, William	3/9/94	Pittsburgh, PA	R, S
Scherlis, William	11/16/94	Pittsburgh, PA	R, S
Schwartz, Jacob and Susan Goldman	12/16/94	New York, NY	R, S
Sears, Alan	10/1/93	McLean, VA	R, S
Simon, Herbert	3/8/94	Pittsburgh, PA	R, S

·Simpson, Robert	12/13/94	telephone	S
Squires, Stephen	6/17/93	Arlington, VA	R, S
Squires, Stephen	6/17/94	Arlington, VA	S
Squires, Stephen	7/12/94	Arlington, VA	R, S
Squires, Stephen	12/21/94	Arlington, VA	S
Swartout, William	8/8/94	Los Angeles, CA	S
Thorpe, Charles	11/17/94	Pittsburgh, PA	S
Toole, John	7/11/94	Arlington, VA	R, S
Treitel, Richard	7/28/94	Mountain View, CA	R, S
Uncapher, Keith	8/8/94	Los Angeles, CA	S
Wactlar, Howard	3/9/94	Pittsburgh, PA	S
Winograd, Terry	7/28/94	Palo Alto, CA	R, S

Interviewers: R = Roland; S = Shiman

Unpublished Primary Sources

The most important archival records for this project were the DARPA documents in the hands of the agency or its former employees. Before describing these, a word is in order about the DARPA documents we did not see. We were promised access to all unclassified documents at the outset. When we interviewed DARPA's Deputy Director Duane Adams in 1994, he observed quite reasonably that it made no sense for DARPA to deny us access, because we could always request documents under the Freedom of Information Act. But when we wrote to Adams to follow up on that apparent offer, we received no reply. Many documents we asked to see were never delivered to us. Many files we asked to exam were never opened to us.

Nevertheless, by pursuing the records that were made available to us, we were able to patch together a reasonably thorough documentary basis for this study. For a federal agency nominally subject to the regulations governing the management of official records, DARPA indulges an exceptionally insouciant and informal style of records management. We found our DARPA records in four different provenances. The records of the DARPA front office, that is the director's office and the support offices directly attached, appear to be in good order. As best we could tell, they are inventoried, retired to the Washington Federal Records Center, and pruned according to a schedule negotiated with the National Archives and Records Service. We were able to identify records in these collections from standard inventories, though we saw little of what we requested access to.

Some of the records of the SC program were maintained in the "cold room" at DARPA headquarters in Arlington, Virginia. This appeared to be an on-site records storage facility available to the program offices such as IPTO, ISTO, and CSTO. Again, we were allowed to see inventories of these records and to request access to them. We saw most of these files that we asked to see, but the inventories are cryptic and irregular, precluding any confidence on our part that we saw and requested the right records.

The computing research offices at DARPA appear to be among the worst recordkeepers—"cowboys," one of the records management personnel told us. Their behavior is culturally driven, at least in part. They see themselves as leading-edge technologists; documenting the past is a waste of time. Many of ISTO's records, for example, were lost when DARPA moved from Rosslyn to Ballston, Virginia, in the early 1990s. Apparently ISTO had a room full, practically floor to ceiling, of boxed records. After the move was announced, the computing personnel were repeatedly alerted to get their records packed and ready for shipping, but they paid no attention until it was too late. Some boxes were apparently saved, but the great bulk was abandoned or thrown out. The "cold room" records appear to be the ragged survivors of this "Great Records Massacre."

Some SC records, mostly those generated by Stephen Squires, were being held in storage at DynCorp, a company that contracted support services to DARPA and other federal agencies. Though these records belong to the government and thus to the taxpayer, Squires maintained them as his own private archive, from which he one day hoped to write his own book. Officially, we were allowed access to these records only when Squires was present. Shiman saw a few of the boxes under these conditions, and he returned surreptitiously at a later date to examine the boxes in Squires's absence. He viewed about a dozen of the sixty to seventy boxes in the room, concluding that they were interesting, but not essential to our project. As with many of these collections, their bulk outweighed their usefulness.

William Scherlis collected and stored at Dyncorp a set of files on the Software and Intelligent Systems Technology Office. It contained about 15,000 items, ranging from contractors' reports, notes, and proceedings of conferences, to articles from journals and magazines. Apparently program managers would sweep their shelves periodically, especially when leaving the agency. The collection was well-catalogued. It was apparently shipped to the Charles Babbage Institute at the University of Minnesota after SC ended.

ARPA order (AO) files were sent from DARPA to DynCorp for our use. These are especially useful for tracing the contracting history of the various programs and include some progress reports, plans, and an occasional nugget such as a hand-written note or observation. They also give overall numbers showing what had been spent on the SC program as a whole for a given year. Like many DARPA documents, however, they must be used cautiously, since funds often migrated from project-to-project without being fully recorded in the files made available to us. When we requested an ARPA order under the Freedom of Information Act after our project was completed, we received a version with some financial data blacked out.

Many of the most useful documents came into our hands from the cellars, attics, and garages of former DARPA employees. In spite of the federal regulations governing disposition of official records, many employees simply emptied their file cabinets when they left the agency, taking their records with them. Most of these employees were willing to share their files with us, though Lynn Conway insisted that she had signed an agreement when she left DARPA promis ing not to disclose any information about her service in the SC program; she brought several stacks of documents to our interview but refused to let us see any of them. In contrast, Ronald Ohlander made available to us his complete e-mail files from his years at DARPA. He had kept his own electronic record of this correspondence and printed it out for us from his server in Marina del Rey, California. More than one interviewee told us that the real record of what went on in this and other DARPA programs appeared not in the official documents, but in the e-mail correspondence. Ohlander's e-mail confirms that observation in part, but it did not produce any smoking guns. It is likely that e-mail records are becoming even more important as an increasing percentage of business is conducted on line.

Whenever possible in the text, we have identified the DARPA documents on which we have relied and the source from which they came. Some of the documents, however, came to hand in photocopied form with doubtful or unknown provenance. We have nonetheless used these where we are confident of their authenticity.

Two other archives warrant special mention. The Charles Babbage Institute of the University of Minnesota is a center for the history of information processing. Among its many useful collections for the project are the archives of the Norberg and O'Neill history, *Transforming Computer Technology*. Other useful records are scattered through its holdings, including materials on Computer Professionals for Social Responsibility, networking, and government computer activities.

Equally useful are the archives at the Massachusetts Institute of Technology, especially the records of the Laboratory for Computer Science, collection no. 268.

Many of the secondary works reviewed for this study are listed in "Strategic Computing: A Bibliography," prepared by Philip Shiman on 31 March 1993. The works relied upon are cited in the notes.

Following publication, the records collected for this study will be offered to the Babbage Institute for use by other researchers.

Index

Printed in the United States
by Baker & Taylor Publisher Services

Printed in the United States
by Baker & Taylor Publisher Services